Studies in Logic
Logic and Argumentation
Volume 109

The Cognitive Dimension of Social Argumentation
Proceedings of the 4th
European Conference on Argumentation
Volume I

Volume 101
The Logic of Partitions. With Two Major Applications
David Ellerman

Volume 102
Bounded Reasoning Volume 1: Classical Propositional Logic
Marcello D'Agostino, Dov Gabbay, Costanza Larese, Sanjay Modgil

Volume 103
The Fertile Debate. Affective Exploration of a Controversy
Claire Polo

Volume 104
Argument, Sex and Logic
Dov Gabbay, Gadi Rozenberg and Lydia Rivlin

Volume 105
Logic as a Tool. A Guide to Formal Logical Reasoning
Valentin Goranko

Volume 106
New Directions in Term Logic
George Englebretsen, ed

Volume 107
Non-commutative Algebras. Pseudo-BCK Algebreas versus m-pseudo-BCK Algebras
Afrodita Iorgulescu

Volume 108
Semitopology: decentralised collaborative action via topology, algebra, and logic
Murdoch J. Gabbay

Volume 109
The Cognitive Dimension of Social Argumentation. Proceedings of the 4th European Conference on Argumentation, Volume I. Fabio Paglieri, Alessandro Ansani and Marco Marini, eds.

Volume 110
The Cognitive Dimension of Social Argumentation. Proceedings of the 4th European Conference on Argumentation, Volume II. Fabio Paglieri, Alessandro Ansani and Marco Marini, eds.

Volume 111
The Cognitive Dimension of Social Argumentation. Proceedings of the 4th European Conference on Argumentation, Volume III. Fabio Paglieri, Alessandro Ansani and Marco Marini, eds.

Studies in Logic Series Editor
Dov Gabbay   dov.gabbay@kcl.ac.uk

The Cognitive Dimension of Social Argumentation

Proceedings of the 4th
European Conference on Argumentation
Volume I

Edited by

Fabio Paglieri

Alessandro Ansani

and

Marco Marini

© Individual author and College Publications, 2024
All rights reserved.

ISBN 978-1-84890-471-2

College Publications
Scientific Director: Dov Gabbay
Managing Director: Jane Spurr

http://www.collegepublications.co.uk

Cover prepared by Laraine Welch

All rights reserved. No part of this publication may be reproduced, stored in a retrieval system or transmitted in any form, or by any means, electronic, mechanical, photocopying, recording or otherwise without prior permission, in writing, from the publisher.

THE COGNITIVE DIMENSION OF SOCIAL ARGUMENTATION

PROCEEDINGS OF THE 4TH EUROPEAN CONFERENCE ON ARGUMENTATION

VOLUME 1

EDITED BY
FABIO PAGLIERI
ALESSANDRO ANSANI
MARCO MARINI

TABLE OF CONTENTS

INTRODUCTION — V
Fabio Paglieri, Alessandro Ansani, Marco Marini

CAN ARGUMENTS CHANGE MINDS? — 1
Catarina Dutilh Novaes

ARGUING WITH ARGUMENTS ARGUMENT QUALITY, ARGUMENTATIVE NORMS, AND THE STRENGTHS OF THE EPISTEMIC THEORY — 25
Harvey Siegel

A PARTICULARIST THEORY OF ARGUMENTATION BY ANALOGY — 61
José Alhambra

A MINIMALIST APPROACH TO ARGUMENTATION BY ANALOGY. COMMENTARY ON JOSÉ ALHAMBRA'S A PARTICULARIST THEORY OF ARGUMENTATION BY ANALOGY — 79
Jean H.M. Wagemans

COMMUNICATIVE ACTIVISM AS AN 'ALTERNATIVE' ARGUMENTATIVE STRATEGY TO CONFRONT CONTEXTS AFFECTED BY HATE SPEECH — 83
Álvaro Domínguez Armas

COMMENTARY ON "COMMUNICATIVE ACTIVISM AS AN 'ALTERNATIVE' ARGUMENTATIVE STRATEGY TO CONFRONT CONTEXTS AFFECTED BY HATE SPEECH BY ÁLVARO DOMINIGUEZ ARMAS — 103
Sara Greco

USING ARGUMENTATION SCHEMES TO MODEL LEGAL REASONING — 107
Trevor Bench-Capon, Katie Atkinson

CONTEXTUAL AND EMOTIONAL MODULATION OF 127
SOURCE-CASE SELECTION IN ANALOGICAL ARGUMENTS
MARCELLO GUARINI

DISTANT SIMILARITY AS AN ADVANTAGE OF (SOME) 145
ARGUMENTS FROM ANALOGY? COMMENTARY ON
MARCELLO GUARINI'S 'CONTEXTUAL AND EMOTIONAL
MODULATION OF SOURCE CASE SELECTION IN
ANALOGICAL ARGUMENTS'
MARCIN KOSZOWY

DEVELOPING REQUIREMENTS FOR RECONSTRUCTING 151
VISUAL ARGUMENTS USING META-VISUAL DISPUTES
BITA HESHMATI

MAPPING ARGUMENTS IN FAVOR OF THE PRESUMPTION 171
OF INNOCENCE
MICHAEL J. HOPPMANN

WHAT JUSTIFIES THE PRESUMPTION OF INNOCENCE? 197
COMMENTARY ON HOPPMANN'S "MAPPING ARGUMENTS
IN FAVOR OF THE PRESUMPTION OF INNOCENCE
PETAR BODLOVIĆ

DEVELOPING AN INVITATIONAL APPROACH TO PRAGMA- 213
DIALECTICS
BROOKE HUBSCH

REFRAMING IN DISPUTE MEDIATION: AN UMBRELLA 231
TERM ENCOMPASSING FOUR ARGUMENTATIVE
PHENOMENA
CHIARA JERMINI-MARTINEZ SORIA

COMMENT ON PAPER: REFRAMING IN DISPUTE 249
MEDIATION: AN UMBRELLA TERM ENCOMPASSING FOUR
ARGUMENTATIVE PHENOMENA
ELENA MUSI

THE CONCEPT OF ARGUMENT: REVISITED AND REENGINEERED MARCIN LEWIŃSKI	251
COMMENTARY ON LEWIŃSKI, "THE CONCEPT OF ARGUMENT: REVISITED AND REENGINEERED" HARVEY SIEGEL	273
ELICITING ARGUMENTATION IN ENGINEERING-SCIENCE EDUCATION. MAKING JUSTIFIED RESEARCH-AND-DESIGN DECISIONS IN CONCEPTUAL MODELLING MARIANA OROZCO	277
COMMENTARY ON "ELICITING ARGUMENTATION IN ENGINEERING-SCIENCE EDUCATION. MAKING JUSTIFIED RESEARCH-AND-DESIGN DECISIONS IN CONCEPTUAL MODELLING" BY MARIANA OROZCO AND MIEKE BOON JAN ALBERT VAN LAAR	297
THE INTERACTION BETWEEN ARGUMENTATIVE NORMS: THE CASE OF MUNĀẒARA RAHMI ORUÇ	301
ARGUMENTATIVE VIRTUE AND DIALECTICAL OBLIGATIONS WENQI OUYANG	321
ARGUMENTATIVE VIRTUES: BACK TO BASICS FABIO PAGLIERI	339
COMMENTARY ON FABIO PAGLIERI'S "ARGUMENTATIVE VIRTUES: BACK TO BASICS" DANIEL COHEN	357
ARGUMENTATIVE PATTERNS INITIATED BY CLOSED-LIST QUESTIONS IN ACCOUNTABILITY DIALOGUES. A CORPUS STUDY OF FINANCIAL CONFERENCE CALLS ANDREA ROCCI, OLENA YASKORSKA-SHAH, GIULIA D'AGOSTINO & COSTANZA LUCCHINI	361

DILEMMATIC ARGUMENTATION: A PRAGMA-   385
DIALECTICAL APPROACH TO A CLASSICAL TOPOS
CHARLOTTE VAN DER VOORT

THE CONCEPT OF ARGUMENTATION IN CHINESE   409
WRITING EDUCATION AND THE MODERN
TRANSFORMATION OF CHINESE THINKING AND
REASONING
HAILONG WANG

COMMENTARY ON: THE CONCEPT OF ARGUMENTATION   421
IN CHINESE WRITING EDUCATION AND THE MODERN
TRANSFORMATION OF CHINESE THINKING AND
REASONING BY HAILONG WANG
FRANK ZENKER

INTRODUCTION

FABIO PAGLIERI
Istituto di Scienze e Tecnologie della Cognizione, Consiglio Nazionale delle Ricerche (ISTC-CNR), Roma, Italy
fabio.paglieri@istc.cnr.it

ALESSANDRO ANSANI
Centre of Excellence in Music, Mind, Body and Brain, University of Jyväskylä, Finland
alessandro.a.ansani@jyu.fi

MARCO MARINI
Istituto di Scienze e Tecnologie della Cognizione, Consiglio Nazionale delle Ricerche (ISTC-CNR), Roma, Italy
marco.marini@istc.cnr.it

The European Conference on Argumentation (ECA) is an academic initiative launched in 2013 by a group of Europe-based argumentation scholars, with the aim of inaugurating a series of biennial international conferences on this thriving area of studies (for details, see https://ecargument.org/), to complement other large-scale events on similar or related topics: the conference of the International Society for the Study of Argumentation (ISSA), the conference of the Ontario Society for the Study of Argumentation (OSSA), the conference on Computational Models of Argument (COMMA), the Rhetoric in Society conference and, more recently, the events organized by the Argumentation Network of the Americas (ANA). The first edition of ECA took place in Lisbon (PT) in 2015, followed by a second one in Fribourg (CH) in 2017 and a third one in Groningen (NL) in 2019: then the Covid-19 pandemic struck and the next edition of ECA had to be postponed to 2022, when it took place in Rome, from September 28 to September 30.

The conference lasted two days and a half, with a very intense and diverse programme, including 3 keynote talks, 1 plenary panel, 16 long papers with invited commentators, and as many as 118 regular papers (14 of which were presented as part of 3 thematic panels). Most of these contributions are collected in written form in these three volumes, as follows:

- *Volume 1* includes 2 of the 3 keynotes presented at the conference, authored by Catarina Dutilh Novaes and Harvey Siegel, followed by 16 long papers, ordered alphabetically by first author's surname: the majority of those (9 out of 16) are accompanied by their respective commentary.
- *Volume 2* includes 30 regular papers, ordered alphabetically by first author's surname, from Aikin & Casey to Licato et al.
- *Volume 3* includes the remaining 30 regular papers, again ordered alphabetically by first author's surname, from Liga to Zemplén & Tanács..

Even though these proceedings cover only a selection of what was presented at the ECA conference in Rome, they provide a faithful approximation of the breadth and depth of the ongoing discussion in argumentation scholarship. They also attest the interdisciplinary character of this field: this has been the hallmark of argumentation studies since their inception, yet the disciplines brought to bear on this subject matter have steadily increased over the years; whereas philosophy and linguistics were always partners in this endeavor, nowadays they are supported also by computer science and experimental psychology, as well as communication and media studies in a broader sense. At the same time, the study of argument from a philosophical perspective is no longer regarded as a specialistic niche for philosophers with a grudge against deductive logic as a model of human reasoning, but it is taking back its place as a central concern for philosophical inquiry in general, as it was at the dawn of the discipline (Aristotle's work is an obvious example).

These are welcome developments in the natural evolution of argumentation studies, which the ECA initiative has always intended to promote and nourish: thus, we expect to see more of the same in future editions of ECA, starting with ECA 2025 in Warsaw, Poland, on "Argumentation in the digital society".

CAN ARGUMENTS CHANGE MINDS?[1]

CATARINA DUTILH NOVAES
Department of Philosophy, Vrije Universiteit Amsterdam
c.dutilhnovaes@vu.nl

Abstract

Can arguments change minds? Philosophers like to think that they can. However, a wealth of empirical evidence suggests that arguments are not very efficient tools to change minds. What to make of the different assessments of the mind-changing potential of arguments? To address this issue, we must take into account the broader contexts in which arguments occur, in particular the propagation of messages across networks of attention, and the choices that epistemic agents must make between alternative potential sources of content and information, which are very much influenced by perceptions of reliability and trustworthiness. Arguments can change minds, but only under conducive, favorable socio-epistemic conditions.

1. Introduction

Can arguments change minds? Philosophers like to think that they can: by engaging in the (presumably rational) process of carefully considering reasons in favor or against a given position or view, we update our beliefs accordingly.[2] According to this optimistic view, famously defended by John Stuart Mill in particular, we not only *do* change our mind when exposed to (compelling) arguments (a descriptive claim), but we also *improve* our overall epistemic position by the careful considerations of reasons (an evaluative claim).

However, a wealth of empirical and anecdotal evidence seems to suggest that arguments are in fact not very efficient tools to change minds

---

[1] This is a reprint of the article which appeared in the Proceedings of the Aristotelian Society CXXIII(2), 2023.
[2] In this paper, I speak of 'changing minds' in a rather loose way, but the concept can also be treated more systematically. There are different formal frameworks that offer an account of what it means to change one's mind, such as Bayesian probabilities and various belief revision theories. For our purposes here, the differences between them are immaterial, as they all deal with how agents update their beliefs in view of incoming information.

(Gordon-Smith 2019) (McIntyre 2021). For example, the well-documented phenomenon of polarization (Isenberg 1986) (Sunstein 2002) suggests that, when exposed to arguments supporting positions different from their prior views, people in fact often (though perhaps not always) become even more convinced of their prior views rather than being swayed by arguments (Olsson 2013). Frequently, argumentative encounters look rather like games where participants want to 'score points' (Cohen 1995) (Dutilh Novaes 2021) rather than engage in painstaking consideration of different views for the sake of epistemic improvement.

What to make of these different assessments of the mind-changing potential of arguments? To address this issue, it seems that we need to look beyond the content and quality[3] of arguments alone: we must also take into account the broader contexts in which they occur, in particular the propagation of messages across networks of communication, and the choices that epistemic agents must make between alternative potential sources of content and information. These choices are very much influenced by perceptions of reliability and trustworthiness, which means that the source of the argument may be even more decisive than its content or quality when it comes to how persuasive it will be for a given person. (In this respect, argumentation would be more akin to testimony than one might expect, as I argued elsewhere (Dutilh Novaes 2020b).) In a nutshell: arguments may well be able to change minds, but only under conducive, favorable socio-epistemic conditions.

In this paper, I deploy a *three-tiered model of epistemic exchange* that I've been developing over the past years (Dutilh Novaes 2020b) to (hopefully) shed light on the mechanisms involved in these processes, and on the conditions under which arguments can change minds. I start with an 'optimistic' assessment of the power of argumentation to change minds, in particular in John Stuart Mill's formulation, and its shortcomings as discussed in the literature (at least as an accurate *description* of the phenomena in question). I then offer a brief description of the three-tiered model and of its relevance for the issue at hand. In Part 4, I discuss two real-life examples of people who had epistemic breakthroughs which involved (at least to some extent) engagement with arguments, but only against the background of favorable socio-epistemic conditions. In part 5, I clarify a few pending issues. I then close with some concluding remarks.

---

3 I understand the quality of an argument as pertaining to familiar criteria for argument quality such as validity and soundness. Argument quality can also be accounted for probabilistically.

2. The Millian conception of argumentation and its limitations

Mill is one of the main exponents of the view that interpersonal argumentative situations involving people who truly disagree with each other have the potential to change minds (primarily for the better, he thinks).4 In On Liberty (1859) (Mill 1999), he notes that, when our ideas are challenged by those who disagree with us, we are forced to evaluate critically our own beliefs.

> [Man] is capable of rectifying his mistakes, by discussion and experience. Not by experience alone. There must be discussion, to show how experience is to be interpreted. Wrong opinions and practices gradually yield to fact and argument; but facts and arguments, to produce any effect on the mind, must be brought before it. (Mill 1999) (p. 41)

This process is often described as a *free exchange of ideas*, and according to Mill, it is beneficial *even* when we are right and our interlocutors are wrong. The expected result is that the remaining beliefs, those that have survived critical challenges, will be better justified than those held before such encounters. As Mill puts it, "both teachers and learners go to sleep at their post, as soon as there is no enemy in the field." (Mill 1999) (p. 83) Dissenters thus force us to stay epistemically alert instead of becoming too comfortable with existing, entrenched beliefs—what Mill describes as 'dead dogma' (Simpson 2021).

But for this process to be successful, dissenters must be permitted to voice their opinions and criticism freely, and indeed Mill's forceful defense of free speech is one of his most celebrated positions. One of his main arguments for free speech is epistemic: he emphasizes the role played by the free exchange of ideas in facilitating the growth of knowledge in a society. The more dissenting views and arguments in favor or against each of them are exchanged, the more likely it is that the 'better' ones will prevail (Halliday and McCabe 2019).

However, it is not sufficient that dissenters be given the opportunity to voice their opinions freely; it is also of crucial importance that receivers of these opinions and arguments be willing to engage in good faith and with

---

4 I have defended this view myself (Dutilh Novaes, 2020a) but with the important caveat that the beneficial epistemic effect of interpersonal argumentation will come about only against the background of specific circumstances that ensure good faith exchange of ideas (for example, within a community of mathematicians). See below for a discussion of circumstances where argumentative exchanges reliably lead to epistemic improvement.

an open mind.⁵ Mill pays much attention to the structural conditions for the free exchange of ideas (in particular, that there should be no state-sanctioned censorship of any kind), but he does not seem to take sufficiently into account our well-documented tendencies to avoid engaging with dissenting views altogether, or to explain away contrary evidence so as to preserve prior beliefs (a point that will be further discussed shortly).

More recently, Alvin Goldman articulated a similar account of the social epistemology of argumentation (Goldman 1994) (Goldman 2004). The starting point for Goldman is the recognition of a situation of epistemic division of labor, where different members of an epistemic community know different things, and so can benefit from exchanging these different epistemic resources with each other. Moreover, given our inescapable fallibility, these exchanges with other knowers may help expose our own mistaken beliefs (as also noted by Mill). A third feature of our socio-epistemic situation is that people sometimes have incentives to deceive and mislead, so a certain amount of epistemic vigilance is needed. It is against these background conditions that argumentation becomes a valuable tool in the pursuit of truth and avoidance of error, according to Goldman.

> Norms of good argumentation are substantially dedicated to the promotion of truthful speech and the exposure of falsehood, whether intentional or unintentional. [...] Norms of good argumentation are part of a practice to encourage the exchange of truths through sincere, non-negligent, and mutually corrective speech. (Goldman 1994) (p. 30)

But does argumentation indeed *reliably* succeed in promoting truth and avoiding error in social epistemic contexts, as suggested by Mill and Goldman? Do we readily revise our beliefs when exposed to (good) arguments that contradict them? Do we really "gradually yield to fact and argument", as claimed by Mill? It seems that Mill and Goldman are overly optimistic regarding the power of arguments to change minds. In fact, argumentation appears to be a rather inefficient way to change minds in many real-life situations (Gordon-Smith 2019).

The truth is that people typically avoid revising their views about firmly entrenched beliefs (a point famously made by Quine (Quine 1951)). When confronted with arguments or evidence that contradict these beliefs, they tend either to ignore the evidence, explain it away (as we know from

---

5 There is also the important issue (to be discussed shortly) of whether dissenting voices will attract attention at all, for example if they belong to marginalized groups.

the literature on confirmation bias (Nickerson 1998)), or to discredit the source of the argument as unreliable.[6] These tendencies are exemplified by so-called science deniers such as flat-earthers (McIntyre 2021), but also in scientific practice where entrenched paradigms often resist a fair amount of counter-evidence before a 'scientific revolution' takes place (Kuhn and Hacking 2012). In particular, arguments that threaten core beliefs, feelings of belonging, and identities (e.g., political beliefs) seem to trigger various forms of motivated reasoning whereby one ignores or rejects those arguments without engaging substantially with their content (Taber and Lodge 2006) (Kahan 2017). Engaging (or not) in argumentation is often a means to express and cement social identities rather than to come closer to the truth (Talisse 2019) (Hannon 2019).

Moreover, when choosing among a vast supply of options, there is a tendency to gravitate towards content and sources that confirm one's existing opinions, in so-called 'echo chambers' and 'epistemic bubbles' (Nguyen 2020). Conversations with like-minded people may reinforce prior beliefs and even drive people to more extreme versions of those beliefs (Olsson 2013). This means that the mere availability of dissenting opinions is not sufficient to ensure that knowers remain epistemically alert and consider all sides of a question. There is always the option of ignoring (i.e., not engaging with) these dissenters and the substance of their arguments, especially if they are perceived as untrustworthy (Dutilh Novaes 2020b). This is the familiar phenomenon of polarization: instead of bringing parties closer together, argumentation and deliberation may have the opposite effect of drawing them further apart (Sunstein 2002).

Another obstacle is the fact that the absence of government-sanctioned censure (as proposed by Mill) is no guarantee that all relevant voices will be truly *heard*. Dissenting views defended by marginalized social groups will tend to attract less attention than those with powerful proponents; the so-called free exchange of ideas is one where power differentials significantly affect the spread and uptake of views. This is the familiar problem of *inclusion* in democratic societies (Young 2000), which has serious political as well as epistemic consequences. More often than not, it is not the force or quality of an argument alone that determines its uptake; the social position of its proponents is a decisive factor in how much it will spread and be viewed as persuasive.

To be sure, there *are* some contexts where the exchange of reasons in argumentative interactions does seem to lead reliably to people changing their minds and to epistemic improvement (Mercier 2018) ((Dutilh Novaes

---

[6] But see (Mercier 2020) and (Coppock 2022), who argue that epistemic agents do regularly, and competently, update their beliefs in view of new information, including on value-laden matters such as politics.

2020a), Chapters 8 and 9).[7] The literature on group problem-solving has established that, for what this literature describes as 'intellective problems', that is, those that have a unique answer within a given theoretical framework (e.g., a mathematical or logical problem), group discussion among peers has a clear beneficial, truth-conducive effect (Laughlin 2011). Indeed, in specialized contexts such as in science or mathematics, argumentative 'friction' is a quintessential way to produce knowledge (Longino 1990) (Lakatos 1976). But this is less obviously the case for so-called 'judgmental problems', that is, those pertaining to values and judgments that do not have a unique correct answer (Laughlin 2011). (The intellective vs. judgmental distinction is one of degrees.) Importantly, in real-life situations, we are very often confronted with judgmental problems, and for these kinds of problems the literature on group problem-solving suggests that argumentation does not reliably lead to better outcomes (which is not surprising, given that for judgmental problems there is often no consensus on what counts as conclusive evidence). In fact, many of them are instances of *deep disagreements* (Fogelin 1985) that may not be amenable to being solved by means of reasoning and argumentation.

These observations suggest that we are not 'proper Millians' when it comes to argumentation and dissent. The epistemic alertness that Mill believed would be the natural, almost automatic consequence of being exposed to dissenting opinions and arguments often fails to come about. The Millian account is thus descriptively inaccurate, or at the very least incomplete. One may retort that the Millian account is still *normatively* correct; but given that it appears to be highly idealized, it is arguably not suitable to offer *prescriptive* recommendations (in the sense of (Bell, Raiffa, and Tversky 1988)) for concrete human agents.[8] Instead, we need a more realistic approach to the (social) epistemology of argumentation, one which takes into account not only the cognitive limitations of individual knowers but also the social complexities of these processes.

---

[7] The concept of 'epistemic improvement' presupposes that there are suitable metrics that allow us to measure progress. One natural metric is simply what epistemologists call accuracy, which roughly corresponds to Goldman's 'pursuit of truth and avoidance of error' (veritism). But more fine-grained metrics may be considered, for example epistemic improvement in terms of understanding (Grimm, Baumberger, and Ammon 2016).

[8] I don't think that the Millian story is fully convincing as a normative account either, but a thorough discussion of this point goes beyond the scope of this paper. See (Fantl 2018) for a critique of the Millian idea that engaging with dissenters is always rational/desirable.

3. The three-tiered model of epistemic exchange

We've just seen that the free exchange of ideas is hindered by various factors such as structural power relations and cognitive and social tendencies, so much so that there is no guarantee that wrong opinions and practices will "gradually yield to fact and argument". To address some of the limitations of the Millian conception of argumentation, I've been developing a three-tiered model of epistemic exchange, which presents a more realistic account of epistemic exchange through argumentation by considering the costs, obstacles, and risks of engaging in argumentative exchanges (C. Dutilh Novaes 2020b). While it is a model of social epistemic processes in general, the key idea is that argumentation truly consists in an *exchange*, where resources flow in both directions (from arguer to receiver but also from receiver to arguer), and thus is a specific kind of epistemic exchange.

This model was inspired by a theoretical framework known as *Social Exchange Theory* (SET) (Dutilh Novaes 2020b). This is a framework developed by sociologists and social psychologists that seeks to explain human social behavior in terms of processes of exchange, involving costs and rewards, and against the background of social networks and power structures (Cook 2013). It was originally developed in the late 1950s and early 1960s under the influence of research in economics (rational choice theory), psychology (behaviorism), and anthropological work by Malinowski, Mauss, and Lévi-Strauss. SET is an influential and empirically robust framework, which has been used to investigate a wide range of social phenomena (such as romantic relationships, business interactions, trust in public institutions, among many others). In particular, and relevant for our purposes, it has been extensively used to investigate interpersonal communication (Roloff 2015). The SET models are neither purely descriptive—as they rely on certain idealized assumptions such as that agents seek to maximize rewards and minimize costs—nor purely normative, given that they incorporate experimental findings as well as extensive observational data. Moreover, SET combines a first-person perspective, which explains and predicts choices that individuals make between different potential exchange partners, with a third-person perspective, which focuses on structural features of these exchange networks.

The three-tiered model of epistemic exchange adapts insights and results from SET to exchanges that are specifically epistemic, that is, when epistemic resources such as knowledge, evidence, information etc.

are involved (possibly alongside other kinds of resources).[9] The model allows for a meticulous account of the conditions under which successful epistemic exchange may occur or fail to occur. Crucially, there seem to be two preliminary stages that determine whether specific agents will be in a position to engage in fruitful epistemic exchange: the *networks* that determine which sources and which epistemic resources an agent is exposed to; and the *contrastive choices* that agents must make regarding which contents and sources to engage with (among those she is exposed to). Thus seen, the three tiers for epistemic exchange are:

1. Attention/exposure. The first tier pertains to whether people are potential exchange partners of each other, given the relevant opportunity structures for epistemic engagement within a network. In simpler terms: who is in an agent's network of potential contacts? Who is in a position to attract the attention of others? It may be that potential lines of communication are cut, say in the case of structural censorship or epistemic bubbles. But it may also be that so many signals are being broadcast that many different sources are competing for the receiver's attention (Gershberg and Illing 2022), in a so-called 'attention economy' (Franck 2019).[10]

2. Choosing whom to engage with. The tier comprises the choices that agents make against the background of possibilities for exchange, as determined by the relevant opportunity structures. Typically, there will be a number of options for a given agent—for example, the various newspapers that I can read on any given day, among those that I have access to. Given limitations of time and attention, contrastive choices will have to be made. Among those sources that have caught my initial attention, who do I view as worthy of consideration as an exchange partner? At this point, considerations of *trustworthiness* (Hawley 2019) and *expertise* (Goldman, 2018) come into play, as well as the perceived value of the content being offered by different potential exchange partners. In particular, trusting someone will often entail *not* trusting someone else, especially when their respective messages conflict (C. Dutilh Novaes 2020b).

3. Engagement with content. It is only in the third tier that engagement with *content* properly speaking occurs; this is when the actual epistemic exchange takes place. At this point, the receiver will reflectively (and perhaps critically) engage with the argument being offered, seeking to understand its substance and evaluate its cogency. In case of a positive evaluation, this may lead to a change in view for the receiver (though even at this point the receiver may still balk at revising her beliefs). It may also

---

9 See (C. Dutilh Novaes 2020b) for further details on how the three-tiered model emerges from SET.
10 See (C. Dutilh Novaes and de Ridder 2021) for a discussion on scarcity vs. overabundance of information in epistemic environments.

lead to a mutually beneficial exchange where both arguer and addressee improve their respective epistemic stances, as posited by Mill, and in some cases even go on to create new epistemic resources together (as in Lakatos' 'proofs and refutations' model of mathematical practice (Lakatos 1976)).

Figures 1 to 3 represent the three tiers.[11] For simplicity, a main agent is depicted with other agent around her, but the model in fact focuses on complex networks of agents who are interconnected to different degrees. The topologies of such networks crucially determine how these socio-epistemic processes come to unfold.[12]

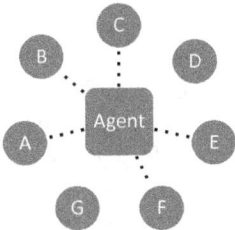

Figure 1. Attention: Agent does not 'see' sources D and G; the other sources catch her attention (dotted lines).

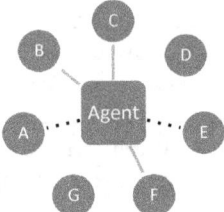

Figure 2. Contrastive choices: Agent deems B, C and F as worth exchanging with (grey lines), but not A and E.

---

11 The model can also be understood in terms of set containment: at a given point in time, the set of people I actually engage in epistemic exchanges with is a subset of those who I deem worth exchanging with (above a certain threshold), which in turn is a subset of those who, due to our respective positions in the network, are potential exchange partners for me.

12 Notice that there are a number of interesting structural similarities between the three-tiered model that I present here and the network epistemology research program, as developed by Zollman (Zollman 2013), Olsson (Olsson 2013), O'Connor and Weatherall (O'Connor and Weatherall 2019), among others. For reason of space, I do not develop this point further here, which will remain a topic for future research.

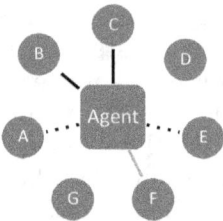

Figure 3. Engagement: Agent eventually engages substantively with B and C (black lines), but not with F.

Millian conceptions of argumentation tend to focus primarily on tier 3— the 'force' of an argument alone should suffice to change minds—and to downplay some of the structural obstacles to a truly free and equal exchange of ideas[13]. Indeed, tiers 1 and 2 crucially determine if and when someone will seriously engage with the epistemic resources being offered by someone else at all. Just as the original SET models, the three-tiered model is neither purely normative nor purely descriptive. It is not purely normative because it does not consider ideal or idealized agents: instead, it considers agents with limited cognitive resources, and who are susceptible to what Neil Levy describes as 'bad beliefs', that is, beliefs that are incompatible with expert opinion and our best available evidence (Levy 2021). Moreover, the model is empirically robust as it draws on decades of SET's experimental and observational findings pertaining to exchanges more generally. However, the model is not purely descriptive or predictive either, as it seeks to explain the mechanisms that lead different people to engage in epistemic exchanges with some sources but not with others; this is done on the basis of a few foundational principles such as reciprocity and fairness, and by highlighting in particular the roles of attention and trust in such processes. As such, the model is perhaps best understood as an explanatory model, in the sense that it seeks to represent some of the causes of the target phenomenon and the mechanisms responsible for bringing it about (Ivani and Dutilh Novaes 2022). (It may also lead to prescriptive recommendations on how to facilitate certain types of epistemic exchanges.)

The three-tiered model offers an explanation for why arguments often fail to change minds, as it highlights some of the necessary conditions for this to occur. First, a suitable relation of attention and exposure must

---

[13] Mill's own emphasis on freedom of speech is aimed at creating a maximally inclusive informational environment, and thus at increased exposure to various views (phenomena belonging to tier 1). Mills mentions some factors as having a central role in making exchanges more likely to succeed, such as the importance of education. However, he does not provide a detailed analysis of the conditions under which successful epistemic exchange may occur or fail to occur; in particular, they may fail even in contexts where (religious or otherwise) persecution is not present.

emerge between sender and receiver—which is far from obvious, especially in highly saturated informational environments such as the ones we currently inhabit (Gershberg and Illing 2022). Secondly, an agent must make choices regarding whom to engage with, among the different possibilities: this is where considerations of trustworthiness—understood as related to both competence and benevolence (Dutilh Novaes 2020b) (Dutilh Novaes 2023)—arise. If I already suspect that a given source does not hold benevolent attitudes towards me, should I really spend my precious time and energy engaging with their arguments? Maybe not (Köymen and Dutilh Novaes forthcoming). For example, the refusal to engage with scientific arguments supporting the efficacy and safety of vaccines on the part of so-called 'anti-vaxxers' is often justified by the (not entirely unreasonable) suspicion that spurious interests are involved (e.g., the 'evil Big Pharma' narrative (Dutilh Novaes 2020b) (Ivani and Dutilh Novaes 2022)). Finally, the exchange itself requires that agents with very diverse epistemic backgrounds find enough common ground and suitable means of communication rather than talking past each other, which is far from obvious especially in situations of ideological/political disagreement (Talisse 2019). If the (potential) exchange fails at any of these three levels, then arguments will not prompt a change of mind.

4. Real-life examples

Despite all these challenges, the conclusion that arguments *never* change minds is also unwarranted: arguments sometimes *do* change minds. The question then becomes, under which conditions is this (more) likely to happen? The three-tiered model provides suitable conceptual tools to address this question. Instead of discussing it in the abstract or with toy examples, I here present two recent concrete examples of people who underwent radical epistemic transformations where arguments (presumably) played a significant role: Megan Phelps-Roper, formerly a prominent member of the Westboro Baptist Church,[14] and Derek Black, formerly a prominent advocate of white supremacy.[15]

---

[14] My discussion of Megan Phelps-Roper's trajectory draws primarily on the 2015 New Yorker profile of her:
https://www.newyorker.com/magazine/2015/11/23/conversion-via-twitter-westboro-baptist-church-megan-phelps-roper. She also wrote a memoir, tellingly titled Unfollow: A Journey from Hatred to Hope (Phelps-Roper 2020).
[15] My discussion of Derek Black's trajectory relies primarily on an interview for the New York Times podcast 'The Daily' (transcript here):
https://www.nytimes.com/2017/08/22/podcasts/the-daily-transcript-derek-black.html). There is also a book narrating Black's journey: Rising Out of Hatred:

The Westboro Baptist Church is a hyper-Calvinist congregation based in Topeka, Kansas (USA), often described as a hate group. It is known for engaging in inflammatory homophobic pickets as well as hate speech against atheists, Jews, Muslims, transgender people, and numerous Christian denominations. Megan Phelps-Roper is a granddaughter of founder Fred Phelps, and was raised to be a prominent member of the group. As such, she grew up immersed in their stern ideology, and from early on participated in pickets at funerals of gay men (with signs featuring slogans such as 'GOD HATES F*GS') and later of soldiers killed at war (as Westboro members believe that the wars that the USA has been involved in in recent decades are God's punishment for the country's tolerance of homosexuality).

Despite the extreme positions of Westboro members, their children, including Megan, usually attended Topeka public schools. At school, she was presumably exposed to other, more tolerant worldviews, but this did not substantially affect her own conviction in the Westboro belief system. Many members had received higher education and some, including Megan's mother, worked as lawyers. Thus, they did not exactly live in an epistemic bubble in the sense of not being exposed to alternative belief systems; in fact, they believed that Westboro members could best preach to the 'wicked' by living among them. The thought was: if you really knew the 'truth' in your heart, exposure to the world of the wicked would not affect your devotion. In practice, however, there was no room at all for epistemic autonomy or dissent: the supreme value that was instilled in children was that of complete obedience.

In 2009, Megan joined Twitter to further spread Westboro's views. Some of her homophobic tweets were picked up on and re-tweeted (in the spirit of mockery) by large accounts, which resulted in her receiving many angry replies but also gaining a significant number of followers. She thereby came to be in contact with a wider range of critics, to whom she diligently replied citing biblical passages (along with pop culture references and emojis). She was used to giving interviews to journalists, but on Twitter she could engage with many people directly, with no journalistic filter.

But Megan had by then also started having doubts about some of the Westboro teachings. In particular, around the time she joined Twitter, Westboro was preparing for the end of the world. There were very specific predictions on how Westboro members would lead a hundred and forty-four thousand Jews who repented for killing Jesus through the wilderness of Israel, until Christ would finally come to save them all. Megan felt there was no proper scriptural support for many of these predictions, and turned to Twitter for answers. More specifically, she started following and

---

The Awakening of a Former White Nationalist, written by journalist Eli Saslow (Saslow 2018).

engaging with Jewish Twitter users, in particular with a Jerusalem-based web designer called David Abitbol.

And thus, even if through bitter debate, she began to forge deeper connections with other Twitter users. Until then, interactions with 'the wicked' had remained superficial and fleeting, such as with counter-protesters at pickets. On Twitter, however, she got involved in extended debates with specific people (such as Abitbol) with whom she developed fierce but friendly patterns of interaction. To her surprise, for the first time in her life she started *caring* about what people outside of Westboro—namely, some of her Twitter acquaintances—thought of her; the connections with some of her Twitter interlocutors became increasingly meaningful. (In fact, she ended up marrying one of them years later.)

And so, as a result of some small seeds of doubt concerning Westboro's preaching (as well as concerns pertaining to changes in how the church was run and the role of women therein), but mostly through her Twitter connections and interactions, Megan embarked in a long and painful process of questioning everything she had been brought up to believe. About three years after joining Twitter, she started seriously considering leaving the church. That would, of course, entail tremendous social and emotional costs; she would basically lose all contact with her immediate and extended family. She eventually made the consequential decision to leave the church (together with her younger sister Grace) on November 2012, and began connecting again, including in offline environments, with some of her Twitter contacts such as Abitbol. She has since become an advocate for dialogue between groups with conflicting views, and has spoken on multiple venues about her experiences (including the inevitable TED talk).

If we are to believe her own account of the process, arguments played an important role in Megan's (slow but profound) 'epistemic breakthrough' of coming to realize that she could no longer endorse the Westboro belief system. There was much deliberation involved, both with herself and with many of her Twitter contacts (some of whom were also knowledgeable on sources she considered authoritative, in particular the Bible). Through these processes (which at times resembled Socratic dialogues), inconsistencies and discrepancies in the Westboro doctrines became apparent to her, leading to a thorough revision of her own convictions. However, two necessary conditions had to be in place for these arguments to do their work: naturally, she had to be exposed to them (on Twitter, she could be exposed to a wide range of sources and interlocutors); but more importantly, these arguments were coming from people she had grown to respect and care about. She had had exposure to ideas that clashed with the Westboro doctrines before (e.g., at school), but this time the sources of these ideas were people she had forged deeper connections with. This time, she paid more attention and engaged in earnest with the substance of their arguments. What is perhaps remarkable about Megan's trajectory is the fact that the process of recalibration of attributions of respect and trust to different people (away from Westboro members and towards 'outsiders') happened primarily by means of online interactions rather than face-to-face. (Her *New Yorker* profile describes the process as 'conversion via Twitter'.) Online connections can become 'real' connections after all, and may offer a much wider net of potential epistemic exchange partners.[16]

Derek Black's trajectory bears interesting similarities to Megan Phelps-Roper's, but in his case the 'conversion' took place primarily through face-to-face interactions rather than online. Derek is the son of Don Black, prominent white supremacist and founder of Stormfront, one of the most influential white supremacist online communities in the US. His godfather is David Duke, one of the most visible and influential Ku Klux Klan leaders in recent decades (as shown for example in the 2018 Spike Lee movie *BlacKkKlansman*). Both Duke and Don Black are Ku Klux Klan Grand Wizards. Derek was raised to be the 'crown prince' of white supremacy in the United States, and from early on was deeply involved in promoting this worldview, including producing a radio show with his father.

Different from Megan Phelps-Roper, Derek was homeschooled, and so had limited exposure to worldviews other than his family's white supremacist beliefs during his youth. His whole socio-emotional world while growing up consisted of people espousing the same ideology. In 2010, at age 21, he decided to enroll at the New College of Florida in Sarasota, a four-hour drive away from home; this was the first time he left the insular world of white supremacism he had grown up in. He began to live what might be described as a 'double life': recording the radio show with his father in the morning, then attending classes and socializing with students who were (left-leaning) social justice advocates during the rest of the day. Initially, his identity as a white supremacist had not been revealed.

But inevitably, after a few months a fellow student discovered and exposed his identity, and his racist beliefs and ongoing activism became public knowledge at the college. Unsurprisingly, this led to him becoming ostracized among students. The one exception was a small group of Jewish students who began to invite him to their Shabbat dinners. (By then, Derek had already had a brief relationship with a Jewish woman, which had come to an end when his white supremacy persona became public knowledge.) Perhaps because these were the only people still willing to socialize with him, he became a regular at their dinners.

While some of the dinner-goers did not seek to confront Derek about his beliefs openly, others engaged in heated intellectual discussions with him. Here is his own account of these discussions:

---

16 See (Lewiński and Dutilh Novaes Forthcoming) for an account of online communication drawing on the three-tiered model of epistemic exchange.

"I would say, "This is what I believe about I.Q. differences, I have 12 different studies that have been published over the years, here's the journal that's put this stuff together, I believe that this is true, that race predicts I.Q. and that there are I.Q. differences in races." And they would come back with 150 more recent, more well researched studies and explain to me how statistics works and we would go back and forth until I would come to the end of that argument and I'd say, Yes that makes sense, that does not hold together and I'll remove that from my ideological toolbox but everything else is still there. And we did that over a year or two on one thing after another until I got to a point where I didn't believe it anymore." [17]

These conversations went on for years, during which Derek gradually moved away from the white supremacist ideology he had grown up with. Eventually, in 2013, Derek wrote a public statement to the Southern Poverty Law Center, publicly renouncing his previous views. He had much to lose socially and emotionally by distancing himself from white supremacy, including his close relationship with his family: changing one's mind can be not only cognitively but also socially costly (an aspect also explored in (Gordon-Smith 2019)). As with Megan Phelps-Roper, Derek's epistemic breakthrough did not happen overnight: it was the result of a long process where his original beliefs were gradually dispelled, at least partially through the force of arguments (that is, if we are to believe his own account of this process). However, once again the fact that arguments came from people whom Derek had come to respect on a personal level (despite the ethnicities of some of them being considered as 'inferior' according to the white supremacist worldview he had espoused until then) was a crucial element in the process. He truly listened and engaged with the substance of their arguments because of this favorable interpersonal setting, which in turn was facilitated by his vulnerability and the fact that these were the only people still willing to interact with him on campus. (Like Megan, Derek also ended up in a long-term romantic relationship with one of the people who challenged his beliefs early on.)

Until he went to New College, Derek's exposure to other worldviews had been limited (a tier 1 phenomenon), and he had been raised to trust only those who espoused similar ideas as his family's (a tier 2 phenomenon). In Nguyen's (2020) terms, he was both in an epistemic bubble and in an echo chamber, whereas Megan Phelps-Roper was

---

17 Source: https://www.nytimes.com/2017/08/22/podcasts/the-daily-transcript-derek-black.html

primarily in an echo chamber but not as much in an epistemic bubble.[18] The rewiring of circuits of attention and trust prompted by his experiences on campus is what enabled arguments to do their mind-changing work on Derek.

Naturally, arguments can also change minds on specific issues. The two cases described here correspond to complete overhauls of whole belief systems, but arguments can also, and likely more easily, cause localized revisions (which may require some accommodations but not as radically as in these two cases). The point here is that, if *even* in these two extreme cases arguments appear to have changed the minds of Megan and Derek, then *a fortiori* arguments can prompt more localized belief revisions as well.

A topic I've investigated in previous work is the change in public opinion regarding the folk character of 'Black Pete' in the Netherlands (Zwarte Piet) (Dutilh Novaes et al. 2020). Black Pete is the assistant of St. Nicholas at the hugely popular St. Nicholas festivities in early December, and was traditionally portrayed in highly racialized ways (blackface, curly hair, thick red lips). In recent years, there has been a significant shift in public opinion regarding the purported racist nature of the character; while until some 10 years ago, the character was viewed by 95% of the population in a positive light, currently at least one third of the population (and rising) came to see it as unacceptable. This has led to important changes in how the character is portrayed, most significantly a sharp decline in the use of blackface makeup. Arguments seem to have played an important role in this shift in public opinion, in particular by challenging what Charles Mills has aptly described as 'white ignorance': a deforming epistemic outlook that renders people classified as white oblivious to the pervasiveness of racism (Mills 2015).

5. Clarifications

Before concluding, a few clarifications on the picture sketched so far are warranted. Firstly, from an egalitarian-progressive perspective, Megan's and Derek's 'conversions' are viewed as positive because they came to renounce what many of us take to be wrong and problematic worldviews. They attained what *we* take to be significantly improved epistemic states.[19] But the general mechanisms described by the three-tiered

---

18 "An epistemic bubble is a social epistemic structure in which other relevant voices have been left out, perhaps accidentally. An echo chamber is a social epistemic structure from which other relevant voices have been actively excluded and discredited." (Nguyen 2020) (p. 141)

19 At least, I am assuming that most readers of this piece will reject homophobia, racism, and white supremacy.

model—pathways of attention, trust, and engagement—do not favor specific ideologies (Dutilh Novaes 2023). Indeed, the spread of 'unsavory' positions such as vaccine rejection and various conspiracy theories follows similar patterns. In particular, propensity to espouse conspiracy theories seems to be strongly associated with distrust towards established institutions such as governments, the press, and the scientific establishment (van Prooijen, Spadaro, and Wang 2022). In the end, whether we come to espouse 'good' or 'bad beliefs' (in Levy's terminology) is to a significant extent a result of the epistemic environments we find ourselves in, including our attributions of credibility and trustworthiness to different sources (Dutilh Novaes, 2023). The belief-forming processes leading to 'good' or to 'bad' beliefs are, perhaps surprisingly, not fundamentally different (Levy 2021). Bad beliefs can also be supported by arguments—'bad' arguments perhaps (though not necessarily!), but arguments nevertheless, and they too can change minds if the conditions are suitable. (Notice that this is also a thorny point for the optimistic Millian who maintains that truth will eventually prevail, provided that all views can be openly expressed and discussed.)

A second clarification pertains to whether there are jointly *sufficient* conditions for arguments to change minds. What the three-tiered model describes are *necessary* conditions pertaining to attention and attributions of credibility. But even if these are in place, there is no guarantee that arguments will indeed change minds; people may, and indeed often do, still stick to their prior beliefs, especially beliefs that are thoroughly enmeshed with their ways of living.[20] While the idea of a fool-proof method to change minds by means of high-quality arguments may seem appealing, in practice arguments alone cannot *force* an epistemic update to occur, try as they might.[21]

---

20 Compare Sally Haslanger's notion of cultural technē, understood as a collection of social meanings "that provides a 'stage-setting' for action and is a constituent part of the local social-regulation system" (Haslanger 2021) (p. 23). A cultural technē 'gone wrong' will organize social structures in unjust ways, and for Haslanger this is exactly what ideology is. To dismantle a cultural technē 'gone wrong', rational arguments by themselves will have little to no effect; instead, the cultural technē in question must first be 'disrupted' to open up possibilities for contestation. In the cases of Megan and Derek, the disruption in question was caused by inhabiting different discursive and affective environments (Twitter for Megan, college for Derek). But at a broader, societal level, more significant disruptions seem necessary, following Haslanger's notion of cultural technē. They may however still partially involve arguments, in for example what is known in the Marxist tradition as consciousness raising.
21 This is a point related to what some authors identify as the intrinsically coercive nature of arguments (Nozick 1981) (Casey 2020), which is however not always effective. As Wittgenstein pointed out, even a correct mathematical proof may fail to persuade, despite 'the hardness of the logical must' (see (Wright 1990)). A related

Relatedly, even when a change of mind apparently prompted by arguments occurs, it may well be that the efficacious causes are ultimately non-epistemic factors such as social dynamics (e.g., a desire to belong to a certain group) or economic incentives. In other words, we cannot be sure that the arguments were persuasive for the 'right' (rational) reasons, i.e., pertaining to their quality *qua* arguments.[22] Indeed, given the human propensity for *rationalization* (Cushman 2020), it is often not transparent to the agent herself what exactly prompted a change of mind.

Finally, Megan and Derek were (presumably) swayed by arguments only because they previously accepted the basic rules of the language-game of argumentation.[23] Megan was skilled at the practice of arguing in support of religious beliefs on the basis of careful scriptural analysis, and came to respect the scriptural knowledge of some of her Twitter interlocutors. Derek referred to 'scientific' studies himself to support his views (e.g., that there are racial IQ differences), but then came to realize that there were much *better* scientific studies supporting opposite views. They were thus receptive to the very practice of supporting positions with arguments and evidence; that is, there was at least a certain degree of meta-level agreement on the 'rules of the game' between them and their interlocutors. Had this not been the case—for example, if they thought that everything was really a matter of 'what feels right to me', or that all opinions are equally valid—then arguments might not have had much grip on them.

In sum, the cases just discussed still do not offer full reassurance to the optimistic Millian that, under the right conditions, good arguments will indeed change minds for the better. 'Bad' arguments may also change minds for the worse; good arguments may fail to prompt a change of mind, even under the right circumstances; even when it looks like a change of mind was caused by engagement with high-quality arguments, we cannot be sure that the actual causes for the change were truly argumentative; and arguments by themselves will have little to no effect in cases where people reject the very idea of updating their beliefs in view of arguments and evidence. Still, the cases discussed offer at least a plausible account of the mechanisms through which good arguments may change minds, and thus partially vindicate Mill's view that engaging with dissenters may allow for the correction of errors.

---

question, not addressed here, is whether it is ethically acceptable to try to change someone's mind (be it by arguments or other interventions); does it not constitute a problematic infringement of someone's intellectual autonomy?

22 Thanks to James Owen Weatherall for raising these two worries.

23 I owe this point to Harvey Siegel.

6. Conclusions

We started with the view that arguments can change minds by the force of reason alone. In practice, however, and certainly in situations where values and political views play a significant role, arguments do not seem to be particularly suitable to change minds; on the contrary, people typically either outright refuse to engage with or else are not moved by arguments that clash with their deep-seated beliefs. But in some circumstances, arguments may in fact succeed in changing minds. I've argued that two important but often underappreciated factors, attention and trust, need to be taken into account to explain the persuasiveness (or lack thereof) of arguments in specific situations. Arguments can only change minds if they catch the receiver's attention, and if the receiver chooses to give them careful consideration, which in turn is significantly (but not completely) determined by attributions of credibility and trustworthiness to the source. If these conditions are in place, then it may well happen (though again, no guarantee!) that arguments will change someone's mind.[24]

We've looked into the real-life cases of Megan Phelps-Roper and Derek Black. In both cases, they came to renounce the worldviews they were brought up with thanks in part to argumentative engagement (at least if we are to believe their own accounts of these processes). However, they had first both come to respect the *sources* of these arguments, and this is why they engaged with their substance in earnest rather than dismissing them outright. Also, in both cases, it was a lengthy process: arguments need time to truly change minds. It was the cumulative effect of many such argumentative interactions that eventually led them to a complete abandonment of their original positions. How representative these two cases are on a broader scale is difficult to establish; but since the main claim of this paper is merely an existential one—arguments can *sometimes* change minds—they offer support to this modest claim and help illustrate the mechanisms involved.

These two cases also show that changing minds through arguments is costly (Casey 2020). It can be costly for the person who changes their mind, as it may entail the loss of their most meaningful social and affective connections; and it is costly for those trying to change minds through arguments, as they must invest significant resources (time, energy) to catch the receiver's attention and to gain enough of their trust so that they will engage in earnest with the substance of the arguments. Moreover, the

---

[24] McIntyre (McIntyre 2021) presents a very similar picture of how science deniers (sometimes) change their minds in view of argument and evidence: "All of these stories are basically the same. They happen within the context of a trusting, personal relationship. As I've said all along, facts and evidence can matter, but they have to be presented by the right person in the right context." (p. 73)

argumentative process itself can be slow and require many iterations. Thus, a plausible conclusion to be drawn is that arguments are not very efficient tools to change minds (as opposed perhaps to e.g., narratives or propagandistic discourse). Still, we need not go as far as concluding that arguments are pointless and futile, as some like to say; in the right circumstances at least, they may in fact change minds for the 'right' reasons.

Acknowledgments: This research was generously supported by the European Research Council with grant ERC-2017-CoG 771074 for the project 'The Social Epistemology of Argumentation'. Thanks to the SEA team for their invaluable contributions over the years. I am grateful for the feedback received from various audiences: the 'Epistemic Breakthroughs' workshop in St. Andrews, the European Conference on Argumentation in Rome, the Brazilian Society for Analytic Philosophy Conference in Rio de Janeiro, and seminars at Ruhr University Bochum and VU Amsterdam.

References

Bell, David E., Howard Raiffa, and Amos Tversky. 1988. "Descriptive, Normative, and Prescriptive Interactions in Decision Making." In Decision Making: Descriptive, Normative, and Prescriptive Interactions, by David E. Bell, Howard Raiffa, and Amos Tversky, 9–30. Cambridge: Cambridge University Press.
Casey, John. 2020. "Adversariality and Argumentation." *Informal Logic* 40: 77–108.
Cohen, Daniel H. 1995. "Argument Is War...and War Is Hell: Philosophy, Education, and Metaphors for Argumentation." *Informal Logic* 17: 177–88.
Cook, Karen S. et al. 2013. "Social Exchange Theory." In John DeLamater & Amanda Ward (eds.) *Handbook of Social Psychology*. Berlin: Springer, 61–88.
Coppock, Alexander. 2022. *Persuasion in Parallel: How Information Changes Minds about Politics*. Chicago Studies in American Politics. Chicago; London: The University of Chicago Press.
Cushman, Fiery. 2020. "Rationalization Is Rational." *Behavioral and Brain Sciences*.
Dutilh Novaes, Catarina. 2020a. *The Dialogical Roots of Deduction*. Cambridge: Cambridge University Press.
———. 2020b. "The Role of Trust in Argumentation." *Informal Logic* 40: 205–36.
———. 2021. "Who's Afraid of Adversariality? Conflict and Cooperation in Argumentation." *Topoi* 40 (5): 873–86. https://doi.org/10.1007/s11245-020-09736.
———. 2023. "The (Higher-Order) Evidential Significance of Attention and Trust—Comments on Levy's Bad Beliefs." *Philosophical Psychology*, no. Online First. https://doi.org/10.1080/09515089.2023.2174845.

Dutilh Novaes, Catarina, and Jeroen de Ridder. 2021. "Is Fake News Old News?" In *The Epistemology of Fake News*, by S. Bernecker, A.K. Flowerree, and T. Grundmann. Oxford: Oxford University Press.
Dutilh Novaes, Catarina, Emily Sullivan, Thirza Lagewaard, and Mark Alfano. 2020. "Changing Minds through Argumentation: Black Pete as a Case Study." In *Reason to Dissent: Proceedings of the 3rd European Conference on Argumentation*, 2:243–60. Studies in Logics : Logic and Argumentation 86. London: College Publications.
Fantl, Jeremy. 2018. *The Limitations of the Open Mind*. Oxford: Oxford University Press.
Fogelin, Robert. 1985. "The Logic of Deep Disagreements." *Informal Logic* 7: 3–11.
Franck, Georg. 2019. "The Economy of Attention." *Journal of Sociology* 55: 8–19.
Gershberg, Zachary, and Sean D. Illing. 2022. *The Paradox of Democracy: Free Speech, Open Media, and Perilous Persuasion*. Chicago: University of Chicago Press.
Goldman, Alvin I. 1994. "Argumentation and Social Epistemology." *Journal of Philosophy* 91: 27–49.
———. 2004. "An Epistemological Approach to Argumentation." *Informal Logic* 23..
Gordon-Smith, Eleanor. 2019. *Stop Being Reasonable: How We Really Change Minds*. New York , NY: Public Affairs.
Grimm, Stephen, Christoph Baumberger, and Sabine Ammon. 2016. Explaining Understanding: New Perspectives from Epistemology and Philosophy of Science. London: Routledge.
Halliday, Daniel, and Helen McCabe. 2019. "John Stuart Mill on Free Speech." In *The Routledge Handbook of Applied Epistemology*, by D. Coady and J. Chase, 72–87. London: Routledge.
Hannon, Michael. 2019. "Empathetic Understanding and Deliberative Democracy." *Philosophy and Phenomenological Research* Early View: 1–20.
Haslanger, Sally. 2021. "Political Epistemology and Social Critique." In *Oxford Studies in Political Philosophy Volume 7*, 23–65. Oxford University Press. https://doi.org/10.1093/oso/9780192897480.003.0002.
Hawley, Katherine. 2019. *How To Be Trustworthy*. Oxford: Oxford University Press.
Isenberg, Daniel J. 1986. "Group Polarization: A Critical Review and a Meta-Analysis." *Journal of Personality and Social Psychology* 50: 1141–51.
Ivani, Silvia, and Catarina Dutilh Novaes. 2022. "Public Engagement and Argumentation in Science." *European Journal for Philosophy of Science* 12 (3): 54. https://doi.org/10.1007/s13194-022-00480-y.
Kahan, Dan M. 2017. "Misconceptions, Misinformation, and the Logic of Identity-Protective Cognition." New Haven: Yale Law School.
Köymen, Bahar, and Catarina Dutilh Novaes. forthcoming. "Reasoning and Trust: A Developmental Perspective." In *Why and How We Give and Ask for Reasons*. Oxford University Press.
Kuhn, Thomas S., and Ian Hacking. 2012. *The Structure of Scientific Revolutions*. Fourth edition. Chicago ; London: The University of Chicago Press.
Lakatos, Imre. 1976. *Proofs and Refutations: The Logic of Mathematical Discovery*. Cambridge: Cambridge University Press.
Laughlin, Patrick R. 2011. *Group Problem Solving*. Princeton, NJ: Princeton University Press.

Levy, Neil. 2021. *Bad Beliefs: Why They Happen to Good People*. New product. New York: Oxford University Press.
Lewiński, Marcin, and Catarina Dutilh Novaes. Forthcoming. "The Many-to-Many Model: Communication, Attention, and Trust in Online Conversations." In *Conversations Online*. Oxford University Press.
Longino, Helen. 1990. *Science as Social Knowledge: Values and Objectivity in Scientific Inquiry*. Princeton: Princeton University Press.
McIntyre, Lee C. 2021. *How to Talk to a Science Denier: Conversations with Flat Earthers, Climate Deniers, and Others Who Defy Reason*. Cambridge, Massachusetts: The MIT Press.
Mercier, Hugo. 2018. "Reasoning and Argumentation." In *Routledge International Handbook of Thinking and Reasoning*, by L. Ball and V. Thomson, 401–14. New York, NY: Routledge.
———. 2020. *Not Born Yesterday*. Princeton, NJ: Princeton University Press.
Mill, John Stuart. 1999. *On Liberty*. Peterborough: Broadview Press.
Mills, Charles. 2015. "Global White Ignorance." In *Routledge International Handbook of Ignorance Studies*, by M. Gross and L. McGoey, 217–27. London: Routledge.
Nguyen, C. Thi. 2020. "Echo Chambers and Epistemic Bubbles." *Episteme* 17: 141–61.
Nickerson, Raymond S. 1998. "Confirmation Bias: A Ubiquitous Phenomenon in Many Guises." *Review of General Psychology*, 175–220.
Nozick, Robert. 1981. *Philosophical Explanations*. Cambridge, MA: Harvard University Press.
O'Connor, Cailin, and James O. Weatherall. 2019. *The Misinformation Age*. New Yaven, CO: Yale University Press.
Olsson, Erik J. 2013. "A Bayesian Simulation Model of Group Deliberation and Polarization." In *Bayesian Argumentation: The Practical Side of Probability*, 113–333. Dordrecht: Springer.
Phelps-Roper, Megan. 2020. *Unfollow: A Memoir of Loving and Leaving Extremism*.
Prooijen, Jan-Willem van, Giuliana Spadaro, and Haiyan Wang. 2022. "Suspicion of Institutions: How Distrust and Conspiracy Theories Deteriorate Social Relationships." *Current Opinion in Psychology* 43 (February): 65–69. https://doi.org/10.1016/j.copsyc.2021.06.013.
Quine, Willard. 1951. "Main Trends in Recent Philosophy: Two Dogmas of Empiricism." *The Philosophical Review* 60 (1): 20. https://doi.org/10.2307/2181906.
Roloff, Michael E. 2015. "Social Exchange Theories." In *International Encyclopedia of Interpersonal Communication*. London: Wiley.
Saslow, Eli. 2018. *Rising out of Hatred: The Awakening of a Former White Nationalist*. First edition. New York: Doubleday.
Simpson, Robert Mark. 2021. "'Lost, Enfeebled, and Deprived of Its Vital Effect': Mill's Exaggerated View of the Relation Between Conflict and Vitality." *Aristotelian Society Supplementary Volume* 95 (1): 97–114. https://doi.org/10.1093/arisup/akab006.
Sunstein, Cass R. 2002. "The Law of Group Polarization." *Journal of Political Philosophy* 10: 175–95.

Taber, Charles S., and Milton Lodge. 2006. "Motivated Skepticism in the Evaluation of Political Beliefs." *American Journal of Political Science* 50: 755–69.

Talisse, Robert. 2019. *Overdoing Democracy*. Oxford: Oxford University Press.

Wright, Crispin. 1990. "Wittgenstein on Mathematical Proof." *Royal Institute of Philosophy Supplement* 28: 79–99.

Young, Iris Marion 2000. *Inclusion and Democracy*. Oxford: Oxford University Press.

Zollman, Kevin. 2013. "Network Epistemology: Communication in Epistemic Communities." *Philosophy Compass* 8: 15–27.

ARGUING WITH ARGUMENTS: ARGUMENT QUALITY, ARGUMENTATIVE NORMS, AND THE STRENGTHS OF THE EPISTEMIC THEORY

HARVEY SIEGEL
University of Miami
hsiegel@miami.edu

> The whole concept of argument... rests upon the ideal of rationality – of discussion not in order to move or persuade, but rather to test assumptions critically by a review of *reasons* logically pertinent to them.
> --Israel Scheffler

Abstract
'Argument' has multiple meanings and referents in contemporary argumentation theory. Theorists are well aware of this, but often fail to acknowledge it in their theories. In what follows I distinguish several senses of 'argument', and argue that some highly visible theories are largely correct about some senses of the term but not others. In doing so I hope to show that apparent theoretical rivals are better seen as collaborators or partners, rather than rivals, in the multi-disciplinary effort to understand 'argument'[1], arguments, and argumentation in all their varieties. I argue as well for a pluralistic approach to argument evaluation and argumentative norms, since arguments and argumentation can be legitimately evaluated along several dimensions, but that epistemic norms enjoy conceptual priority.

---

1 Scheffler 1989, p. 22, emphasis in original.

1. Some Examples of Arguments

A. All people are mortal.
Socrates is a person.
Therefore, Socrates is mortal. (Found written on a Philosophy 101 classroom whiteboard)

B. God is the being greater than which none can be conceived.
Existence in reality is greater than existence in the understanding alone.
Therefore, a being that exists in the understanding alone is not God.
God exists in the understanding.
Therefore, God exists in reality. (Ditto)

C. Euclid's argument (proof) that there is no highest prime number.

D. Democrats: The US Constitution guarantees the right of privacy, which includes the right to end a pregnancy by means of abortion, free of government interference.
Republicans: No it doesn't. Moreover, the fetus' right to life outweighs any right to privacy, if there is one.

E. HS: A and B are arguments.
DC: No, they're not. Arguments require arguers.
HS: Yes they are. No they don't. Quarrels require quarrelers; arguments don't require arguers[2]. (With thanks to Daniel Cohen)

F. AMS: The aim of argument is the resolution of disagreement and the attainment of consensus.
BS: No it isn't. (With thanks to Michael Gilbert)

G. That's a good argument. It sure persuaded me!

H. Jack: Let's see the Spielberg remake of 'West Side Story' tonight.
Jill: No, I don't like musicals. Let's go out for a Rijsttafel instead.
Jack: Good idea, I haven't had one in ages. Maybe we could have a koffie verkeerd afterwards.
Jill: Sounds like a plan!

I. Sophia: NATO and the EU should move immediately to accept Ukraine's applications for membership.

---

[2] Biro and Siegel 2015, pp. 30-32.

Jens: That is a terrible idea; it would increase the chance of war, which would result in unnecessary death, destruction, and misery for the people of Ukraine.
Sophia: In the short term, you may be right. But in the longer term, the benefits of NATO and EU membership outweigh the costs, not only for Ukraine but for Western Europe more broadly.
Jens: You're right. Let's go to the rally this afternoon and register our support for Ukraine's entry into NATO and the EU. (Written before the Russian invasion)

These few examples of arguments are not all of a piece, and no doubt many more yet different types could be added. Examples A, B and C are exemplars of the (1) *abstract propositional* sense of 'argument'. A and B are familiar to philosophy students and are perfectly analyzable in terms of abstract propositions [3] (or the sentences that express them) and logical/inferential or epistemological relations, sometimes referred to as 'premise-conclusion (or 'reason-conclusion') complexes'. C is perhaps best understood as a sub-category of such complexes, namely mathematical proofs, again understood in terms of abstract propositions and inferential relationships. These examples also exemplify a second, (2) *speech act* sense of 'argument' when they are spoken or otherwise enacted.

Examples D, E and F exemplify arguments involving *disagreements*: All three register disagreements by way of speech acts, and set the stage for attempts at persuasion and dispute resolution. As such they are examples of the second, speech act sense of 'argument'. They also exemplify a third, (3) *social/dialogical/communicative* sense of 'argument', which is often better understood as the social phenomenon of *argumentation*. Arguments involving speech acts needn't be social, dialogical, or communicative – I might speak to and argue with myself, for example. Nevertheless, senses (2) and (3) frequently go together. Even so, it is useful to distinguish them for analytical purposes, especially because it is crucial theoretically to distinguish *arguments* – entities, composed either of sentences or propositions, or of speech acts – from the social, dialogical phenomenon of *argumentation*. E and F are disagreements involving the nature of argument, and as such rightly find a home in the conceptual space of argumentation theory. D and E contain arguments in the first, abstract propositional sense of 'argument', while F does not.

Example G is not itself an argument, but rather a report of the speaker's reaction to and evaluation of one. Example H is a conversation, or dialogical exchange, in which the parties are negotiating a mutually agreeable plan for the evening. It does not seem to be an argument, even in the social/dialogical sense of that term, although exchanges like it are sometimes so regarded and labeled (perhaps because Jack and Jill both

---

3 Goddu 2015 offers a sustained case for treating arguments as abstract objects.

offer reasons for their preferences). Example I is a successful resolution of a difference of opinion concerning an issue of political moment, now sadly outdated by the Russian invasion of Ukraine. It exemplifies all three of the senses of 'argument' mentioned thus far: it contains arguments in the abstract propositional sense; it records specific speech acts; and it exemplifies the social/communicative sense of 'argument' – that is, it is an instance of argumentation – in that Sophia and Jens exchange reasons in order to communicate with one another in an attempt to resolve a difference of opinion concerning an issue of moment to them both.

It goes without saying that nothing much turns on the examples themselves. They are intended only to illustrate the different ways in which 'argument' is used, both by ordinary speakers of English and by scholars studying arguments and argumentation. Such theorists use the word 'argument' to refer to all these things and more, despite the substantial ways in which they differ. The multiple senses of 'argument' in play in the scholarly world of argumentation theory, I submit, has led to both theoretical problems and seemingly intractable disagreements. They also have led to a confusing array of criteria of argument evaluation: logical validity, logical or epistemological cogency, persuasive force or effect, the achievement of consensus, the 'satisfying-ness' of an argumentative exchange, and so on. In what follows I will suggest that an explicit recognition of the ambiguity of 'argument' in the literature, and a delineation of the domains in which the different senses of 'argument' rightly play a role, will help us resolve such disagreements. They will also help us see more clearly the strengths and weaknesses of some familiar theories of argumentation, in particular the Pragma-Dialectical Theory and Virtue Argumentation Theory, both of which are right about some senses of 'argument' but go wrong when they either ignore or extend themselves to other senses of the term, or understand that term in overly broad ways.[4] Equally helpful is an explicit recognition of the multiple, legitimate criteria of argument quality: arguments and argumentative exchanges, I will argue, can be evaluated along several dimensions, all of which are perfectly legitimate, though not of equal priority from the point of view of argumentation theory.

2. The Nature of Argument

Argumentation theorists theorize about all sorts of things, including the nature of argument itself. Here an often cited contribution is Daniel O'Keefe's (1977, p. 121) distinction between *argument$_1$* ("a kind of

---

[4] In a longer version of this paper, forthcoming in *Informal Logic*, I discuss a third important theory – Christopher Tindale's Rhetorical Theory – which here, for reasons of space, I set aside.

utterance or a sort of communicative act" – the *products* of argumentative episodes) and *argument₂* ("a particular kind of interaction" – the *processes* by which those products are produced during such episodes). This is an important distinction, to be sure – but notice that both disjuncts fall in the domain of argument*ation*, the social communicative activity, not argument in the (abstract propositional) sense illustrated by examples A and B above. Charles Willard similarly held that argumentation is best understood in social communicative terms, arguing that an argument is "a kind of interaction in which two or more people maintain what they construe to be incompatible positions." (1983, p. 21) The more fundamental distinction is that between *arguments* as constituted by sentences, or the propositions they express, and the inferential or epistemic relations obtaining among them[5], or the speech acts by which they are expressed, on the one hand, and *argumentation*, the interpersonal activity in which reasons for beliefs, opinions, and policy proposals are considered, discussed, and exchanged and, in the good case[6], defended and criticized by way of arguments (in the abstract propositional sense), on the other. Both disjuncts of O'Keefe's distinction, as well as Willard's construal, fall on the argumentation side of the ledger.

But 'argument' is itself multiply ambiguous, as we have seen. Most fundamentally, *arguments are what arguers traffic in when arguing*. If I'm (sincerely) arguing with you concerning a candidate belief, proposition or viewpoint about which we disagree, I'm giving you *reasons* which I believe, hope and intend will make the case for my preferred attitude (belief, acceptance, rejection, doubt, hope, etc.) toward that proposition or viewpoint, and/or make the case against your preferred attitude. If you're arguing with me, you're doing the same thing with respect to your own preferred attitude and/or mine. If we're not giving each other such potentially epistemically forceful reasons, we're not *arguing* – though we may of course be cajoling, bullying, persuading, inquiring, joking, quarreling, or many other things. Arguments engage this epistemic task

---

5 As J. Anthony Blair put it, "an argument is a proposition and a reason for it" (2004, 137). More expansively, he writes: "I propose that we conceive a set of one or more propositions to be an argument (understanding 'proposition' in a broad sense) just when all but one of them constitute a reason for the remaining one. And a set of propositions are a reason for an belief, attitude or decision, just when the former support the latter to some degree. What constitutes support is an epistemological question, understanding epistemology in a broad way, so as to be the theory of the justification of attitudes and various kinds of normative propositions as well as of beliefs." (141-2, emphases in original) It would be hard to find a clearer statement of the epistemic view defended here, though I recommend keeping open the possibility that a reason can fail to provide support but still be a reason, albeit a bad one that offers no support.
6 A 'bad' case is one that John Biro and I have termed 'quarrels', i.e., disagreements in which no reasons or arguments are advanced.

of advancing/challenging the cases made for and against the standpoints at issue and discussed in argumentative dialogues or dialectical exchanges, and, as John Biro and I have argued for decades, their ability to accomplish this task is the mark of an argument's epistemic quality: an argument is good, epistemically, to the extent its premises/reasons warrant belief in its conclusion/standpoint, that is, to the extent it *renders belief rational*.[7] But as already noted, 'argument' is also used to refer to the actions performed to make such a case, or to effect the exchange of reasons that occurs during our argumentative interaction. A useful way to understand this ambiguity, we have suggested, is in terms of *abstract structures* vs. *sequences of events*: 'argument' is used to refer both to the abstract propositional structure advanced or challenged by arguers in the course of their argumentative activities, and to the acts of arguing, usually speech acts, in which arguers engage when arguing. (Biro and Siegel 2006a, p. 92)[8]

It is perhaps regrettable that 'argument' can be and is often used to refer to both of these, and more besides. Catarina Dutihl Novaes' excellent *Stanford Encyclopedia of Philosophy* entry on 'Argument and Argumentation' begins:

> An argument can be defined as a complex symbolic structure where some parts, known as the premises, offer support to another part, the conclusion. Alternatively, an argument can be viewed as a complex speech act consisting of one or more acts of premising (which assert propositions in favor of the conclusion), an act of concluding, and a stated or implicit marker ("hence", "therefore") that indicates that the conclusion follows from the premises...
>
> Argumentation can be defined as the communicative activity of producing and exchanging reasons in order to support claims or defend/challenge positions, especially in situations of doubt or disagreement. (Dutilh Novaes 2021a, pp. 2-3)

thus using 'argument' to refer to (1) the abstract structures exemplified by examples A-C above, and (2) the speech act of arguing exemplified by someone articulating, in speech or some other way, such an abstract structure (e.g., Aristotle uttering A, or Anselm writing B). *Argumentation* in turn picks out the communicative, dialectical/dialogical social practices and activities exemplified by examples D-F and H-I, as well as A-C if

---

7 For an early statement, see Biro and Siegel 1992, p. 92. The view is developed further in Biro and Siegel 2006a, 2006b, and 2015, and Siegel and Biro 1997, 2008, 2010, and 2021.

8 We also distinguish arguments from the uses to which they are put by arguers. Cf. Biro and Siegel 2015, 30-32.

articulated during the course of such activities. D and E include arguments (in the abstract propositional sense), and also satisfy Dutilh Novaes' 'speech act' characterization of the second sort of argument she delineates, while F does not contain any argument of either sort, and so is a disagreement or quarrel, but not an argument in either sense.

While the fact that 'argument' is used to refer to the two quite different things Dutilh Novaes delineates – roughly, abstract structures and complex speech acts articulating and utilizing such structures – may be regrettable, since the risk of equivocation is high, this is how the word is used in English, and theorists of argument and/or argumentation should be on guard to avoid such equivocation. The situation is complicated by the fact that the second, 'complex speech act' sense of 'argument' is typically manifested in the social activity of *arguing*, which, as already noted, itself is what argumentation theorists usually call *argumentation* – the latter is simply the social, communicative activity that utilizes such speech acts in arguing.[9] More troublesome still is the fact that extended episodes of argumentative interaction are also referred to as arguments, as in "Over the past two weeks my students have had a rip-roaring argument about the mind-body problem." 'Argument' is thus (at least) quadruply ambiguous: we use the term in the abstract propositional sense, the speech act sense, the social communicative sense, and the extended argumentative episode sense. There is little to distinguish the second, 'speech act' sense of 'argument' from argumentation, other than that the former can be manifested in the absence of a dialogical partner: they both pick out complex speech acts involving arguments (in the abstract propositional sense), and those social, communicative practices might not involve arguments in that sense at all, as in example F, but rather quarrels, fights, or disagreements, none of which need involve arguments in that sense. There are thus four senses of 'argument' that are best distinguished: (a) arguments in the abstract propositional sense, (b) arguments in the complex speech act sense, which may or may not constitute instances of argumentation, (c) arguments as communicative activities involving arguments in either or both of the first two senses, which activities themselves constitute instances of argumentation, and (d) extended episodes of argumentative interaction. These should all be distinguished from disagreements or quarrels, which are communicative activities not involving arguments at all.

I argue next that two important argumentation theories – the Pragma-Dialectical Theory of the Amsterdam School (henceforth PD) and Virtue

---

9 For some reservations concerning the sense of 'argumentation' just mentioned, see Biro and Siegel 2006b, p. 1, note 1, and Siegel and Biro 2021, p. 184, note 4, which also discusses what in our view is Douglas Walton's unfortunate treatment of the terms 'argument' and 'argumentation'.

Argumentation Theory (henceforth VAT) – either fail to honor these distinctions, or insightfully treat one sense of the term but illicitly extend their analyses to other senses of it, and as a result err in important but remediable ways. I will also suggest that the usual measures of argument quality – validity, cogency, dispute resolution, the achievement of consensus, the psychological reaction of arguers to argumentative episodes in which they are engaged, etc. – are all reasonable and legitimate ways to evaluate arguments in one or more of that term's senses, but that they are not of equal priority.

3. Pragma-Dialectics and The Dialectical Conception of Arguing

The argument/argumentation distinction looms large in the argumentation literature, and some theories address both; when they do, one of them sometimes dominates and takes precedence in the expressed theory, and that theory's treatment of the other suffers. The Amsterdam School is a case in point. As John Biro and I have argued *ad nauseam*, PD takes arguments to be fundamentally dialogical or dialectical exchanges, and although it incorporates epistemic-evaluative terms like 'validity', 'rational', 'fallacy', and the like, it reconceives these terms so that they apply to dialectical 'moves' that do/do not conform to the theory's rules for conducting critical discussions, rather than to arguments in the abstract propositional sense. As a result, the theory declares hoped-for argumentative outcomes – the resolution of a difference of opinion and the achieving of consensus – to be rational, even though a given outcome might nevertheless enjoy no epistemic support from its allegedly justifying reasons, and so is not rational in that primary and standard sense of the term.

In his response to our (and others') criticisms, Frans van Eemeren underlines PD's insistence that arguments are to be understood as interpersonal dialectical exchanges, thus failing to address the problem posed: arguments (in the sense of dialectical exchanges) that result in resolutions the theory deems rational are not. Van Eemeren is clear that for PD, argument quality involves 'normatively ideal argumentative discourse' (2012, p. 440[10]) rather than the usual marker of epistemic quality, namely, the ability of the premises/reasons offered to increase the justificatory status of a standpoint or the achieved resolution, outcome or conclusion of the discussion. In his most explicit comments on criticisms of PD launched from the epistemic perspective, he writes:

---

10 Unless otherwise indicated, all references in this section are to this paper.

Basically, the criticisms of the epistemic dimension of pragma-dialectics boil down to the accusation that following the pragma-dialectical discussion procedure correctly may in some cases lead to the acceptance of standpoints that are not epistemically tenable – which generally means that they are not to be considered true. Leaving aside that it is sometimes hard to tell with certainty that a standpoint which is accepted is untrue, this accusation misses the point. As argumentation theorists, pragma-dialecticians are out for the best method for resolving differences of opinion on the merits and determining whether the standpoints at issue are acceptable on reasonable grounds. This means that they want to develop adequate ('problem-valid') testing procedures for checking the quality of the premises used in argumentative discourse and the way in which they are used in defending standpoints. (451-2, notes and references deleted)

I will address van Eemeren's subsequent text below. First, let me note some initial worries.

3.1. Truth and Certainty

Although it is correct that standpoints that are not epistemically tenable are not to be considered true, 'epistemic tenability' is not equivalent to 'true', and the focus on truth here is misleading. Such tenability is first and foremost a mark of *justificatory status*: a belief, resolution or standpoint is epistemically tenable to the extent that it enjoys some substantial measure of evidential or reasoned support. Such support is typically truth-indicative, in that if I have good reason for thinking that p, I have good reason for thinking that p is true. That is, to believe that p just is to believe that p is true. This is what it is to believe something, and this is one way that truth enters into the epistemic theory.[11] A second way that truth enters the picture on the epistemic view is that good arguments, according to that view, afford epistemic *improvement* and so the opportunity for gains in knowledge or justified belief – and since truth is a condition of knowledge (insofar as one cannot know something that is false), a good argument may afford its recipient gains in knowledge, which entails gains in true beliefs. Of course it is sometimes hard to tell whether or not a candidate belief or standpoint is true, as van Eemeren rightly notes. Equally, sometimes it is easy. (For example, it is clearly and unproblematically true (skeptical worries aside, which are not entertained

---

11 For further discussion of the relations obtaining among knowledge, belief, justification and truth, see Siegel 1998.

in van Eemeren's discussion) that it's now sunny outside my study window, that I'm now discussing van Eemeren 2012, that van Eemeren prefers PD to the epistemic theory, and that his middle initial is 'H'.) The epistemic theory takes good arguments to be vehicles for epistemic improvement, and such improvements sometimes involve gains in true beliefs. Epistemic improvement includes as well – indeed primarily – gains in justificatory status, which is what a good argument delivers to its conclusion/standpoint. For this reason van Eemeren's focus on truth is misplaced. It is the *justificatory support* offered to candidate beliefs/standpoints/conclusions by premises, reasons or evidence that renders such standpoints *worthy* of belief that is the chief preoccupation of the epistemic theory. And while strong support or high justificatory status is, as is often said, a 'fallible indicator' of truth, it is indeed fallible, as even strongly justified beliefs and standpoints can nevertheless be false, just as unjustified beliefs can nevertheless be true. The epistemic theory endorses fallibilism as strongly as van Eemeren does. His mention of *certainty* ('it is sometimes hard to tell with certainty') is disappointingly straw-mannish, as neither truth nor evidential/reasoned support require it, and it is no part of the epistemic theory, which rejects it as firmly as PD does. (The epistemic theory is characterized further below.)

3.2. 'On the Merits'

van Eemeren writes that "pragma-dialecticians are out for the best method for resolving differences of opinion on the merits". (452) What are these 'merits'? Two pages earlier he explains that one minor change made to PD over the years is "the addition of the qualification 'on the merits' to 'resolving a difference of opinion' (which is exactly what 'resolving in a reasonable way' in pragma-dialectics means)." (450) And what is it to resolve a difference of opinion in a reasonable way? On the PD view, the reasonableness of a particular discussion rule – the rule that licenses a particular argumentative move – is a function of the rule's conducing to the resolution of the relevant difference of opinion in ways that are acceptable to the discussants. Thus 'the merits' in 'resolving a difference of opinion on the merits' involve the efficacy of dispute resolution (problem validity) in accordance with rules governing procedures the parties accept (conventional validity). They are not *epistemic* merits, such that the strong support offered to them by their premises, reasons or evidence renders their standpoints, opinions or conclusions better justified, epistemically speaking. In this respect the addition of 'on the merits' to PD's account of the resolution of differences of opinion does little to defend PD from the criticisms that Biro and I, along with other defenders of the epistemic view, have leveled against its account of reasonableness. For PD, 'reasonable' refers to dialectical rules and the moves they sanction, and 'justified' refers to resolutions reached by such moves in accordance with such problem- and conventional-valid rules. Justification in the sense of evidential or reasoned support for the standpoint at issue plays no role – for PD, if a resolution is achieved in accordance with such rules, the standpoint is 'justified'. This is manifestly not what 'justified' means, epistemically speaking.[12]

3.3. 'Justificationism' and Positive Support

van Eemeren protests that Biro and I misunderstand PD's rejection, following Popper and 'critical rationalism', of 'positive justification', pointing out that PD has always allowed for 'pro' as well as 'contra' argumentation. (451)[13] He insists that PD's rejection of justification amounts only to the rejection of the possibility that standpoints can be legitimized "definitively". (*ibid.*) We agree, if 'definitively' means that a dispute's resolution is *certain* or that the dispute can never be reopened – since we all endorse fallibilism, we agree that a dispute can always be reopened or a seemingly justified standpoint challenged anew, on the basis of new evidence or a new evaluation of previously considered evidence or arguments.[14] As we argued (in Siegel and Biro 2008), this is not what 'justificationism' means in the critical rationalist literature. This is crucial, since van Eemeren adds in a footnote here that "Notions such as 'pro argumentation' and 'justificatory force' are in pragma-dialectics understood in a dialectical fashion and acquire a non-justificationist meaning." (*ibid.*, note 27) Here van Eemeren seems to want to have it both ways, both allowing for the positive support that pro argumentation might provide, but embracing critical rationalism, which rejects such positive support. If 'pro argumentation' offers positive support for a standpoint, or positive reason to embrace a belief, conclusion, or resolution, then he is rejecting critical rationalism, which endorses no such thing. (cf. Siegel and Biro 2008, 195-199) If by 'non-justificationist meaning' he means simply that pro argumentation can provide support, although non-'definitive' support, then the new meaning proposed is just the same old meaning, while emphasizing the fallible nature of human judgment. I hope that van Eemeren really does endorse the idea that arguments can provide positive

---

12 Eugen Octav Popa (2016, 196 ff.) makes a related objection in terms of circularity. Thanks to José Ángel Gascón for suggesting this reference.
13 It is perhaps worth noting that we anticipate van Eemeren's complaint that we misunderstand PD's rejection of justificationism, and respond to the complaint, in Siegel and Biro 2008, pp. 201-202.
14 It is disappointing that van Eemeren has not seriously engaged our discussion of Popper, justificationism, and critical rationalism. He repeats his earlier usage of 'definitively', but fails to address the ambiguity we have pointed out. (Siegel and Biro 2008, p. 195; see also Siegel and Biro 2010, pp. 461-2)

support, even if they cannot provide 'definitive' support. If this is actually the PD view, it is difficult to see how it differs from the epistemic view, which says the same thing. But that depends on the 'non-justificationist meaning' PD gives to the notion of positive support, which is problematic in the ways pointed out above and in the just-mentioned papers.[15]

Van Eemeren continues by offering "three crucial points [that] need to be born in mind, however, when considering how they deal with argumentative reality":

> First, *in the pragma-dialectical view, argumentation theory is neither a theory of proof nor a general theory of reasoning or argument, but a theory of using argument to convince others by a reasonable discussion of the acceptability of the standpoints at issue...* (452, emphasis added)

Leaving aside the provocative phrase 'argumentative reality' – suggesting as it does that arguments that aren't dialectical aren't real[16] – the italicized text is remarkable in a couple of ways. For one thing, the sense of 'argument' in play is ambiguous in that it suggests all of the first three senses of the term delineated above. For another, it suggests that, in the PD view, argumentation theory is not, even in part, a theory of argument. This will surprise many theorists, including some of the biggest names in the history of the field, who took themselves to be offering just such a theory. Perhaps more surprising still, it suggests that, in the pragma-dialectical view, anyone who isn't a PD theorist isn't engaged in argumentation theory at all! For who besides PD theorists theorize about 'using argument to convince others by a reasonable discussion of the acceptability of the standpoint at issue', with the particular meanings those terms have been given in PD? Are there really no other topics or issues about which argumentation theorists, operating in that capacity, might address? Are there no other approaches to the field that argumentation theorists might legitimately take? The passage suggests that the pragma-dialectical view restricts the entire domain of argumentation theory to its own concerns.

But the biggest problem lurking for PD here is its implicit conceding of the main criticism of it launched by the epistemic theory: that arguments

---

15 I leave aside the obvious sense in which the redefinition of epistemic terminology in PD terms invites charges of equivocation and straw man.

16 I am here speculating on what van Eemeren means by 'argumentative reality', since he doesn't clarify the term in the text. It would be good to know which arguments fail the 'reality test' (so to speak), and what such reality comes to. For example, are examples A and B above arguments occurring in 'argumentative reality', even though they are composed of sentences on a whiteboard and so are not part of any dialectical exchange?

judged to be good by PD lights may nevertheless be bad, epistemically speaking, in that they do not enhance the justificatory status of the belief, conclusion or standpoint at issue. This is because 'reasonable discussion of the acceptability of the standpoints at issue' amounts for PD simply to following the PD rules, which can be followed without any resulting epistemic or justificatory gain. Problem-validity does not help secure such gain in justificatory status, since it involves simply the possibility of resolution of differences of opinion, whether epistemically rational or not; nor does conventional validity, which is simply a measure of the acceptability to the participants of the recommended dialectical procedure, which acceptability again does nothing to secure such gain.

Van Eemeren emphasizes the importance of persuasion and agreement on the PD view, noting that "getting the truth of a standpoint accepted by others who are in doubt" (452) is independent of the justificatory status of the standpoint. He is right about this, and importantly so; that is one reason that argumentation theorists are interested in rhetoric and persuasion, and have been since at least the days of Aristotle. The epistemic view does not deny this; it simply insists on distinguishing between an argument's ability to secure the epistemic status of its conclusion, on the one hand, and its ability to persuade interlocutors of that status, on the other hand.[17] To insist that the latter is the mark of argument quality is to conflate these two independent measures of argument quality, two different ways that arguments can be good (or not). More on this below.

Van Eemeren's second crucial point is that PD, allegedly unlike the epistemic view, is interested in more than truth claims, including

> standpoints involving acceptability claims of a somewhat different nature, such as evaluative standpoints expressing ethical or aesthetic judgments and prescriptive standpoints advocating the performance of a certain action or the choice of a certain policy option.... This means that in the pragma-dialectical view a theory of argumentation needs to have a scope that extends dealing with the truth-related issues which are the primary interest of epistemologists. (*ibid.*)

I have already noted van Eemeren's misleading focus on, and claims regarding, truth. Here the false suggestion is that the epistemic theory cannot speak to the quality of reasons that may be offered in support of ethical or aesthetic judgments, or of particular actions or policies. Its falsity is manifest, as people regularly offer reasons for such judgments, actions and policies, and those reasons are manifestly evaluable in terms of their ability to support their judgments, actions and policies. (E.g., 'It's

---

17 See especially Biro and Siegel 2006b.

wrong to do action *A* because it'll cause unnecessary suffering'; 'Shakespeare is better than Donald Trump in characterizing the qualities and subtleties of basic human emotions'; 'Don't perform action *A* because it'll frustrate your subsequent effort to secure goal *G*; do *B* instead'; 'The government should do what it can to eliminate hunger, rather than concentrate on economic growth at the expense of the poor'.) The epistemic theory's scope is as broad as the many domains in which reasons can be offered, challenged, or evaluated, and in which justificatory considerations can be raised and addressed.[18]

Van Eemeren's third crucial point centers on his denial of the epistemic theory's main criticism of PD, namely, that problem-validity and conventional validity do not ensure or amount to justificatory status. In response to Biro's and my complaint that "discussants may share, and rely on, unjustified beliefs, and they may accept, and use, problematic rules of inference and reasoning" – which van Eemeren claims is a caricature of PD – he writes that "if problem-validity is properly understood, this is not possible – at least if there is a better – i.e., more problem-valid – alternative available." (453, quoting Siegel and Biro 2010, p. 458) Why is it not possible? Because "starting points and rules are reasonable only if they have been subjected to and passed critical tests." (*ibid.*, quoting Botting 2010, p. 423)[19] And what are those critical tests? They are those that establish the problem- and conventional-validity of dialectical moves sanctioned by those very starting points and rules. Despite van Eemeren's complaint here, it is manifestly possible for discussants to share and rely on unjustified beliefs and use problematic rules of inference and reasoning – which he surprisingly but quietly concedes.[20]

---

18 In a footnote at this point, van Eemeren suggests that "the 'justified beliefs' involved in dealing with evaluative and prescriptive issues can as a rule be better treated in terms of intersubjective acceptability than in terms of objective truth." (2012, p. 452, note 33) He is right about this, if the focus of argumentation theory is restricted to the overcoming of disagreement and the achieving of consensus, although such intersubjective acceptability, as we have urged, is not in itself a mark of epistemic quality. For consideration of the breadth of the domains in which reasons can and often do play a justificatory role, see my 1988, 1997, and 2017, and the many references to Israel Scheffler's work, in Scheffler 1989 and elsewhere, contained therein.

19 I resist the urge to reply to Botting's paper here. Instead I strongly recommend Christoph Lumer's powerful, indeed devastating, reply, Lumer 2012.

20 Van Eemeren grudgingly concedes this: "A consequence [of the requirement of intersubjective agreement] may be that… 'good' arguments and standpoints are eventually rejected and 'bad' arguments and standpoints accepted." (2012, p. 452, note 31) He continues that "This happens only on reasonable grounds however if the arguers have complied with all the required testing procedures. A 'better' result can only be achieved if the problem-validity of the testing methods for establishing truth etc. are first improved and the tests are made acceptable to would-be discussants", thus again making problem- and conventional-validity the ultimate arbiters of epistemic quality which, as we have seen, they are not and cannot be.

3.4. Is Dialectic All There Is to Argumentation?

Perhaps the fundamental point that divides PD and the epistemic theory is the centrality to argumentation theory of *dialectic*. The basic counter-objection van Eemeren poses to the epistemic theory is that it is insufficiently communicative, discursive, dialogical or dialectical: it insists on considering arguments in non-dialectical terms, and it doesn't embrace PD's redefinition of epistemic terms into dialectical ones. As we have seen, PD, on the other hand, embraces those redefinitions and, most importantly here, insists upon a dialectical understanding of arguments: "As argumentation theorists, pragma-dialecticians are out for the best method for resolving differences of opinion on the merits and determining whether the standpoints at issue are acceptable on reasonable grounds" (452); argumentation is "not to be studied as a structure of logical derivations, psychological attitudes or epistemic beliefs, but as a complex of linguistic (and sometimes also non-linguistic) acts with a specific communicative function in a discursive context."[21] While epistemic theorists recognize the value of linguistic and dialectical analysis, they insist that that cannot be the whole of argumentation theory, precisely because within such dialectical exchanges lurk the substance of actual arguments (in the abstract propositional sense), and their justificatory force is a central mark of argument quality.

In effect, the dispute between PD and the epistemic theory comes down to this: PD, wanting argumentation theory to facilitate high quality 'real world' argumentation – that is, communicative, discursive efforts to resolve differences of opinion in 'argumentative reality', in accordance with PD rules that honor and ensure problem- and conventional validity – insists on a dialectical approach, and further on understanding argument quality in dialectical terms, such that the quality of an argumentative exchange consists entirely in its specifically dialectical quality. What is needed, on the PD view, is an account of argument quality "that does justice to dialectical considerations." (van Eemeren and Grootendorst 2004, p. 50) The epistemic theory, by contrast, is interested in the

---

21 Van Eemeren and Houtlosser 2003, p. 388. For another explicit statement of the centrality for PD of dialectical considerations, including where participants might best start a critical discussion, the importance of "the particularities" of actual discussions, and the centrality of "dealing with argumentative discourse", see van Eemeren 2012, p. 453, note 36. For yet further evidence of PD's understanding of reasonableness and argument normativity in strictly dialectical terms, see Biro and Siegel 2006b, pp. 5-10, and the several passages from van Eemeren and Grootendorst 2004 quoted and discussed therein, as well as the italicized quoted passage from p. 452 above.

epistemic, justificatory-force-enhancing quality of the actual arguments embedded in dialectical exchanges, as well as arguments occurring in non-dialectical contexts. PD aims to guide such exchanges; the epistemic theory aims to determine the specifically epistemic improvements those exchanges might bring. Biro and I have argued that both are important dimensions of argumentation theory, and that the two theories should be seen as partners rather than rivals. (Biro and Siegel 2006b) [22] Van Eemeren, in insisting on a wholly dialectical, discursive, communicative approach that concentrates on 'argumentative reality', rejects the very essence of argumentation, as we see it, by denying that argument quality is a function of an argument's ability to enhance the justificatory status of its conclusion. Van Eemeren denies that PD denies this, but this is because he understands the key epistemic phrase 'justificatory status' in non-epistemic terms.[23]

The justificatory force of premises/reasons in securing the epistemic propriety of conclusions/standpoints cannot be captured by dialogical/dialectical rules, however 'reasonable' or helpful in the resolution of differences of opinion they might be. Epistemic quality is simply not a function of dialogical or dialectical rules – p's (propositional) justificatory status, and the (doxastic) justificatory status of a subject S's belief that p, are not functions of the dialectical features of an exchange, or of S persuading her interlocutor or being persuaded by her that p in accordance with dialectical rules, or of S and her interlocutor achieving a consensus concerning p in accordance with such rules. Rather, it is strictly a matter of the objective support p enjoys from the reasons and evidence that provide (or not) such support. Dialectical rules simply cannot determine epistemic propriety, tenability, or justificatory status.[24] For this we must look not to rules such as those put forward by PD, but rather to

---

[22] Gascón 2017 similarly and insightfully argues for the compatibility of PD and VAT. The view here defended – that the several theories of argumentation discussed are compatible in particular ways, and have their strengths with respect to particular senses of 'argument' – suggests the prospect of a broad compatibility among apparently conflicting theories. This suggestion leads naturally to a call for theory integration, such as that made by Tony Blair in 'A Time for Argument Theory Integration', reprinted in his 2012, ch. 15. Thanks to Chris Tindale for suggesting the connection between Blair's call for integration and the present effort, which recommends acknowledging the strengths of the several theories with respect to one or another of the senses of 'argument' delineated above, while cautioning against their extension to other senses of the term, where they fare less well.

[23] As do Garssen and van Laar 2010. For detailed discussion see Siegel and Biro 2010.

[24] For further arguments for this conclusion, with special reference to Habermas, see Siegel 2018. As Blair succinctly puts it, "Dialectic thus presupposes reason-giving as a tool or move, and reason-giving presupposes the possibility of reasons supporting propositions, namely arguments." (2004, 142)

the usual criteria of epistemic quality – strength of evidence; degree of support offered to the conclusion/belief in question by that evidence; probability of the conclusion, given the evidence; strength and security of the inferential link between premises/reasons/evidence and conclusion; consideration of the total evidence, including counter-evidence; fair evaluation of the evidence; etc. – that have long been the business of epistemologists to theorize, codify and explore. In insisting that arguments be evaluated in strictly dialectical terms, PD comes dangerously close to throwing the baby out with the bathwater. We counsel partnership rather than rivalry, which preserves the baby – arguments evaluated epistemically – along with PD's dialectical evaluation. Dialogue, discourse, dialectic and persuasion are important loci of argumentation theory's concerns, to be sure. But they are not, and cannot be, the whole story.

Let me conclude this section by declaring once again that in my view resolving differences of opinion in reasonable ways is a good thing, and that providing a theory of it is a laudable one. PD has made an important contribution, one that has set the agenda for much of the argumentation theory community for decades. Biro's and my criticism is simply that it does not capture the epistemic normativity of arguments (in the abstract propositional sense), and so of argumentation, which does its business by way of such arguments.[25]

4. Virtuous Argumentation and Virtue as a Criterion of Argument Quality

Virtue Argumentation Theory (henceforth VAT) holds that the *character of the arguer* is a key determinant of argument quality. More specifically, VAT offers an *agent*-based assessment of argument quality, such that an argument (in the social communicative sense) is good insofar as its participating arguers manifest *argumentative virtues* – e.g., open-mindedness, intellectual humility, willingness to listen and to take unfamiliar positions seriously, etc. – in their arguing. Such agent-based assessment, which focuses on "arguers, rather than (just) arguments" (Aberdein and Cohen 2016, 340)[26], insists upon "the importance of agents

---

[25] I want to thank van Eemeren for his helpfulness over the years, both personally and professionally, and to acknowledge my debt to him, despite our scholarly differences.

[26] Gilbert 1995 offers a clear and in many respects compelling defense of shifting the focus of evaluation from argument to arguer in critical reasoning courses. Gilbert's view also aligns strongly with Christopher Tindale's, discussed in the longer version of this paper, which Tindale 2021 explicitly acknowledges.

to the normative evaluation of arguments." (*ibid.*, 339) It is offered as an alternative to argument evaluation that centers on the properties of arguments conceived independently of the arguers who engage in them.[27] Inspired by philosophical enthusiasm for and recent advances in virtue theories in ethics and epistemology, VAT takes its place alongside those more familiar virtue theories, extending the reach of virtue theories into the domain of argumentation. (Aberdein 2010, 169-170)

Of special relevance here is VAT's conception of arguments: "Arguments are dynamic, multi-agent events". (Aberdein and Cohen 2016, 339) As such, VAT, like PD (as we have seen) and Christopher Tindale's rhetorical theory (AA, which is discussed in the longer version of this paper), focuses on arguments in the social, communicative sense; it downplays or ignores arguments in the abstract propositional sense, and, again like PD and AA, emphasizes "the dialectical nature of argumentation" (Aberdein 2010: 165), holding that "Argument, unlike knowledge, is intrinsically dialectical" (*ibid.*, 175). We have already seen some difficulties with this claim. First, it is argument*ation*, rather than argument, that might be intrinsically dialectical; otherwise, the argument concerning Socrates' mortality, used as an example of a valid argument in logic instruction for centuries, doesn't count as an argument. More importantly, at least philosophically, it rules out famous arguments – Anselm's Ontological Argument, Aquinas' Cosmological Argument, Moore's Open Question Argument, Searle's Chinese Room Argument, and other famous arguments that are routinely rendered on classroom whiteboards and studied as such in philosophy classes for their structure and the degree of support their premises afford their conclusions – *as arguments*: The 'intrinsically dialectical' conception of argument entails that the Ontological Argument, as rendered in example B above, is not an argument! This untoward result, by itself, should raise serious doubts about this view of argument and argument evaluation. But there are other problems with the 'intrinsically dialectical' conception of argument in general, and with VAT in particular, that I explore next.

Virtue theories of argumentation face what seems an overwhelming initial difficulty, that of explaining how the virtues and vices of arguers could have anything to do with the quality of arguments. Can't a vicious arguer produce a good argument? Can't a virtuous arguer produce a bad one? There seems to be a fundamental distinction to be drawn between the character traits of arguers and the quality of the arguments they produce. It is difficult to see how a virtue theory of argumentation might shed light on the normative evaluation of arguments. It seems to be committed to something like the view that an argument is good – epistemically good,

---

[27] This is an overgeneralization, as one prominent advocate, Andrew Aberdein, defends VAT as relevant to the assessment of argument quality in the abstract propositional sense as well. His view is taken up below.

such that its premises/reasons provide support for its conclusion – *because* it has been argued for virtuously. This seems at best a *non sequitur*, since the quality of an argument (in the abstract propositional sense) hinges entirely on the support for the conclusion offered by its premises. It seems also to conflate *arguments* – in the primary sense of the term (as argued above and below), abstract objects whose premises support (or not) their conclusions – and *argumentation* – the social, communicative activity of giving, analyzing, criticizing and evaluating arguments, which can be evaluated in terms of their epistemic strength, their rhetorical or persuasive force or effect, their ability to bring about consensus, their aesthetic properties, or along yet other dimensions.[28] My diagnosis is that some advocates of VAT, such as Daniel Cohen, do not mean to be, or take themselves to be, offering a theory concerning the evaluation of arguments in the abstract propositional sense at all. Rather, theirs is an account of argument quality concerning arguments in the social, communicative, dialogical/dialectical sense. Other advocates, in particular Aberdein, do take themselves to be offering a theory concerning the evaluation of arguments in the structural, abstract propositional sense as well in the social communicative sense. Aberdein's case is taken up next. In any case, distinguishing these senses of 'argument' enables us to see that VAT's contribution to argument evaluation (in the social communicative sense) is important and largely correct, but that it goes astray when applied to argument evaluation in the abstract propositional or speech act senses.

Aberdein (2014, 78) notes correctly that VAT seems more amenable to "rhetorical and consensus approaches" to argument evaluation, but defends its appropriateness to "the epistemological approach" as well. His discussion in this paper centers on the critique of VAT offered by Tracy Bowell and Justine Kingsbury (2013), in which they suggest that VAT, in committing itself to an agent-based approach to argument evaluation, runs the risk of committing the *ad hominem* fallacy, since it urges that argument quality depends upon features of the arguer rather than the argument, whereas textbook discussions of the fallacy typically hold that all such evaluations are fallacious. Aberdein's discussion of the *ad hominem*, in particular his distinctions among various forms of it, is sophisticated and telling. He is right that some *ad hominem* arguments

---

28 Which of these should 'wear the pants', as John Austin memorably if politically incorrectly put it, is itself a hotly contested issue among argumentation theorists. I trust it is clear that I am here plumping for the legitimacy of them all but the primacy of the first. More on this below.

are in fact epistemically strong[29], and that VAT can escape the charge.[30] That said, Bowell and Kingsbury's discussion is clear and compelling in several respects. In particular, they urge, in agreement with Aberdein and Cohen,[31] that whether or not VAT offers a plausible account of argument goodness, "there is much to be gained by identifying the virtues of the good arguer and those of the good evaluator of arguments, and by considering the ways in which these virtues can be developed in ourselves and in others" (23), thus acknowledging the benefits of VAT's focus on arguers. Even critics of VAT as a measure of argument quality endorse the value of developing accounts of argumentative virtue, both pedagogically and as an independent dimension on which to assess such quality.[32]

Bowell and Kingsbury characterize good arguments thus: "A good argument is an argument that provides, via its premises, sufficient justification for believing its conclusion to be true or highly probable, or for accepting that the course of action it advises is one that certainly or highly probably should be taken." (2013, 23) Aberdein suggests that 'sufficient justification' in their characterization can be understood "in terms of the virtues of the arguer (and, perhaps, those of the respondent)" (2014, 78), thus rendering it compatible with VAT. Is this right? Can 'sufficient justification' of the sort that is rendered to conclusions by suitable premises be afforded by arguers' virtues?[33] It is hard to see how, since a virtuous arguer can offer reasons or premises that, although arrived at virtuously, fail to afford any such justification – though open-minded, intellectually humble, willing and able to take unusual positions seriously and to modify her own position, etc., she might nevertheless reason badly, such that her reasons/premises afford little or no justification to her conclusion. Aberdein responds by arguing that the 'traditional intuition' that "arguments should be evaluated on their own merits, and not on the basis of who puts them forward" (2018b, 127, quoting Gascón 2018, 163) is in fact compatible with an agent-based approach to argument evaluation. That is, he urges that (1) '*arguments*

---

29 Cf. Biro and Siegel 1992, 88-9; Siegel and Biro 1997, 285-9; Aberdein 2014, 82-3.
30 It is worth noting that Bowell and Kingsbury agree that in some cases information about the arguer can be relevant to the evaluation of an argument – "there are legitimate ad hominem arguments". (25; cf. 31) There is in fact less disagreement here than meets the eye.
31 Aberdein 2010, 171-2; Aberdein and Cohen 2016, 342.
32 Cf., for example, José Ángel Gascón's discussion of "the fostering of argumentative virtues in education", 2016, 448.
33 Aberdein is clear that he wants his account to deal with cogency, a mark of argument quality that has traditionally centered on arguments in the abstract propositional sense. That is, he wants cogency to be itself determined by the virtues of participating arguers. Cf. Aberdein 2010, 171; 2014; 2018b, 124; and especially 2018a.

should be evaluated on their own merits, and not on the basis of who puts them forward' is actually compatible with (2) *'the quality of an argument in the abstract propositional sense is a function of the virtues of its arguer'*. This seems suspiciously like a contradiction, as it seems to entail that argument quality both is and is not independent of who puts the argument forward. Aberdein suggests that the contradiction is only apparent, and that an arguer's argument will afford sufficient justification if it is made *while arguing virtuously*:

> A crucial qualification is that a virtuous arguer can put forward a bad argument, but not *qua* virtuous arguer, not when they are arguing virtuously. Likewise, a vicious arguer can put forward a good argument, but only by arguing as a virtuous arguer would argue. The foundation of a virtuistic analysis of argument quality will not be whether the arguers are *actually* virtuous – perhaps an impossible question to answer – but whether they are arguing as virtuous arguers would argue. What this standard actually comprises may not be so very different from what more conventional accounts of good argument propose (at least, it won't be any laxer). Good arguments1 should still have true premises and conclusions that follow from them with certainty or high likelihood; good arguments2 should still be chiefly composed of good arguments1. But this will be because that is how a virtuous arguer is overwhelmingly likely to argue.... So we are not presented with two evaluative strategies—evaluate arguments on their own merits; evaluate arguments on the basis of who puts them forward – nor am I proposing that we should abandon the former and embrace the latter. Rather, when properly understood, these are two differently incomplete descriptions of the same strategy: evaluate arguments on their own merits as manifest in the actions of the arguers who put them forward (and are otherwise engaged in them). (Aberdein 2018b, 127-8, emphases in original)[34]

Whether or not the contradiction is genuine or spurious can perhaps be set aside here. For Aberdein's proposal, even if correct, does not help VAT overcome its difficulty concerning the 'traditional intuition'. It may be true that good arguments share two features: they are good because of their epistemic merits – for example, because they "have true premises and conclusions that follow from them with certainty or high likelihood" – and that when they are advanced in argumentative

---

[34] 'Argument1' and 'argument2' here refer to O'Keefe's (1977) distinction, briefly discussed above. Aberdein (forthcoming, 5-6) repeats the argument just quoted, but does not offer any additional consideration that would deflect or challenge the criticism offered in the text immediately following this note.

exchanges those merits are "manifest in the actions of the arguers who put them forward (and are otherwise engaged in them)." But their being so manifest is not what makes them good; rather, their goodness is strictly a reflection of their epistemic merits. That the merits are manifest in virtuous exchanges is only derivatively (if at all) a mark of an argument's quality. The manifestation is in effect an epiphenomenon of the epistemic features of the argument: it is those features – not the fact that they are reflected in virtuous argumentative exchanges – that make the argument good. Those features are independent of who (if anyone) puts them forward, and of how they are put forward, and remain so even if the argument is never put forward, virtuously or otherwise.[35]

In short: Aberdein is clearly correct that arguers' status as virtuous arguers is a function of the argumentative virtues being manifest in their argumentative efforts. But that is because those efforts comport with epistemic criteria of argument quality. If they did not so comport, they would not be argumentative virtues. When it comes to argument evaluation as determined by the degree of support offered to a conclusion by its reasons/premises, it is that support (or its absence) that determines an argument's quality. The virtues manifested might reflect that independently established quality, but do not determine it.

Because argument quality can be measured along several distinct axes, as I've been arguing and argue further below, Fabio Paglieri's (2015) delightful and insightful critique of Bowell and Kingsbury's cogency-based account of that quality does not quite succeed: he is right that cogency is not the only measure of such quality, but not that it should be simply ignored by virtue argumentation theorists. It is one legitimate measure, and, if the epistemic theory is correct, the central one, though the others just mentioned are also legitimate. Paglieri is clear that his critique depends on "people's intuitions" of such quality (69; cf. 70), and in particular the "Cohen reaction" – "*Really? That's your example of a good argument?!*" – to "uninformative, trivial, pedantic" examples of valid arguments, like that concerning Socrates' mortality. (69, citing Cohen 2013, 479) But people's intuitions vary widely here – mine, for instance, are quite different than Cohen's – and that variance supports the 'multiple measures of quality' position defended here, rather than the 'cogency is irrelevant to argument quality' view he attributes to 'radical' (74) virtue theorists. As José Ángel Gascón suggests, we can and should evaluate arguments in terms of both cogency and argumentative virtues, and VAT shines on the latter measure. (2016, 445-6)

---

35 Goddu 2016 offers a compelling version of a closely related objection.

Bowell and Kingsbury argue that

> The fact that a good argument can be put forward by an argumentationally unvirtuous arguer suggests that in those cases in which a good argument is put forward by a virtuous arguer, the goodness of the argument is not constituted by the virtues displayed by the arguer. (30-1)

This is exactly right, and is a variant of a point Biro and I have made repeatedly: people are all too often persuaded by bad arguments, and fail to be persuaded by good ones. Consequently, the quality of an argument cannot be a matter of its persuasive effect.[36] In the same way, vicious arguers can put forward good arguments, and virtuous arguers bad ones. Consequently, the quality of an argument cannot be a matter of the virtues/vices of the arguer. David Godden makes this point forcefully: "The problem [for VAT] is that neither [an arguer's] capacity for virtue nor his exercise of it on some occasion provides any support for his claims." (2016, 354; cf. also Goddu 2016, 442-3) More expansively, he writes:

> The goodness of a reason is a function of whether, and the extent to which, it supports a claim.
> *Thus, support for claims originates in, and is explained by, the way reasons act, not the way reasoners act...* While argumentative virtues might well prescribe the ways that we should go about working with reasons (and hence engage in argumentative practices), virtues neither constitute the reasons themselves nor are they the features on the basis of which the goodness of reasons are determined. (355, emphasis in original)[37]

That is, while VAT informs argumentative practice, and in that way determines arguers' quality *qua* arguers, it does not determine argument quality so long as that is conceived in terms of the strength of epistemic support conferred upon conclusions by their premises or the reasons offered on their behalf. As Gascón puts the point (2016, 2018): "a theory of argumentative virtue should not focus on argument appraisal, as has been assumed, but on those traits that make an individual achieve excellence in argumentative practices. An agent-based approach in argumentation should be developed, not in order to find better grounds for argument appraisal, but to gain insight into argumentative habits and excellence." (2016, 441)

---

36 Most recently, Biro and Siegel 2015, 31.
37 Godden 2016 offers additional powerful reasons for doubting that argument quality can be a function of the virtues/vices of the arguer advancing it.

4.1. Is Virtue All There Is to Argumentation?

As Bowell and Kingsbury, Gascón, and Godden urge, VAT provides important criteria for evaluating argumentative practices, moves and behavior, and important norms governing such practices. It is uncontroversially better to argue virtuously than viciously, and VAT provides important insight into the nature and desirability of argumentative virtue. All this is salutary and to be applauded. But it goes too far when it holds, as Aberdein does, that argumentative virtues and vices can determine the epistemic quality of arguments in the abstract propositional (or premise/reason-conclusion) sense. Here the character of the arguer is irrelevant to the epistemic strength of the argument: a vicious arguer can put forward an excellent argument (in that sense), and a virtuous arguer a terrible one. Argument evaluation in terms of arguers' virtues is one important way in which argumentation – the social, dialectical, communicative phenomenon – can be evaluated. But it is not the only way, and it should not be thought that arguments, in all the senses of that term rehearsed above, can or should be evaluated only in terms of virtue, vice, and character. Arguments can be evaluated along many dimensions; VAT insightfully emphasizes one such dimension.[38]

This completes my survey of the strengths and weaknesses of PD and VAT, and in particular their common defect of treating the social, communicative sense of 'argument' as the only one, or the central one, or the only one worthy of argumentation theorists' attention.[39] I next take up

---

38 I don't think that prominent virtue argumentation theorists, in particular Cohen, actually think that VAT offers the only way to evaluate arguments. Insofar, their claims concerning the 'intrinsic dialecticality' of arguments, and the relevance of virtue theory to all argument evaluation, are probably best seen as enthusiastic overstatements. I hope I'm right about this! For a recent presentation of Cohen's view, see his 2022.

39 Christoph Lumer (personal communication) forcefully reminds me that epistemic theories do not address only the epistemic strengths/weaknesses of arguments construed as abstract objects; they also offer alternative views of dialectical and rhetorical argumentative moves and procedures, and, as such, they constitute rivals to PD and AA. In principle they could do this for VAT as well. My position might therefore be unduly conciliatory to the theories treated thus far: even on their own terms, they are inferior to their epistemic rivals. (For example, Feldman 1994 and Lumer 2005b develop accounts of epistemically rational persuasion, and Lumer 1988 and Goldman 1999 develop accounts of epistemically rational argumentative discussion rules.) I happily concede Lumer's point. My conciliatory attitude is meant only to acknowledge the obvious point that arguments can be studied and evaluated independently of their epistemic strengths/weaknesses, for example in terms of their abilities to foster agreement/consensus or to persuade, independently of their specifically epistemic qualities. As Lumer insists and as argued above and below, insofar as such approaches ignore the epistemic, they do not, strictly speaking, address

issues concerning *criteria of argument evaluation* and *norms governing argumentative interactions*. This will put us in a better position to appreciate the strengths of the epistemic theory.

5. Criteria of Argument Evaluation, Norms of Argumentation, and Priorities among Norms[40]

By what criteria should arguments and episodes of argumentation be evaluated? Which norms rightly govern argumentative moves or episodes? With respect to both these related but distinct questions, we face a plethora of candidates that can be proposed for either arguments or argumentative exchanges, episodes, or moves. Arguments can be evaluated in terms of logical validity, logical soundness, epistemological cogency, and more generally in terms of their ability to increase the *belief-worthiness*[41] of their conclusions. Arguments in the social communicative sense – argumentative episodes, exchanges, and moves – can likewise be evaluated in such epistemic terms, since the epistemic view of argumentation complements the epistemic view of arguments: high quality argumentative practices can promote epistemic improvement, e.g., more justified beliefs.[42] But those practices can also be evaluated in terms of dialectical propriety, rhetorical force, persuasive power, persuasive effect, dispute resolution, arguers' virtues, arguers' goals/purposes[43], etc., and also in terms of their aesthetic properties, e.g. beauty or elegance[44],

---

arguments, since arguments fundamentally involve case-making and are thus, first and foremost, epistemic objects. While Lumer and I do not agree on everything, his contributions to epistemic argumentation theory have been fundamental, and I gratefully acknowledge them here.
40 This section was developed in response to an invitation to speak at a workshop in Siena in April 2022 on 'Norms of public argumentation: Select theoretical perspectives and applications', organized as part of the EU research network project 'APPLY: European Network for Argumentation and Public Policy Analysis'. I am grateful to the workshop organizers, Jan Albert van Laar, Christoph Lumer, and Frank Zenker, for their invitation, and to them and the other participants for their instructive comments and suggestions.
41 What Biro and I have earlier called epistemic seriousness. See Biro and Siegel 1992, p. 92; Siegel and Biro 1997, p. 278.
42 See here Dutilh Novaes 2021, pp 19-22, both for a clear articulation of the idea that argumentative practices can further epistemically worthy ends, and for reservations concerning their impact on 'real-life' exchanges.
43 See here especially the pioneering work of Douglas Walton and his co-authors. For references and discussion, see Siegel and Biro 2021.
44 As in 'That's a beautiful (or elegant) argument.' Thanks here to Dan Cohen.

and in terms of the psychological upshots of such exchanges, including their evocation of particular responses.[45] How should we best navigate these many proposed criteria and norms?

In pursuing this question, we should acknowledge a preliminary point: 'What makes an argument (in the abstract propositional sense) good?' is a quite different question from several other similar-sounding questions: 'What makes an argument (in either the speech act or the social/dialectical/communicative sense) good'?, 'How should one argue?', 'How should argumentative exchanges, episodes, or moves be conducted'?, and 'How should argumentative exchanges, episodes, or moves be evaluated'? Many theories, including PD, VAT, and coalescent, multi-modal theories[46] speak to one or more of the latter questions, while the epistemic theory, which addresses all these questions, first and foremost targets the first. The distinction between the first question and the latter set of questions – in effect, the distinction between argument (in the abstract propositional sense) evaluation and argument (in the social, communicative, dialectical sense) evaluation – is fundamental, but not always acknowledged. Acknowledging it allows us to resolve outstanding debates concerning both criteria of argument evaluation and norms of argumentative interaction, because evaluating arguments (in the abstract propositional sense) is related to but distinct from evaluating argumentative practices in their social, communicative contexts.

The two sorts of evaluation are connected, since, as urged above, arguments are what arguers traffic in when arguing. Whatever else an arguer is doing or trying to do when arguing – persuading her interlocutor (or a broader audience) of the acceptability or otherwise of a belief/conclusion/standpoint, trying to score a debate victory, demonstrating her intelligence, amusing her audience, or whatever – if she is (sincerely) arguing, she is offering reasons for the conclusion or standpoint that she believes, hopes and intends provide justificatory support for it. Insofar, the standard epistemic criteria of argument quality noted above remain in play. But they needn't be the only criteria in play, since, depending on the argument (in the social communicative sense) and its context, other criteria, such as dialectical appropriateness, persuasive effect, or the virtuousness of the arguers may also be appropriate.

We might, for example, rightly praise or criticize an argument for its persuasive effect: 'Extending the child tax credit in the USA is so important for struggling families and their children, thank goodness its Democratic defenders' arguments succeeded in persuading enough Republicans to secure its passage, thereby bringing the USA more in line

---

45 As in 'That was an engaging (or satisfying) argument.' Thanks here too to Dan Cohen.
46 In addition to Tindale 2021, see the pioneering work of the 'father' of coalescent, multi-modal argumentation theory Michael Gilbert 1994, 1995, 1997, 2007, 2011.

with its more socially responsible and civilized European peers.' If we're recalcitrant Republicans, we might view that passage negatively, thereby ruing the persuasive effect of those arguments, but we'd still give them high marks for persuasive effect if they indeed persuaded a sufficient number of Republicans to secure the legislation's passage. So persuasive effect is one criterion by which arguments (in the social, communicative sense) can be evaluated.

Similar remarks apply to the other purported criteria of argument quality just mentioned. For example, we might evaluate arguments (in the latter sense) in terms of *dialectical propriety*, as PD does. We might also evaluate argumentative exchanges, or the individual moves within them, in terms of the *virtues* displayed by the arguers during the exchange, or, more deeply, in terms of the virtuousness or viciousness of their *characters*, as VAT does. Here argumentative virtue serves as a criterion by which such exchanges can be evaluated. Other dimensions of quality already mentioned, e.g., the elegance of an argument or the degree to which the participants find it satisfying, are also legitimate. All are sensible criteria by which arguments (in the social communicative sense) can be assessed.[47]

5.1. Criteria versus Norms

Criteria are those considerations that we consider when evaluating finished products: in the cases of interest here, either arguments or argumentative exchanges. They are *product-oriented*, purporting to tell us how to judge or evaluate a given case. *Norms*, on the other hand, are procedural guidelines, intended to guide or govern ongoing activities. Argumentative norms are *process-oriented*: they purport to guide our argumentative behavior, in holding us to account for violations of argumentative propriety.[48] So they are especially relevant to argumentative practice. If I insult you during our exchange, for example, I violate the argumentative norm of *civility*.[49] Other norms fall into

---

[47] Michael Gilbert puts the point well, in his discussion of flaws of typical critical reasoning courses: "... the tools that have been used by Informal Logic" should not 'be abandoned', but rather 'put into perspective as one way of examining one aspect of an argument.'" (1995, 134, emphases Gilbert's; cf. Gilbert 1994, 1997.)

[48] I utilize the familiar product/process distinction for convenience here, but the distinction, as it applies to 'argument', is problematic, as Geoff Goddu (2011) argues. The crucial thing, as Goddu notes, is "to distinguish acts of arguing from arguments-as-objects" (87), without holding that the latter are necessarily or inevitably the products of the former, before discussing the criteria and norms that govern their respective evaluations. Thanks here to Goddu for helpful comments.

[49] Civility is of course not just an argumentative norm; it governs communicative efforts generally. But it is as applicable in argumentative contexts as it is in other communicative contexts.

distinguishable categories (e.g., logical, epistemic, linguistic, dialectical, rhetorical) and include items as diverse as *epistemological* advice concerning the justificatory force of candidate reasons and evidence (e.g., Offer reasons that you think support the conclusion that you think are themselves belief-worthy; Reason cogently, in accordance with standard criteria of epistemic quality; Consider fairly and open-mindedly counter-evidence and counter-argument; Be on the lookout for confirmation bias and other psychology-of-reasoning flaws), *dialectical* advice (e.g., Obey the PD rules/code of conduct), and *rhetorical* advice (e.g., Take into account your audience and their existing beliefs and shape your interventions accordingly; Be mindful of the background beliefs and attitudes shared by your audience; Respond sensitively to their objections, concerns, and interventions).

We must be careful not to insist on too sharp a distinction between evaluating finished argumentative products in terms of criteria, and guiding ongoing argumentative practice by way of norms. For one thing, we sometimes use 'norm' to refer to what I'm here calling 'criterion', even when we're evaluating finished products; 'norm' is itself ambiguous in this way. Perhaps more importantly, ongoing argumentative activities end, resulting in finished argumentative products, i.e., arguments (in the social, communicative sense); and even before they end, we can evaluate slices or segments of them as if they were finished products. Consider again the norm of civility. As an argumentative exchange proceeds, the norm tells us to argue civilly. Once the exchange is completed, or before then if we want to evaluate a specific segment of it, we can evaluate the exchange (in whole or in part) by checking its civility. So civility can serve as both norm and criterion. Similar remarks apply to other norms. Still, the distinction between the evaluation of finished argumentative products in terms of product-oriented norms or criteria, on the one hand, and guidelines for conducting ongoing argumentative exchanges by way of process-oriented norms, on the other hand, is helpful.

5.2. Priorities among Compatible Norms

All the sorts of norms considered thus far – epistemic, dialectical, rhetorical, virtue-theoretic, etc. norms – are compatible. All can be utilized and appealed to, depending on the type of evaluation in play: We can ask, of a given argument:

Is the abstract propositional structure logically valid? Epistemically strong? Do its premises/reasons provide probative support to its conclusion?

Is the argumentation dialectically kosher?

Is the argumentation rhetorically effective?

Is the argumentation virtuous?

Is the argumentation beautiful? elegant?

Was the argumentation engaging or satisfying to the participating arguers?

All of these are legitimate avenues of argument evaluation. No doubt there are others.

However, although the norms are complementary, they are not of equal priority. On the epistemic view that Biro and I have defended, epistemic norms are of highest priority. This is because *arguments are what arguers traffic in when arguing*. The other senses of 'argument' are derivative of this one. Recall: If I'm arguing (sincerely) with you concerning a candidate belief, proposition or viewpoint about which we disagree, I'm giving you reasons which I believe, hope and intend will make the case for my preferred attitude (e.g., belief, acceptance, rejection, doubt, hope, etc.) toward that proposition or viewpoint, and/or make the case against your preferred attitude. The basic phenomena of arguments and arguing involve *making cases* for/against particular claims or propositions. And this is fundamentally an epistemic matter: If the case is well made, we have good reason, *ceteris paribus*, to embrace the claim, proposition or attitude in question; if not, not. If you're trafficking in arguments, you're engaged in making cases, or in challenging or evaluating them. If you're so engaged, the quality of the case under consideration is paramount. If you're not so engaged, the other norms deployed (dialectical, rhetorical, etc.) aren't being applied to *arguments*, in any of the senses considered thus far. So the most fundamental sense of 'argument' is the abstract propositional one, and the most fundamental sort of argument evaluation is epistemic.

What sort of priority is this? There is a case to be made that argumentation is *causally* prior to arguments (in the abstract propositional sense), in the sense that the latter are, as Ralph Johnson puts it, "the distillate of the practice of argumentation." (2000, 168) I am happy to grant the plausibility of this causal priority claim, although I think it needs qualification in various ways. My priority claim is not causal, though. It is rather a claim concerning *conceptual* priority, based on my claim, argued for throughout, that arguments are most fundamentally abstract propositional structures that *make cases* for their conclusions. This is why their epistemic evaluation – how strong is the case made? – is the most fundamental sort of argument evaluation. The other sorts of evaluation (dialectical, rhetorical, virtuosic, etc.) are dependent on the case-making nature of arguments, since if no case is being made, such evaluations are not evaluations *of arguments*. For an argument to be dialectically, rhetorically, or virtuosically good (or not), it must first of all be (or contain) an argument in the primary, abstract propositional sense of the term. If so, an evaluation of its case-making strength (or weakness) is preeminent among the many dimensions along which arguments can be evaluated.[50]

6. Conclusion: The Strengths of the Epistemic Theory

In some respects the opening motto from Scheffler captures the core of the epistemic theory of argument/argumentation, emphasizing as it does reasons and rationality, as Biro and I have emphasized from the beginning. In other respects, though, it embodies the ambiguity I have been belaboring at the core of argumentation theory: is an argument a set of abstract propositions, to be evaluated in terms of the logical or epistemic relationships obtaining among its members, or is it rather a social, dialectical, communicative activity, to be understood in social terms and evaluated in terms of its dialectical or rhetorical effects, or the purposes or virtues of arguers? The answer to this question, of course, is that it rests on a false dichotomy, and that 'argument' picks out both.[51] Nevertheless, keeping them separate and clear has proved difficult; as we have seen, some theories have failed to do so, and their theories suffer from the failure. In particular, PD and VAT run into trouble in this way.[52] Once we draw the distinctions above, between the several senses of 'argument', product v. process argumentative criteria and norms, and epistemic v. social, communicative, rhetorical, virtuosic, and other evaluations, things fall into place. The just-mentioned theories treat argumentative practices and their normative evaluations insightfully and well. But their attempted prioritizations of the social, communicative, and rhetorical dimensions of argumentation, and, indeed, their attempts to extend their social, communicative argument evaluations into the epistemic evaluations of arguments generally, fail. In large part this is because they either fail to draw those distinctions, or fail to recognize both their importance, and that their own analyses concern only the social, communicative sense of 'argument'.

I should note for the record that there is not just one epistemic theory. I have emphasized the 'BS' version of that theory – the one Biro and I have

---

50 I am indebted to José Gascón and Barbara Stengel for helpful discussion of the sort of priority in play here.

51 Biro and I have urged this more than once; see esp. Biro and Siegel 2006b.
52 There are obviously many other theories and approaches than those I've considered here. I apologize for not treating them, for reasons of both space and competence. But I hope the examined cases are sufficiently representative of the field to indicate something true and important about the theoretical space in which argumentation theorists work.

been articulating and defending – but there are several approaches that fly under the epistemic banner, as noted earlier.[53,54]

The epistemic theory emphasizes the relationships existing (or not) between premises, reasons, and evidence and the conclusions/targets they putatively support, and conceives arguments primarily as reason-conclusion complexes. Its strengths are several. It keeps a central focus on arguments in the abstract propositional sense, while acknowledging and speaking to the other senses of that term. It maintains and clarifies the argument/argumentation distinction, which is central to argumentation theory, but is mischaracterized by some theorists. It keeps the several distinct questions concerning normative evaluation clear, and answers them in a coordinated way. It clarifies the priority relationships among the several criteria that can be rightfully used to evaluate arguments, in the several senses discussed above. But the most important strength of the epistemic theory is that it captures and explains the most fundamental sense of 'argument': that an argument, in the hands of an arguer, attempts and purports to offer justificatory support to a conclusion; that a good one succeeds and a bad one does not; that it can be evaluated independently of the arguer who advances it, and even in the complete absence of such an arguer, as in the case of examples A and B above; that argument evaluation is most fundamentally an epistemic matter; and that the several other senses of the term are derivative of the fundamental one: arguments are what arguers traffic in when arguing.[55]

---

[53] Fellow epistemic theorists (albeit with varying emphases, account details, and disagreements) include Scott Aikin (2008), Sharon Bailin (1999), Mark Battersby (1989, 2016), Bailin and Battersby (2016, 2022), J. Anthony Blair (2004, 2012), Blair and Ralph Johnson (1993), Patrick Bondy (2015, 2018, 2021), Tracy Bowell and Justine Kingsbury (2013), Richard Feldman (1994, 2005a, 2005b), James Freeman (2005), David Godden (2015, 2016, 2017), Alvin Goldman (1994, 1997, 1999, 2003), Christoph Lumer (1988, 1991, 2005a, 2005b), and Robert Pinto (2001). This is not an exhaustive list; apologies to those I've erroneously left out.

[54] Thanks to Michael Gilbert (2007, 157-8) for christening the Biro/Siegel view 'BS', thus exploiting its rather more déclassé meaning in colloquial American English – 'bullshit' – thereby bringing some welcome humor into argumentation theory. At least it made me laugh!

[55] Presented as a keynote lecture at the ECA 4 conference, Rome, September 2022. I am grateful to Fabio Paglieri and the conference organizers for the invitation, and to Andrew Aberdein, Mark Battersby, John Biro, Tony Blair, Dan Cohen, José Gascón, Geoff Goddu, Christoph Lumer, Barbara Stengel, Christopher Tindale, and John Woods for helpful advice on an earlier draft.

References

Aberdein, A. (2010). Virtue in argument. *Argumentation 24(2)*, 165-179.
----- (2014). In Defence of virtue: The legitimacy of agent-based argument appraisal. *Informal Logic 34(1)*, 77-93
----- (2018a). Inference and virtue. In S. Oswald and D. Maillat, eds, *Argumentation and Inference: Proceedings of the 2nd European Conference on Argumentation, Fribourg 2017, vol. 2* (pp. 1–9). London: College Publications.
----- (2018b). Commentary on Gascón, "Virtuous arguers: Responsible and reliable". In S. Oswald and D. Maillat, eds., *Argumentation and Inference: Proceedings of the 2nd European Conference on Argumentation, Fribourg 2017, vol. 1* (pp. 123–128). London: College Publications.
----- (forthcoming). The Fallacy fallacy: From the owl of minerva to the lark of arete. *Argumentation* (in press).
----- and Cohen, D. H. (2016). Introduction: Virtues and arguments. *Topoi 35(2)*, 339-343.
Aikin, S. F. (2008). Three objections to the epistemic theory of argument rebutted. *Argumentation and Advocacy 44(3)*, 130-142.
Bailin, S. (1999). The Trouble with Percy. *Informal Logic 19(2-3)*, 161-170.
----- and Battersby, M. E. (2016). *Reason in the balance*, 2e. Indianapolis: Hackett.
----- (2022). Inoculating students against conspiracy theories: The case of covid-19. In S. Oswald, M. Lewiński, S. Greco, and S. Villata, eds., *The Pandemic of Argumentation*, Argumentation Library, vol. 43. (pp. 271-289). Dordrecht: Springer.
Battersby, M. E. (1989). Critical thinking as applied epistemology: Relocating critical thinking in the philosophical landscape. *Informal Logic 11(2)*, 91-100.
------- (2016). Enhancing rationality: heuristics, biases, and the critical thinking project. *Informal Logic 36(2)*, 99-120.
Biro, J. and Siegel, H. (1992). Normativity, argumentation, and an epistemic theory of fallacies. In F. van Eemeren, et. al., eds., *Argumentation Illuminated: Selected Papers from the 1990 International Conference on Argumentation* (pp. 85-103). Dordrecht: Foris.
----- (2006a). In defense of the objective epistemic approach to argumentation. *Informal Logic 25(3)*, 91-101.
----- (2006b). Pragma-dialectic versus epistemic theories of arguing and arguments: Rivals or partners? In Houtlosser, P. and van Rees, A. (Eds.), *Considering pragma-dialectics: A festshrift for Frans H. van Eemeren on the occasion of his 60th birthday* (pp. 1-10). Mahuah, NJ: Erlbaum.
----- (2015). Argument and context. *Cogency 7(2)*, 27-41.
Blair, J. A. (2004). Argument and its uses. *Informal Logic 24(2)*, 137-151.
----- (2012). Groundwork in the theory of argumentation: Selected papers of J. Anthony Blair. *Argumentation Library* vol 21.
----- and Johnson, R. H. (1993). Dissent in fallacyland, Part 1: Problems with van Eemeren and Grootendorst. In R. E. McKerrow (Ed.), *Argument and the Postmodern Challenge: Proceedings of the Eighth SCA/AFA Conference on Argumentation* (pp. 188-190). Annandale VA: Speech Communication Association.
Bondy, P. (2015). Virtues, evidence, and ad hominem arguments. *Informal Logic 35(4)*, 450-466.

----- (2018). *Epistemic rationality and epistemic normativity*. NY: Routledge.
----- (2021). The epistemic norm of inference and non-epistemic reasons for belief. *Synthese 198(2)*, 1761-1781.
Botting, D. (2010). A Pragma-dialectical default on the question of truth. *Informal Logic 30(4)*, 413-434.
Bowell, T., and Kingsbury, J. (2013). Virtue and argument: Taking character into account. *Informal Logic 33(1)*, 22-32.
Cohen, D. H. (2013). Virtue, in context. *Informal Logic 33(4)*, 471-485.
----- (2022). You can't judge an argument by its closure. *Informal Logic 42(4)*, 669-684.
Dutilh Novaes, Catarina (2021a). Argument and argumentation. *The Stanford Encyclopedia of Philosophy* (Fall 2021 Edition), Edward N. Zalta (ed.), URL = <htpps://plato.stanford.edu/archives/fall2021/entries/argument/>.
Feldman, R. (1994). Good arguments. In F. F. Schmitt (Ed.), *Socializing epistemology: The social dimensions of knowledge*. NY: Rowman and Littlefield (pp. 159-188).
----- (2005a). Deep disagreement, rational resolutions, and critical thinking. *Informal Logic 25(1)*, 13-23.
----- (2005b). Useful advice and good arguments. *Informal Logic 25(3)*, 277-287.
Freeman, J. (2005). *Acceptable premises: An epistemic approach to an informal logic problem*. Cambridge: Cambridge University Press.
Garssen, B., and van Laar, J. A, (2010). A Pragma-dialectical response to objectivist epistemic challenges. *Informal Logic 30(2)*, 122-141.
Gascón, J. Á. (2016). Virtue and arguers. *Topoi 35(2)*, 441-450.
----- (2017). Brothers in arms: Virtue and pragma-dialectics. *Argumentation 31(4)*, 705-724.
----- (2018). Virtuous arguers: Responsible and reliable. *Argumentation 32(2)*, 155-173.
Gilbert, M. A. (1994). Multi-modal argumentation. *Philosophy of the Social Sciences 24(2)*, 159-177.
----- (1995). Arguments & arguers. *Teaching Philosophy 18(2)*, 125-138.
----- (1997). *Coalescent argumentation*. Mahwah, NJ: Lawrence Erlbaum Associates.
----- (2007). Natural normativity: Argumentation theory as an engaged discipline. *Informal Logic 27(2)*, 149-161.
----- (2011). The kisceral: Reason and intuition in argumentation. *Argumentation 25(2)*, 163-170.
Godden, D. (2015). Argumentation, rationality, and psychology of reasoning. *Informal Logic 35(2)*, 135-166.
----- (2016). On the priority of agent-based argumentative norms. *Topoi 35(2)*, 345-357.
----- (2017). On the norms of visual argument: A case for normative non-revisionism. *Argumentation 31(2)*, 395-431.
Goddu, G. C. (2011). Is 'argument' subject to the product/process ambiguity? *Informal Logic 31(2)*, 75-88.
----- (2015). Towards a foundation for argumentation theory. In F. H. van Eemeren and B. Garssen (Eds.), *Reflections on theoretical issues in argumentation theory*, Argumentation Library 28 (pp. 43-51). Dordrecht: Springer.
----- (2016). What (the hell) is virtue argumentation? In D. Mohammed and M. Lewiński (Eds)., *Argumentation and reasoned action: Proceedings of the 1st*

European conference on argumentation, Lisbon, 2015, Vol. II (pp. 439-448) London: College Publications.

Goldman, A. I. (1994). Argumentation and social epistemology. *Journal of Philosophy 91(1)*, 27-49.

----- (1997). Argumentation and interpersonal justification. *Argumentation 11(2)*, 155-164.

----- (1999). *Knowledge in a social world*. Oxford: Oxford University Press.

----- (2003). An epistemological approach to argumentation. *Informal Logic 23(1)*, 51-63.

Johnson, R. H. (2000). *Manifest rationality: A pragmatic theory of argument*. Mahwah, NJ: Lawrence Erlbaum Associates.

Lumer, C. (1988). The disputation – a special type of cooperative argumentative dialogue. *Argumentation 2(4)*, 441-464.

----- (1991). 'Structure and function of argumentations – An epistemological approach to determining criteria for the validity and adequacy of argumentations.' In van Eemeren, F. H., Grootendorst, R., Blair, J. A., and Willard, C. (Eds.), *Proceedings of the second international conference on argumentation*, vol. 1A, (pp. 98-107).Amsterdam, SICSAT.

----- (2005a). Introduction: The epistemological approach to argumentation – A map. *Informal Logic 25(3)*, 189-212.

----- (2005b). The epistemological theory of argument: How and why? *Informal Logic 25(3)*, 213-243.

----- (2012) The epistemic inferiority of pragma-dialectics – A reply to Botting'. *Informal Logic 32(1)*, 51-82.

O'Keefe, D. (1977). Two concepts of argument. *The Journal of the American Forensic Association 13*, 121-128.

Paglieri, F. (2015). Bogency and goodacies: On argument quality in virtue argumentation theory. *Informal Logic 35(1)*, 65-87.

Pinto, R. C. (2001). Argument, inference and dialectic: *Collected papers on informal logic*. Dordrecht: Kluwer.

Popa, E. O. (2016). Criticism without fundamental principles. *Informal Logic 36(2)*, 192-216.

Reed, C. A., and Norman, T. J. (2004). *Argumentation machines: New frontiers in argument and computation*. Dordrecht: Kluwer.

Scheffler, I. (1989). *Reason and teaching*. Indianapolis: Hackett. First published 1973.

Siegel, H. (1988). *Educating reason: Rationality, critical thinking, and education*. London: Routledge.

----- (1997). *Rationality redeemed?: Further dialogues on an educational ideal*. New York: Routledge.

----- (1998). Knowledge, truth and education. In Carr, D., (Ed.), *Education, knowledge, and truth: Beyond the postmodern impasse* (pp. 19-36). London: Routledge.

----- (2017). *Education's epistemology: Rationality, diversity and critical thinking*. New York: Oxford University Press.

----- (2018). Justice and justification. *Theory and research in education 16(3)*, 308-329.

----- and Biro, J. (1997). Epistemic normativity, argumentation, and fallacies. *Argumentation 11(3)*, 277-292.

----- (2008) Rationality, reasonableness, and critical rationalism: Problems with the pragma-dialectical view. *Argumentation 22(2)*, 191-203.

----- (2010) 'The pragma-dialectician's dilemma: Reply to Garssen and van Laar. *Informal Logic 30(4)*, 457-480.
----- (2021) Walton on argument, arguments, and argumentation. *Journal of Applied Logics 8(1)*, 183-194.
Tindale, C. W. (2021). *The Anthropology of argument: Cultural foundations of rhetoric and reason.* NY: Routledge.
van Eemeren, F. H. (2012). The pragma-dialectical theory under discussion. *Argumentation 26(4)*, 439-457.
----- and Grootendorst, R. (2004). *A systematic theory of argumentation: The pragma-dialectical approach.* Cambridge: Cambridge University Press.
----- and Houtlosser, P. (2003). The development of the pragma-dialectical approach to argumentation. *Argumentation 17(4)*, 387-403.
Willard, C. A. (1983). *Argumentation and the social grounds of knowledge.* Tuscaloosa: University of Alabama Press.

A PARTICULARIST THEORY OF ARGUMENTATION BY ANALOGY

JOSÉ ALHAMBRA
Autonomous University of Madrid
Jose.alhambra@uam.es

Abstract

In this paper I defend what I call a "particularist theory of arguments by analogy". Particularism is opposed to generalism, which is the thesis that arguments by analogy require a universal principle that covers cases compared and guarantees their conclusion. Particularism rejects this claim and holds that arguments by analogy operate case-to-case. Two lines of argument will be explored here. On the one hand, I will contend that analogy consists of a comparison of higher-order relationships. On the other hand, I will argue that arguments by analogy can be evaluated without recourse to any principle, just by looking at the similarities and differences between the cases at hand.

1. Introduction

Analogy is said in many ways. This paper deals with its argumentative use, that is, with arguments in which an analogy plays a non-trivial role in giving support to a claim. By analogy I will understand "a comparison between two objects, or systems of objects, that highlights respects in which they are thought to be similar" (Bartha 2010, p. 1). I will call these objects or systems of objects "source" and "target" cases. As we shall see, my understanding of analogy puts special emphasis on systematicity – broadly understood. It rests on the notion of "parallelism of argumentative relationships", not only because I am concerned with arguments by analogy, but because, as we shall see, I argue that the source and the target in the sort of analogies I am interested in are themselves arguments. It is therefore essential to outline here the theory of argumentation that I am going to work with, namely the so-called "argument dialectics" (see Leal and Marraud 2022).

According to this approach, "arguing is presenting to someone something as a reason for something else" (Marraud 2019, p. 2). This practice of giving –and receiving– reasons for examination can be approached from different perspectives. Standard practice is to take the classical trichotomy as a reference and distinguish three viewpoints: rhetoric, which conceives of argumentative exchanges as communicative processes centered on the purposes of the arguer –usually persuasion; dialectics, which studies argumentation as a procedure subject to a series of conventional rules aimed at achieving goals shared by discussants, and logic or theory of arguments, which studies the products of argumentation, i.e., arguments and their relationships. Here I will adopt a logical approach.

Since arguments are products of argumentation –as distinct from processes of persuasion and procedures of discussion–, and to argue is to exchange reasons with others, we can characterize them as compounds of two elements: the consideration presented as a reason and that for which that consideration is a reason (i.e., the claim). I will refer to the statements that together comprise a single reason as "premises" and to the statements that comprise the claim as "conclusion". To depict an argument, I will use the system of diagrams used in (Leal and Marraud 2022). A single argument (i.e., one that puts forward a single reason) is depicted by two rectangles joined by the connector "therefore":

$$\boxed{R}$$
Therefore
$$\boxed{C}$$

Diagram 1. Single argument.

When we argue, then, we give others reasons to consider in order to defend a particular claim. But when we argue we also make commitments. Whoever presents an argument "R, therefore C" agrees that it is the case that R and that, given R, there is a reason for C. The second commitment can be expressed by resorting to the conditional "if R, then C"[1].

During a discussion, these commitments may be challenged, and this gives rise to a "chaining of reasons" and a "warrant". A chaining arises when we give a reason to justify a consideration that we had previously presented as a reason, and a warrant is a general principle or rule that justifies the conditional associated with an argument, i.e., it tells us that cases like R generally function as reason for cases like C. If someone argues, for example:

[1] You said that you would go to the cinema, so you ought to go to the cinema,

---

1 These conditionals express the relationship between the premises and the conclusion of a particular argument, but they do not add anything to that argument. That is, they are neither a premise nor a warrant in the sense I am about to say, but rather an expression of what the arguer does in presenting something as a reason for something else.

they may be asked to present evidence that this was said or to justify the conditional 'if you said that you would go to the cinema, then you ought to go to the cinema' for example by calling on a principle such as "promises must be kept" (as we shall see, the justification of the relation expressed by the conditional can also be carried out without recourse to a general rule). The warrant may in turn be justified and that results in a chaining-like structure called "backing". Although chainings, warrants, and backings are not part of arguments in the same sense as reasons and claims are −as we can argue without them, but not without a reason or a claim− they play an important role in the logical evaluation of arguments.

A good argument from a logical point of view is one that puts forward a good reason, and a good reason is one that stands up to counterarguments. A counterargument to an argument A is an argument whose conclusion is incompatible with some element or commitment associated with A. If someone presents [1], we can imagine at least three replies: (a) "that's not true, in this recording it can be seen that I didn't say that"; (b) "I said what I said only because you threatened me", or (c) "it's true, I made that promise, but a friend of mine has just had a car accident and I have to go to the hospital". In (a) we give a reason to defend that a premise in [1] is not true; in (b) we mention an exception to the principle "promises must be kept", which justifies the conditional associated with [1], and in (c) we give a stronger reason to do something incompatible with previous action. They are an "objection", a "rebuttal" and a "refutation", respectively. If an argument withstands objections and rebuttals, we say that it is "correct", and that it puts forward a *pro tanto* reason. If it also resists refutations, we say that it is not only correct, but also "conclusive", and that it puts forward a relatively strong (or all-thing-considered) reason. Conversely, if an argument does not resist objections or rebuttals, we say that it is "incorrect", and if it does not resist refutations, we say that, although it is correct, it is not conclusive (i.e., it puts forward a worthwhile reason, but a relatively weak one). Chainings and warrants can be seen as responses to (possible or actual) criticism towards an argument.

Now we can tentatively characterize an argument by analogy as follows: it is one in which a comparison between two objects or systems of objects (i.e., an analogy) is used as a reason for assigning to one of them (target) a property of the other (source). There are two possible readings of the relation between the comparison and the property to be transferred: (1) depending on the elements compared, the property transferred will vary; or (2) depending on the property to be transferred, the comparison will highlight different aspects of the source and the target cases. Although I think (1) and (2) can be seen as complementary approaches, I will stress the latter: in arguing by analogy, we hold that a claim is supported by reasons because the case parallels another case in which it is assumed that a claim is supported by other reasons. So, what we are interested in are comparisons between relationships of a certain kind and not comparisons between objects. My proposal is that this idea may help us design a particularist theory of arguments by analogy. But before going into details, let us look at the debate between what I am calling particularist and generalist theories of analogy.

2. The debate.

"Generalism" and "particularism" comes from the theory of normative reasons. They are opposite answers to the question of what role principles play in moral reasoning. Generalism is the thesis that "the very possibility of moral thought and judgement depends on the provision of a suitable supply of moral principles" (Dancy 2004, p. 7). Particularism rejects this thesis and argues that, although principles may sometimes play an important role in moral reasoning, they are not necessary.

I will apply these labels to the study of arguments by analogy. When I speak of generalism and particularism, I will understand them as opposite answers to the question of whether arguments by analogy need principles to support their claim. In this context, it is generally assumed that there are two types of principles, generalizations ("most things that have x, y, z are W") and universalizations ("all things that have x, y, z are W"). The discussion has been focused on the latter, understood as substantive rather than purely formal principles, so I will do the same here. Generalism argues that principles of this sort, which subsumes the cases compared and guarantees the conclusion of the argument, are required, so that they must be included in the analysis as implicit premises. I call this "plain generalism," and it is defended by authors such as Monroe Beardsley.

The main problem with plain generalism is openly pointed out by Beardsley: "What makes an analogical argument plausible is always a hidden generalization; but when we make that generalization explicit, we can throw away the rest of the analogy" (Beardsley 1975, p. 113). It turns out that generalist recasting explains analogy so well that it makes it disappear. Another problem is that, in practice, arguers rarely state such a principle, and attributing it to them as an implicit premise is problematic, if not outright fallacious. Govier argues that this reconstruction is *ad hoc*, "appearing to be due only to a desire to look at argument through deductivist goggles" (Govier 1989, p. 145), as well as uncharitable, since it holds the arguer responsible for a principle that is generally less plausible than the conclusion of the argument (*see* also van Laar 2014, p. 92). A third problem is strength variations. Two things can be more or less similar to each other, so it seems reasonable that arguments by analogy are more or less strong depending on the relevant similarities and differences between the source and target cases (Guarini

2004, p. 159). But, if we conceive of arguments by analogy as deductions in disguise, variations in strength are, by definition, out of the picture. Finally, this position relies upon "the assumption that particular cases have to be known by having universal generalizations applied to them" (*Ibid* p. 145), and this is problematic to say the least, for it makes the origin of such generalizations a sort of mystery (Wisdom 1991, p. 47-48; Marraud 2020, p. 5).

Other theorists have developed more elaborate analyses that attempt to solve these problems. Fabio Shecaira, for example, proposes the following reconstruction:

1. It is true that a.
2. The most plausible (i.e., the best) reason for believing a is the principle C.
3. Therefore, it is true that C.
4. C implies b.
5. Therefore, it is true that b (Shecaira 2013, p. 429).

As can be seen, arguments by analogy are no longer single arguments, but structures composed of two arguments. The first is an abduction in which source case a is presented as a reason for principle C, and the second is a deduction in which target case b is entailed from principle C. This analysis seems to solve the problem pointed out by Beardsley, since the source information is not superfluous. Moreover, it can accommodate variations in strength, because the first part of the scheme is a non-deductive argument that can lend its conclusion different degrees of support. But a challenge remains: Although there is no doubt that these complex structures may be interesting on their own right, no comparison between source and target cases is made in them, so why should we conceive of them as arguments *by* analogy?

The issue of principle reconstruction is addressed by Bruce Waller. For him, arguments by analogy are based on an appeal to consistency: by resorting to the source case, a principle shared by interlocutors is elicited and from there the target case is deduced, which requires the same treatment for both cases. However, this does not imply that principles are "eternal verities set in stone, awaiting our certain discovery" (Waller 2001, p. 206). On the contrary, they can be refined and modified as particular cases arise that do not conform to previous formulations, something that Waller calls "thoughtful mutual adjustment". This makes principles allegedly underlying arguments by analogy more flexible and sensitive to variations depending on the context and, thus, more plausible. Furthermore, it sheds light on the process by which such principles are reached: there would be no epistemic priority of them over particular cases, but a constant thoughtful mutual adjustment. But again, problems remain: it makes no sense to speak of joint reconstruction of an already shared principle and, at the same time, of rational persuasion (Guarini 2004, pp. 155-156). In addition, once we recognize –as Waller and Shecaira do– that cases can function as reasons for principles, and that this does not in turn presuppose any principle as an implicit premise, a legitimate complain arises: why should these universal principles be necessary in arguments by analogy?

The task for the particularist is then to propose a theory that dispenses with these principles in the explanation of arguments by analogy. In the face of this challenge, two questions arise: if we get rid of principles, (1) on what do we base arguments by analogy? And (2) is there any other role for principles in arguments by analogy?[2]

3. A particularist scheme.

I have said that an analogy is a comparison between two objects or systems of objects that points out features in which they are thought to be similar. This characterisation encompasses two ways of understanding analogies. On the one hand, some contend that an analogy is a sort of aggregation of similarities between objects: the more properties two objects share, the more likely they are to be similar in other respects. On the other hand, others argue that analogy is not a comparison of objects, but of relationships, i.e., it is a parallelism of relations. Here I will adopt this second position. The question, then, is what kind of relationships are compared in an argument by analogy? Let us take a short detour to clarify this issue.

In cognitive sciences, Keith Holyoak and Paul Thagard have also defended a theory of analogy based on the notion of parallelism. They argue that analogical thinking is to "reason and learn about a new situation (the *target* analog) by relating it to a more familiar situation (the *source* analog) that can be viewed as structurally parallel" (Holyoak and Thagard 1997, p. 35). The process by which two situations are compared is called "mapping", namely "the construction of orderly correspondences between the elements of a source analog and those of a target" (Holyoak

---

[2] For reasons of space, I have had to leave out two interesting particularist proposals. First, there is Lilian Bermejo-Luque, who responds to the challenges posed by Waller and Shecaira with her Linguistic-Normative Model of Argumentation. This model combines an adaptation of Toulmin model with the notion of modal qualifiers and makes it possible to defend a deductive yet non-principled scheme of arguments by analogy (Bermejo-Luque 2012, 2014). Secondly, André Juthe proposes a relational model based on a one-to-one correspondence between elements in the source that determine the predicate to be transferred and elements of the target. This allows him to account for arguments by analogy without recourse to universal principles (Juthe 2005, 2016, 2019) and answer the generalist criticism raised by David Botting (Botting 2017, 2022).

and Thagard 1989, p. 195). To understand what these elements are, it is necessary to pay attention to the work of Dedre Gentner, who first proposed a structure-mapping theory of analogy. She makes a twofold syntactic distinction that is relevant here. On the one hand, she distinguishes between "object attributes", predicates that takes one argument (e.g., "María is Spanish"), and "relationships", predicates that takes two or more arguments (e.g., "John is *taller than* Peter"). And on the other hand, she distinguishes between first-order relationships, which takes objects as arguments (as in the previous one), and higher-order relationships, which takes propositions (e.g., "John will be selected for the basketball team *because* he is taller than Peter", or "María has a better CV than John, *so that* she will receive the grant"). According to these authors, what differentiates analogy from mere similarity is that in analogy the comparison operates on higher-order relationships. Take the classic Platonic analogy: Just as a ship needs a captain to direct her course, so a state needs a good leader to set its agenda. Following (Shelley 2004), we can analyse it using this table:

Table I. Representation of the ship-state mapping.

Ship (source)	*State (target)*
Ship	State
Captain	Leader
Course	Agenda
Crew	Citizens
Well-being	Well-being
Need (ship, captain)	Need (state, leader)
Direct (captain, course)	Set (leader, agenda)
Enjoy (crew, well-being)	Enjoy (citizens, well-being)
Because (need, direct)	Because (need, set)
So-that (direct, enjoy)	So-that (set, enjoy)

As can be seen, the source and target cases appear in different columns. With respect to rows, three levels are set corresponding to objects, first-order relationships, and higher-order relationships. In light of this analysis, it is easy to see that Plato's analogy rests not so much on the similarities between objects –a ship and a state bear little resemblance to each other–, but on the parallelism of relationships between the element of both cases. The more relationships and the higher their order, the better the analogy, according to these theories.

So, we have that an analogy is a comparison between two systems of relationships, and that an argument by analogy is one in which that comparison is used as a reason for assigning to one of them a property of the other. On the other hand, I have suggested that in arguing by analogy, we hold that a claim is supported by reasons because the case parallels another case in which it is accepted that a claim is supported by reasons. How can we combine these ideas? My suggestion is that by understanding higher-order relationships as argumentative relationships (*see* Marraud 2007, 2016). By argumentative relationship I mean the relation between the consideration that is presented as a reason and that for which that consideration is a reason. If we take this step, then we have that arguments by analogy are meta-arguments, i.e., arguments about other arguments, since the source and target cases are themselves arguments (*see* Woods and Hudak 1989; Marraud 2007; van Laar 2014, or Stevens 2018)[3].

An example may shed light on this idea. In a report on the decision of the Supreme Court of Virginia (USA) to remove the statue of Confederate General Robert E. Lee, reporter Gregory S. Schneider collects the testimony of Janice Hall Nuckolls, a citizen who lives near the statue. Asked what to do with the base on which the statue stands, she defends that it should be removed too, and argues:

> "The base will be forever linked to the Lee monument, no matter how much paint is on it," she [Janice Hall Nuckolls] said. "Having to start with that would be like being given a canvas to paint but being told to work with the painting that has already been started by someone else. And it's not a good painting". Gregory S. Schneider, "Virginia Supreme Court clears way for Lee statue in Richmond to come down", *The Washington Post*, 02/09/2021[4].

As can be seen, in the first line Nuckolls gives a reason for removing the base of the statue. Then, to show the appropriateness of that reason, she appeals to a parallel case: that of the canvas which has already been painted.

The base will be forever linked to the Lee monument, no matter how much paint is on it
Therefore
It is not a good idea to use the base to build something new on it

Diagram 2. Target argument: monument argument.

---

[3] A question arises here: are all arguments by analogy meta-arguments? Although I am not very interested in the terminological wrangling, I think it could be useful to reserve the notion of arguments by analogy for this type of arguments and to use "arguments by comparison" as a more general category that encompasses other forms of arguing by making comparisons (see Juthe 2005, p. 7, and Stevens 2018, p. 441, note 17).

[4] https://www.washingtonpost.com/local/virginia-politics/lee-statue-richmond-court-removal/2021/09/02/4a2ee794-0bee-11ec-a6dd-296ba7fb2dce_story.html

> A canvas has already been painted by someone else and it is a bad painting
> Therefore
> It is not a good idea to use that canvas to paint something new on it

Diagram 3. Source argument: canvas argument.

What Nuckolls is trying to do here is to defend that the consideration presented for removing the base is a worthwhile reason, because it parallels the consideration presented for not using the canvas –which is supposed to be an obviously worthwhile reason. We have then an argument about other arguments.

If we use Shelley's table, we can appreciate the similarity with Plato's analogy:

Table II. Representation of the monument-canvas mapping.

Canvas (source)	*Monument (target)*
Canvas	Base
Bad painting	Controversial statute
Painter	Sculptor
New painting	New statute
Bound-to (bad painting, canvas)	Bound-to (controversial statute, base)
Not-paint-on (painter, new painting, canvas)	Not-build-on (sculptor, new statute, base)
Because (not-paint-on, bound-to)	Because (not-build-on, bound-to)

Now we can answer the question arisen above: the relationships compared in an argument by analogy are argumentative relationships. Nonetheless, these arguments can also operate on what I will call "inter-argumentative relationships", that is, relations between arguments. Let us consider another example:

> "Well, I have just had a pang of regret, yes: since Cardinal Cañizares said the other day that cells from aborted foetuses are being used to make a vaccine against Covid-19. [...]
> Respected Monsignor, imagine that you have just spiritually assisted a youngster who has been "legally" executed, horror, and that in a hospital bed there is a person whose life depends on the youngster's heart, or his kidneys. Would your eminence authorise the transplantation of his organs? [...] I think so. Does not your eminence not find any similarity between the youngster's corpse and the aborted foetuses? I do, with apologies" (Agapito López Villa, "El

diablo y las vacunas" [The devil and the vaccines], *Hoy*, 21/06/2020 –translation is mine)[5].

Here Agapito López Villa answers to Cardinal Cañizares' complains about the use of stem cells from aborted foetuses in the search for a Covid-19 vaccine. He does not directly state his position but uses a hypothetical case: the situation in which a youngster has been executed and another person needs his organs. To the question of whether it is legitimate to authorise the transplant in such circumstances, he answers that it is. Once that verdict is taken for granted, analogy is posed by the second rhetorical question.

What is interesting here is that the subject of the analogy is not arguments, as in Nuckolls' case, but weighings of reasons for opposite claims. I said before that this is called refutation: a counterargument where two reasons for incompatible claims are weighed against each other and more weight is attributed to one of them, imposing its claim. Connectors such as "but", "however", or "although" are often refutations marks. Using "but", we can depict López Villa's position:

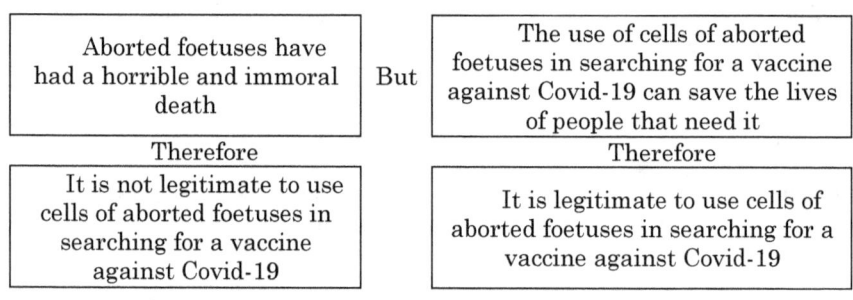

Diagram 4. Foetuses argumentation.

He justifies this weighing of reasons by relying on the hypothetical case of the executed youngster, which can be represented as follows:

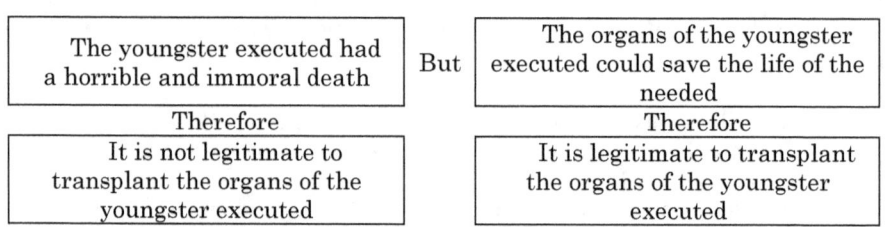

Diagram 5. Youngster executed argumentation.

---

5 https://www.hoy.es/extremadura/diablo-vacunas-20200621112832-nt.html

The following diagram depicts López Villa's argumentation:

The youngster executed had a horrible and immoral death	But	The organs of the youngster executed may save the life of the needed
Therefore		Therefore
It is not legitimate to transplant the organs of the youngster executed		It is legitimate to transplant the organs of the youngster executed
Therefore		
Aborted foetuses have had a horrible and immoral death	But	The use of cells of aborted foetuses in searching for a Covid-19 vaccine may save lives
Therefore		Therefore
It is not legitimate to use cells of aborted foetuses in searching for a Covid-19 vaccine		It is legitimate to use cells of aborted foetuses in searching for a Covid-19 vaccine

Diagram 6. Foetuses-executed argumentation by analogy.

We can appreciate the difference with Plato's and Nuckolls' examples by adding an extra level of rows to Shelley's table (I am interested in the level of relationships marked by the "buts", and not so much in the accuracy of my analysis)

Table III. Representation of the foetuses-youngster executed mapping

Execution (source)	*Foetuses (target)*
Youngster executed	Aborted foetuses
Horrible death	Horrible death
Organs	Cells
Lives	Lives
Transplant	Search
Have (youngster executed, horrible death)	Have (aborted foetuses, horrible death)
Save (organs, lives)	Save (cells, lives)
Be-legitimated (transplant)	Be-legitimated (search)
Not-being-legitimated (transplant)	Not-being-legitimated (search)
Because[1] (not-being-legitimated, have)	
Because[2] (being-legitimated, save)	Because[1] (not-being-legitimated, have)
	Because[2] (being-legitimated, save)
But (Because[1], because[2])	
	But (Because[1], because[2])

I call these varieties "argumentation by parity of reasons" and "argumentation by parity of weighings" (*see* Alhambra 2022). I propose the following schemes:

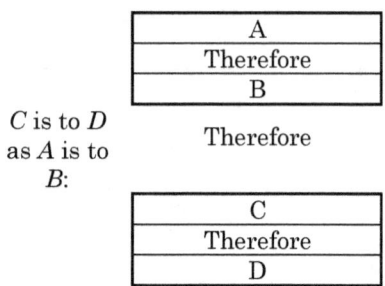

Diagram 7. Scheme for argumentation by parity of reasons.

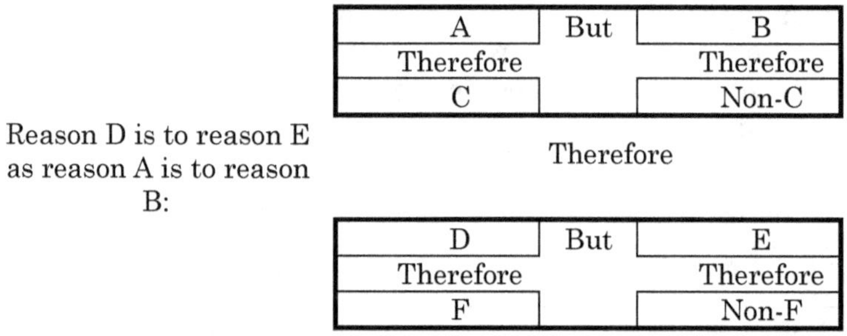

Diagram 8. Scheme for argumentation by parity of weighings.

In both cases we are dealing with arguments about other arguments. In both cases it is argued that a claim is supported by reasons because the case is parallel to another –hypothetical– case in which a claim is supported by reasons. The difference is that while in Nuckolls' case it is argued that the claim is supported by a worthwhile reason, in Lopez Villa's case it is argued that the claim is supported not only by a worthwhile reason, but by one that is stronger than a reason against. In other words, while in Nuckolls' case the property transferred from the source to the target is "to pose a *pro tanto* reason" (assuming the truth or acceptability of what is presented as a reason), in López Villa's case it is "to pose a stronger reason".

This requires some clarifications. As I said, from the perspective of argument dialectics an argument is correct and poses a *pro tanto* reason if it withstands objections and rebuttals, and it is conclusive and poses a relatively strong (or all things consider) reason if it also resists refutations. As we have seen, an objection is an argument whose conclusion is incompatible with some premise of the criticised argument; a rebuttal is an argument whose conclusion attacks the relationship between the premises and the conclusion of the criticised argument, and a refutation is an argument that gives a stronger reason for an incompatible conclusion. So, both "to be a *pro tanto* reason" and "to be a stronger reason" are properties that are defined with respect to an argumentative context. In this sense, argumentation by parity of reasons can be seen as the reverse of rebuttals, and argumentation by parity of weighings as the reverse of refutations. In the first variety, it is argued that the target argument resists rebuttals, because it parallels the source argument which –it is taken for granted– does so. And in the second variety, it is argued that the target argument resists refutations, because it parallels the source argument which –it is taken for granted– does so.

4. Evaluation and principles.

The account that I have just presented is clearly particularist. It does not resort to any universal principle that covers the source and target cases and turns the argument into a deductive one, neither full nor partial. Arguments by analogy are case-to-case arguments based on a parallelism of relationships. We can recognise similarities between ways of arguing and using them as reasons to justify our own arguments. However, someone might reply that what I have just presented is an analysis, but that generalism and particularism are theories of evaluation. The question is not so much whether these arguments are principled, but whether they can be evaluated without recourse to principles of any kind. To answer this question, I will build on Govier's suggestion that, although "the universal claim might be implied by the argument" (Govier 1989, p. 147), "we can evaluate [it] without raising the issue, just by sticking to the cases at hand" (Ibid, p. 148).

Let us start with the basics, what would these principles look like? One thing that sets me apart from many of those who have discussed this issue is that I advocate a meta-argumentative approach. In my view, when we argue by analogy, we contend that a claim is favoured by reasons, because the case is parallel to a case in which we assume that another claim is favoured for other reasons. But, as we have seen, this is precisely the job of the warrant. Take the Nuckolls case. Instead of arguing by analogy, she could have said "because if something is forever linked to a negative idea, it is better not to use it as a starting point for a new thing". In short, arguments by analogy thus understood can be regarded as a substitute for

the warrant. However, there are two readings of this idea, which reproduce the debate between generalists and particularists:
1. Two arguments are analogous because they follow the same principle or rule, or
2. Two arguments can be considered as following the same principle or rule because they are analogous (Marraud 2020, p. 5).

Of course, both positions are open to nuance. Generalists might say that warrants are implicit in some way and, in any case, can be refined and modified on the fly; and particularists might argue that we often –or even mostly– argue using warrants, but that there are argumentative practices that do not require them, and arguments by analogy may be part of those practices (*see* Lamond 2005). The problem is the same with argumentation by parity of weighings, but at a higher level: now we have principles or rules that attribute more strength to one reason than to another. In fact, a particularist might take advantage of this specificity and argue that, although at the level of warrants generalism has some plausibility, a universal principle of the second type is a hard pill to swallow, because the strength of reasons is largely determined by contextual factors (*see* Bader 2016).

The question is then how to evaluate these arguments without resorting to warrant-like principles. My answer is quite simple: by looking for relevant differences between the cases compared. But what is a relevant difference? Well, the same as a relevant similarity but in reverse. In the first section I advanced the idea that depending on the property to be transferred, the comparison will highlight different aspects of the source and the target cases. Something similar happens with differences. If the arguer contends that a claim is favoured by worthwhile reasons, then relevant differences will be those considerations that show either in the source or in the target that this does not really happens. This is done, in argument dialectics, by looking for rebuttals to the reasons posed by the source and target arguments. On the contrary, if the arguer contends that a claim is favoured by relatively strong reasons, then relevant differences will be those considerations which reverse the strength attributed to the reasons weighed, either in the source or in the target case. Here we are looking for refutations or contextual factors that alter the strength of reasons. Thus, the evaluation process would go along these lines:

Table IV. An evaluation process for argumentation by analogy.

1. An analogy is put forward as a reason for a claim.
2. We identify the variety by looking at the arguer's claim.
3. We search for counterarguments:
   3.1. Should it be the first variety, search for rebuttals either to the source or to the target.
   3.2. Should it be the second variety, search for refutations or contextual factors that alter the strength of reasons considered either in the source or in the target.
4. We set the outcome:

4.1. Should it be any counterargument, the argument is incorrect and fails to pose any reason
4.2. Should it be no counterargument, the argument can be considered correct and that it poses a pro tanto reason.

In short, we have a process that does not resort to general principles or rules and therefore avoids generalism problems. This process is case-driven: depending on the property to be transferred in the case at hand, the analogy will highlight some relationships or others, and that gives us the key to look for relevant differences, i.e., adequate counterarguments

But then, in what sense is the universal principle *implied* by the argument? Govier does not make it easy. She contends that the universal principle, which she calls "U-Claim", is neither a background assumption (an example being the principle of non-contradiction in deductive logic), nor an implicit premise. And yet it is somehow implied by the arguer: "the use of an argument by analogy does commit the arguer to some U-claim in the sense that if what she says in her argument is right, then some U-claim must be true" (Govier 1989, p. 148). My position on this is that, if the argument by analogy is good (in the sense I have just outlined), then it can be seen *as a reason for* a general principle or rule covering cases compared. In other words, principles should be seen as a by-product of good analogies, not the other way around. This qualifies Govier's claim in two ways. First, good arguments by analogy do not imply but are reasons for general principles, i.e., they can be defeated or overcome by stronger reasons. And secondly, arguments by analogy need not always lead to a defence of any principle, sometimes we simply move at the level of cases without any intention of generality.

5. Conclusions.

Generalism and particularism give opposite answers to the question of whether arguments by analogy need principles to support their claim. For generalism, these arguments presuppose a principle that attributes the property to be transferred to any object having the features shared by the source and the target cases, which makes the argument a deduction, either full or partial. Particularism rejects this position and argues that arguments by analogy are case-to-case arguments based on the similarities and differences between the source and target cases. Here I have defended a particularist position.

First, I have shown that arguments by analogy operates on a parallelism of relationships. These relationships are argumentative in nature, which turns arguments by analogy into meta-arguments, i.e., arguments about other arguments. I have shown that argumentative relationships may be of different sorts, giving rise to different varieties of arguments by analogy. Here I have distinguished two varieties: argumentation by parity of reasons and argumentation by parity of weighings. Second, I have shown that at this meta-argumentative level principles correspond to Toulminian warrants. Generalists contend that every argument, including arguments by analogy, require a warrant in order to be evaluated, while particularists argues that there are argumentative practices that indeed require warrants, but others that do not, and arguing by analogy may be part of the latter. Finally, I have proposed a procedure for evaluating arguments by analogy without recourse to warrant-like principles, and have argued that, rather than being a requirement of arguments by analogy, they can be seen as a by-product of good analogies.

Acknowledges: This work has been possible thanks to a grant for Research Staff Training of University Autonomous of Madrid (FPI-UAM - for its Spanish acronym) and the research project *Argumentative Practices and Pragmatics of Reasons* (PGC 2018-09594-B-100) of the Spanish Ministry of Science, Innovation and Universities.

References

Alhambra, J. (2022). "Argumentation by Analogy and Weighing of Reasons". Informal Logic, 42(4), 749–785.
Bader, Ralf M. (2016). "Conditions, Modifiers and Holism". In: Weighing Reasons, eds. Errol Lord and Barry Maguire, 27-55. Cambridge: Cambridge University Press.
Beardsley, M. ([1956] 1975). Thinking Straight. Englewood Cliff: Prentice Hall.
Bermejo-Luque, L. (2012). "A Unitary Schema for Arguments by Analogy". Informal Logic, 32 (1): 1-24.
(2014). "Deduction without Dogmas: The Case of Moral Analogical Argumentation". Informal Logic, 34(3): 311-336.
Botting, D. (2017). "The Cumulative Force of Analogies". Logic and Logical Philosophy, 17, 3: 1–37
(2022). "Deductive Analogy Defended (En defensa de la analogía deductiva)". Revista Iberoamericana De Argumentación, (23), 1–22.
Dancy, J. (2004). Ethics without principles. Oxford New York: Clarendon Press Oxford University Press.
Gentner, D. (1983). "Structure-Mapping: A Theoretical Framework for Analogy". Cognitive Science, 7(2), 155-170.
Govier, T. (1989). "Analogies and Missing Premises". Informal Logic, 11(3): 141-152.
(2002). "Should a Priori Analogies Be Regarded as Deductive Arguments?". Informal Logic, 22(2), 155-157.
([1987] 2017). Problems in Argumentation and Evaluation. Windsor: Windsor Studies in Argumentation.

Guarini, M. (2004). "A Defence of Non-deductive Reconstructions of Analogical Arguments". Informal Logic, Vol. 24, No.2, 153-168.
Holyoak, K. J., and Thagard, P. (1997). "The Analogical Mind". American Psychologist, 52(1), 35-44.
Juthe, A. (2005). "Arguments by Analogy". Argumentation 19: 1-27.
(2016). "Argumentation by Analogy: A Systematic Analytical Study of an Argument Scheme", Dissertation, University of Amsterdam, The Netherlands.
(2019) "A Defense of Analogy Inference as Sui Generis", Logic and Logical Philosophy, 1-51.
van Laar, Jan Albert (2014) "Arguments from Parallel Reasoning". In: Henrique Jales Ribeiro (ed.) Systematic Approaches to Argument by Analogy. Amsterdam: Springer, pp. 91-107.
Lamond, G. (2005). Do precedents create rules?. Legal Theory, 11(01): 1-26.
Leal, F., and H. Marraud (2022). How Philosophers Argue. An Adversarial Collaboration on the Russell-Copleston Debate. Switzerland: Springer.
Marraud, H. (2007). "La analogía como transferencia argumentativa". Theoria, 0(59): 167-188.
(2016). «Argumentos e inferencias: teoría de la argumentación y psicología del razonamiento». Cogency, 7(1), 47-68.
(2020). "A Modest Proposal for Classifying the Theories of Argument". Unpublished. Retrieved from: https://www.academia.edu/44808846/A_MODEST_PROPOSAL_FOR_CLASSIFYING_THE_THEORIES_OF_ARGUMENT
2022. "An Unconscious Universal in the Mind is Like an Immaterial Dinner in the Stomach. A Debate on Logical Generalism (1914–1919)". Argumentation, pp. 1-25
Mckeever, S. and M. Ridge (2006). Principled Ethics. Generalism as a Regulative Ideal. Oxford: Oxford University Press.
Perelman, Ch. and Olbrechts-Tyteca, L. ([1958] 1971). The New Rhetoric. A Treatise on Argumentation. Notre Dame: University of Notre Dame Press.
Shecaira, F. P. (2013). "Analogical Arguments in Ethics and Law: A Defence of a Deductivist Analysis". Informal Logic, 33(3): 406-437.
Shelley, C. (2004). "Analogy Counterarguments: A Taxonomy for Critical Thinking". Argumentation, 0(18): 223–238.
Stevens, K. (2018). "Case-to-Case Arguments". Argumentation, 32 (3):431-455.
Toulmin, S. E. (2003 [1958]). The Uses of Argument (updated edition). Cambridge: Cambridge University Press
Waller, Bruce (2001). "Classifying and Analyzing Analogies", Informal Logic, 21(3): 199-218.
Wisdom, J. ([1954] 1991). Proof and Explanation. The Virginia Lectures. Maryland: University Press of America.

Alhambra

A MINIMALIST APPROACH TO ARGUMENTATION BY ANALOGY. COMMENTARY ON JOSÉ ALHAMBRA'S A PARTICULARIST THEORY OF ARGUMENTATION BY ANALOGY

JEAN H.M. WAGEMANS
University of Amsterdam
j.h.m.wagemans@uva.nl

1. Introduction

In his paper titled *A particularist theory of argumentation by analogy*, José Alhambra characterizes the generalist perspective on this type of argumentation as one that necessitates a universal principle that covers cases compared and guarantees their conclusion' (p. 1). Alhambra opposes this perspective by incorporating the distinction between generalism and particularism, which originates from the theory of normative reasons, to the concept of argumentation by analogy. His particularist theory is interesting and relevant. Argumentation by analogy is found in many categorizations of arguments ranging from classical rhetoric to present-day argumentation theory – for a recent overview, see Wagemans (2021b, pp. 584-586). Also, it is a much debated type of argument – for an extensive survey of the literature, see Juthe (2016, pp. 23-68).

Alhambra states that the generalist perspective on arguments from analogy involves describing them as necessitating a universalization in the form of "all things that have x, y, z are W" that "subsumes the cases compared and guarantees the conclusion of the argument, so that they must be included in the analysis as implicit premises'". He then cites the most significant objections to this perspective, including the redundancy of mentioning the source case, the uncharitable attribution of the implicit premise to the arguer, the resulting variation in strength that contradicts the deductive nature of the completed argument, and the enigmatic nature of the generalization's origin. Furthermore, Alhambra demonstrates that the suggested solution, which involves reconstructing the argument by analogy as a combination of abduction from the source case to a general principle and deduction from that principle to the target case, invalidates the reference to this combination as an argument by analogy, as the argument does not depend on a comparison between a source and target case.

I find Alhambra's refutation of the generalist perspective very convincing. As I have explained elsewhere (Wagemans, 2020), the practice of adding premises to arguments can be problematic, especially when it is done to ensure their conclusion. In this commentary, therefore, I shall focus on the subsequent task Alhambra sets for the particularist, namely, 'to propose a theory that dispenses with these principles in the explanation of arguments by analogy'. I shall examine whether Alhambra's proposal indeed dispenses with these principles and, more importantly, whether it saves us from running into similar troubles as the generalist perspective.

2. A remaining issue

According to Alhambra's particularist theory, argumentation by analogy involves the source and target cases being treated as arguments in and of themselves, each with a premise and conclusion. He argues that, consequently, 'the relationships compared in an argument by analogy are argumentative relationships' and that 'arguments by analogy are case-to-case arguments based on a parallelism of relationships'. In this way, Alhambra avoids some of the problems associated with the generalist approach to argumentation by analogy, which relies on a universal principle that connects source and target cases and turns the argument into a deductive one. But it does not seem to save us from all of them. Let me illustrate this point through an example.

Example 1 (original text). *Cycling on the grass is prohibited because walking on the grass is prohibited.*

The example is taken from the legal domain, where the concept of analogy plays a vital role, as laws usually do not cover all cases (Kolb, 2016, p. 152). But how and why did we identify this combination of statements as an argument from analogy in the first place?

To answer this question, I use the Argument Type Identification Procedure (ATIP) (Wagemans, 2021a), a step-by-step method enabling identification in terms of the argument categorization framework of the Periodic Table of Arguments (PTA). Within this framework, an argument type is conceived as a combination of the values of three parameters: form, substance, and lever. When applying the procedure to the example, it can easily be shown the argument has the form 'a is X, because b is X' and the substance 'VV'.

Example 1 (annotated text). *Cycling on the grass (a) is prohibited (X) (V) because walking on the grass (b) is prohibited (X) (V).*

The trouble lies in determining the value of the third parameter, the argument lever, which expresses the underlying mechanism of the argument and is not present in the linguistic material. Formulating such an implicit lever can be challenging: the reader or listener must add something to the discourse, and in doing so, runs the risk of adding something the author or speaker did not intend to communicate or could plausibly deny.

In my view, which could be called a 'minimalist' approach to argument type identification, we are only allowed to name an argument with such a configuration of subjects and predicates an 'argument from analogy' if we can justify in some way or the other that the argument lever can be conceived as 'cycling on the grass (a) is ANALOGOUS to walking on the grass (b)'. But this is not trivial. Since there is no characterization of the relationship between a and b in the linguistic material, we must choose the best fitting candidate among the levers associated with the argument form (see also Wagemans, 2018): Is a ANALOGOUS to b? Or is a SIMILAR to b? Or EQUAL, or PARALLEL? Depending on our choice, the argument receives its name, which usually derives from the lever – see www.periodic-table-of-arguments.org/argument-lever for a permanently updated list of levers and names that can be used as a heuristic tool. In legal argumentation, there are apparently good reasons to opt for ANALOGOUS as the best candidate.

On Alhambra's particularist account, we should analyze this example differently. According to him, argumentation by analogy is to be seen as a parallelism of argumentative relationships. Applied to this example, it would mean that are two extra arguments hidden under the surface. And indeed, we could say that the reason for why walking on the grass is prohibited is that it damages the grass, and that this is the same reason for why cycling on the grass is forbidden. By conceiving the argument in this way, however, we evoke at least two of the problems mentioned in Alhambra's refutation of the generalist perspective: (1) uncharitable attribution of the implicit premise to the arguer and (2) the enigmatic nature of the origin of the addition to the discourse.

3. Conclusion

I fully subscribe to Alhambra's criticisms of the generalist perspective on argumentation by analogy. I am not so sure, however, that his particularist theory solves all the problems he himself associates with the generalist one. While the analyst or addressee of the argument does not have to add a general rule for analytical and evaluative purposes, according to the Alhambra's account, they still must think of source and target as 'meta-arguments' and assume a parallelism between them, thereby going quite far beyond what can be found in the linguistic expression of the argument under scrutiny.

My concern with both the particularist and the generalist view of argumentation by analogy is that they add too much to the discourse to identify the argument type, thus running the risk of providing an uncharitable interpretation and not being able to provide a satisfactory justification of the origin of what is added. As explained above, this may have to do with the fact that when it comes to argument reconstruction and argument type identification, I prefer a 'minimalist' approach.

That said, I think Alhambra and I agree on many points, and I am looking forward to reading the next version of his particularist account, which will hopefully address the remaining issues raised in this commentary.

References

Alhambra, J. (this volume). A particularist theory of argumentation by analogy. In A. Ansani, M. Marini & F. Paglieri (Eds.), ECA 2022 Proceedings London: College Publications.

Juthe, L.J.A. (2016). Argumentation by analogy: A systematic analytical study of an argument scheme. Dissertation University of Amsterdam.

Kolb, R. (2016). The law of treaties. An introduction. Cheltenham: Edward Elgar Publishing.

Wagemans, J.H.M. (2018). Analogy, similarity, and the Periodic Table of Arguments. Studies in Logic, Grammar and Rhetoric, 55 (68), 63-75. DOI: https://doi.org/10.2478/slgr-2018-0028

Wagemans, J.H.M. (2020). Why missing premises can be missed: Evaluating arguments by determining their lever. In J. Cook (Ed.), Proceedings of OSSA 12: Evidence, Persuasion & Diversity. Windsor, ON: OSSA Conference Archive. URL = https://scholar.uwindsor.ca/ossaarchive/OSSA12/Saturday/1

Wagemans, J.H.M. (2021a). Argument Type Identification Procedure (ATIP) – Version 4. Published online December 30, 2021. URL = www.periodic-table-of-arguments.org/argument-type-identification-procedure

Wagemans, J.H.M. (2021b). The Philosophy of Argument. In P. Stalmaszczyk (Ed.), The Cambridge Handbook of the Philosophy of Language (pp. 571-589). Cambridge: Cambridge University Press.

Wagemans, J.H.M. (2023). How to identify an argument type? On the hermeneutics of persuasive discourse. Journal of Pragmatics, 203, 117-129. DOI: https://doi.org/10.1016/j.pragma.2022.11.015

COMMUNICATIVE ACTIVISM AS AN 'ALTERNATIVE' ARGUMENTATIVE STRATEGY TO CONFRONT CONTEXTS AFFECTED BY HATE SPEECH

ÁLVARO DOMÍNGUEZ-ARMAS

NOVA Institute of Philosophy, NOVA University of Lisbon
a.dgueza@campus.fcsh.unl.pt

Abstract
This paper describes communicative activism as an 'alternative' argumentative strategy in arguments contaminated by hate speech. Communicative activism is a category of communicative confrontations carried by social movements that challenge hate speech and aims to bring bystanders to the protests. The paper will first reconstruct the abortion debate in Argentina, identifying the protests of Pañuelos verdes as argumentative contributions defending the legalisation of abortion. It will then describe how hate speech was used to discredit the Pañuelos verdes protests. Finally, three argumentative strategies to confront hate speech are examined: (a) arguing with hate-speakers; (b) advocating for a dialogue with restrictions; (c) opting for 'communicative activism'.

1. Introduction

This paper aims to defend communicative activism as an 'alternative' argumentative move. 'Communicative activism' refers to those confrontative communicative practices that aim to hinder the passage of hate speech and pull third parties into the protests. I will refer to the Pañuelos verdes protests to exemplify communicative activism. To illustrate my argument, I will first reconstruct the debate about the legalisation of abortion in Argentina. There will then be an analysis of how social actors in positions of power used hate speech to prevent the protests of Pañuelos verdes from counting as reasonable contributions to the debate. Finally, I will examine three argumentative strategies to confront the harms of hate speech: (a) to continue arguing with hate speakers; (b) to establish a dialogue with restrictions; (c) opting for 'communicative activism', which is the strategy that I favour.

2. The debate about abortion in Argentina

Since the establishment of the democratic political system in Argentina (1983), the debate about the legalisation of abortion has played an important role in the political sphere (Dulbecco et al., 2021)[1]. In this section, I will reconstruct the participants and positions of the debate in argumentative terms. This paper embraces the definition of argumentation as a communicative activity of producing and exchanging reasons in situations of doubt or disagreement where argumentation functions as a form of disagreement management (Jackson, 2015; Lewiński & Mohammed, 2016).[2]

The simplest reconstruction of the abortion debate in Argentina would be to see it as a two-sided debate (Lewiński, 2019) with participants exchanging reasons about the legalisation (yes-side) or illegalisation of abortion (no-side) between the years 1983 to 2020. Positions in support of the legalisation of abortion, like the one defended by Ginés González García (ex-minister of Health during the presidency of Néstor Kirchner (2003-2007), exemplify their position by creating campaigns to legalise/secure abortion in Argentina. Opponents of the legalisation of abortion, such as Cristina Fernández de Kirchner (ex-president of Argentina 2007-2015) and Mauricio Macri (president of Argentina 2015-2019), defended their standpoint in multiple interviews and parliamentary sessions by characterising abortion as a criminal practice (Bessone, 2020).

However, on closer scrutiny, this reductionist account of the perspectives of the debate is unsustainable. Dulbecco et al. (2021) identify

---

1 Note that the debate about the legalisation of abortion in Argentina was structured by the question: Should abortion be legalized? (Besonne, 2020). Arguing about 'what to do' in society is described in the literature as practical reasoning; the question that structures politics (Fairclough & Fairclough, 2012; Lewiński, 2021). Notice the difference between practical and theoretical reasoning. While the former reasons about what to do, the second reasons about what to believe (Lewiński, 2017, 2021). These questions are the only ones in the abortion debate. For example, Schiappa (2003) describes the abortion debate as involving a conceptual debate about 'what is X?' (where X refers to 'person').

2 This paper focuses on the dialectical perspective of argumentation that relates to the regulation of discussions among people (Wenzel, 1990, 14). The dialectical perspective calls up a procedural sense of argument and considers all the methods people and institutions use in order to bring the processes of arguing to produce good decisions. Thus, the goal of the dialectical perspective is the study of rules, limits, and formats for debate and discussion (Wenzel, 1990).

diverse participants with different perspectives. On the one hand, they identify participants in the Argentinian Parliament who defended the legalisation of abortion because of (i) its immediate effect on the number of deaths caused by practising abortion in unhealthy conditions (the 'health perspective') and (ii) the intrinsic right of women to choose what should be done with their bodies (the 'feminist perspective'). On the other hand, perspectives that defended the criminalisation of abortion because (i) it is a morally reprehensible act from a religious point of view (the 'religious perspective') and (ii) it was already established as an illegal practice in the Penal Code by the law 'Interrupción Voluntaria del Embarazo' (IVE) (Ley 27.610) since 1921 (the 'legal perspective'). [3] Moreover, Dulbecco et al. (2021) identify participants of the debate outside of the Argentine Parliament (the 'social perspectives'). For instance, organizations against the legalisation of abortion which represented religious views (i.e., 'Pañuelos celestes') and social movements in support of the legalisation of abortion (i.e., Pañuelos verdes) (Rosende, 2018). [4] Finally, it included citizens not directly involved in the abortion debate (the 'bystanders').

Therefore, the debate about abortion in Argentina cannot be reduced to a dyadic difference of opinion (yes/no to legalisation of abortion). The work of Dulbecco et al. (2021) instead allows to characterise the debate about abortion in Argentina as a multi-party argumentative discussion (namely a 'polylogue' (Aakhus & Lewiński, 2017)) in which 'multiple players discuss their distinct, and often incompatible, positions across a variety of places' (Lewiński, 2021, 435). The different 'players' identified by Dulbecco et al. (2021) differentiated their 'positions' on the legalisation/illegalisation of abortion in different 'places': politicians advanced their arguments in parliament, social movements on social media, and on the streets.

Let us now focus on how the latter player (the social movement 'Pañuelos verdes') presented its position in the abortion debate in Argentina. Social movements are collective political agents that emerge from the efforts of ordinary people to participate in policymaking by extra-institutional means ('protests'). Studies in political theory characterise social movements as 'agents for social change' whose protests aim to recruit supporters to the social movement by targeting socially oppressive practices against minorities, typically perpetuated by 'social actors in

---

3 The IVE was established in the Penal Code in 1921 and dictated that women who solicited an abortion and professionals involved in the procedure must be punished with one to four years in prison (Bessone, 2020).
4 The 'social perspectives' have a different locus (their voices are not expressed in parliament) from the others mentioned above, but their argumentative positions are similar to the 'feminist perspective' and the 'religious perspective'.

positions of power' (politicians, social and economic elites, media directors) (Barker & Kennedy, 1996; Meyer et al., 2002; Zheng, 2022).[5]

These characteristics are exemplified by the emergence of Pañuelos verdes in Argentina as the *vox populi* of women victims of inequality in different aspects of society.[6] Pañuelos verdes advocated to challenge multiple aspects in which Argentinian women were affected (e.g., gender violence, wages inequality, unequal capacity to access high positions in companies, etc.). Their protests were aimed at the inequality of women in Argentine society and were also directed toward the social actors in positions of power (such as the government, representatives of religious opinions, the media, etc.) (Muzi, 2018).

Pañuelos verdes emerged in 2004 with the campaign 'Campaña Nacional por el Derecho al Aborto Legal Seguro y Gratuito', which led to multiple protests in support of the legalisation of abortion (Bessone, 2020). Their protests had two interrelated fronts. First, the creation of the hashtag #AbortoLegalYa on Twitter. The hashtag presented reasons to argue for the legality of abortions by appealing to the principle of women's right to decide over their bodies and to the medical implications of legalising abortion (the reduction of deaths from unsafe abortion practices) (Bessone, 2020). Secondly, the development of rallies, blockading streets, occupying public spaces, etc. to highlight their perspective in the public debate. All these actions were baptised as the 'pañuelazo' in support of the legalisation of abortion (Dulbecco et al., 2021).

Describing protests in speech act terms is helpful to study how social movements contribute to political argumentation.[7] Austin (1962) once commented that protests express disappointment. Searle (1975) elaborates on this by describing protests as involving 'an expression of disapproval and a petition for change' (Searle, 1975, 22). Recently, Chrisman and Hubbs (2021) describe protests as involving (i) expressions of opposition that negatively evaluates the object of the protest; (ii) prescriptions for redress to social actors in positions of power to change some aspect of the society. The common point between these approaches is that protests involve expressive speech acts; protestors 'express' disapproval towards the object of the protest.

---

[5] To be a 'social actor in positions of power' means to have a certain authority in the social realm. Such authority can have different forms. For instance, politicians and representatives of the government have a de jure authority in society (Dahl, 1961). By contrast, social media has a de facto authority exemplified by its influence on society (Curran, 2002).
[6] They named the social movement Pañuelos verdes because of their attire: a green scarf that made them easily recognisable (Muzi, 2018).
[7] See Jacobs (1989) for a description of how different speech acts contribute to argumentation.

The next section looks at how other participants in the argument (i.e., social actors in positions of power) used hate speech in response to the Pañuelos verdes protests and its effect on the abortion debate.

3. Hate speech in response to protests

Although social movements protest to bring about change in society, protests can have adverse effects (Zheng, 2022). For example, social actors in positions of power may use physical means to suppress protests, e.g. with intense repression by riot police (Blay, 2013) or use hateful communicative practices (hereafter, hate speech) to oppress protests (Young, 1990).

The focus of this paper is how to deal with the harms of hate speech in argumentative exchanges. But how does hate speech affect argumentation? The response to this question is not widely explored in argumentation theory.[8] I will focus on contributions from the philosophy of language to fill this gap.

It should be noted that the definition of hate speech is a subject of debate among scholars. In this paper, I adopt the definition of hate speech as public expressions that spread, incite, or justify discrimination, subordination, and hostility against its victims (Waldron, 2012). Victims of hate speech can be historically oppressed groups (Bettcher, 2014), class, sex or gender (Cortina, 2017), or social movements.

3.1. Hate speech makes protests not count as contributions to the abortion debate

Recent contributions in the philosophy of language address how hate speech harms its victims. Langton (1993, 2018b) develops a pragmatic analysis of hate speech and describes three harms that it constitutes: (i) subordination; (ii) legitimation of discriminatory behaviour, and (iii) silencing of the target. By using hate speech, she says, a system of oppression is enacted: hate speakers rank a given group as inferior with their hate speech and thus legitimise discrimination against them. In

---

[8] The harm that different forms of hate speech do to argumentation is beginning to be a topic of concern for some studies. Bondy (2010) draws on Fricker's (2007) concept of epistemic injustice to describe situations in which arguers are unable to construct an argument because they lack credibility in the context in which they argue. Bowell (2021) investigates cases in which virtuous argumentation might harm the marginalised. Innocenti (2022) deals with situations in which social actors retreat to metadiscussions to derail ground-level discussions by qualifying claims of others by using 'not-all' qualifiers (e.g., stating that 'not all men rape' in discussions of systemic sexism).

doing so, hate speech makes the victim count as an inappropriate speaker for further speech acts.[9] Langton (1993) exemplifies this oppression by describing how pornography makes sexual refusals 'unspeakable' for women by celebrating and promoting the sexual abuse of women. Pornography thus makes sexual refusal not count as the intended act even if women utter the appropriate locution (i.e., 'No') and want their act to be taken as such (Langton, 1993).

Although Langton's (1993) idea that hate speech (i) subordinates, and (ii) legitimises discrimination towards its victims is accepted in the literature (see Kukla, 2018; McGowan, 2019), her idea that hate speech (iii) silences is contested (Kukla, 2014). Kukla (2014) argues that silencing is a borderline case of the harm of hate speech because it transforms the intended act of the victim into no act at all. Kukla (2014) argues that hate speech generally distorts victims' acts by receiving a response different to the one they intended solely because they belong to a certain group.

I find Kukla's (2014) account useful in describing the case of Pañuelos verdes: that social actors in positions of power used hate speech to make the pañuelazo be seen as another kind of act (i.e., as a threat to social security).[10] Let us examine this idea by describing the effects of some derogatory terms (i.e., slurs) and surreptitious messages (i.e., provocative insinuations) used against Pañuelos verdes.

Some Spanish tabloids and representatives of conservative views in the Argentinian Parliament described the pañuelazo as run by 'feminazis' (Amnesty International, 2018; Díez, 2020). 'Feminazi' is a derogatory term that refers to women who defend abortion (Williams, 2015). Lexical items like 'feminazi' are called 'slurs' in the philosophy of language. Slurs convey the derogatory conceptualisation of a group and function to disparage the target (Cepollaro et al., 2019; Jeshion, 2013). Slurs harm their victims because they rank people as low value in society (Jeshion, 2013). Accordingly, I think that social actors in positions of power characterised Pañuelos verdes as unworthy of consideration in society by using the slur 'feminazi'. The term 'feminazi' is originally from Limbaugh's The Way Things Ought to Be (1992). He defines the term 'feminazi' as referring to 'a woman to whom the most important thing in life is seeing to it that as many abortions as possible are performed' (Limbaugh, 1992, 194). Therefore, in referring to Pañuelos verdes as 'feminazis', the pañuelazo

---

9 Langton (2015, 2018b) adopts Witek's (2013) idea that authoritative acts can shift the felicity conditions of other speech acts. We can describe social actors in positions of power as having a pre-established authority (Langton, 2015) (see footnote 5) that allows them to shift the conditions of social movements' acts.

10 Bianchi (2020) exemplifies discursive injustice with the case of a gay activist intending to assert that some behaviour is homophobic and, because of his membership in a low-status group (i.e., a different sexual orientation than the local majority), his performance is taken up as a mere subjective expression of feelings.

were associated with genocidal practices and treated as threats to social security due to the notion of abortion as the 'elimination of a human life' (Vatican News, 2020).

The discrediting of the pañuelazo also occurred through more surreptitious means. For example, the religious journal AICA (2018) reported on the protests of Pañuelos verdes in Argentina in 2018 with the headline 'Una clínica privada de Neuquén que se opone a realizar esta prácitica [abortar] repudió escrache de los "pañuelos verdes" ['A private clinic in Neuquén that opposes the practice [abortion] repudiated the "Pañuelos verdes" attack.]. Although the headline reports a fact, it could also be read as conveying the message that Pañuelos verdes are violent/delinquent and thereby licensing the inference that the pañuelazo pose a social risk because it obliges health centres to perform abortion. Domínguez-Armas and Soria-Ruiz describe sentences like the one in the AICA headline as 'provocative insinuations' (Domínguez-Armas & Soria-Ruiz, 2021).

Insinuations are pragmatic off-record inferences triggered by specific sentences. Insinuations consist of two messages: an unobjectionable literal on-record content and an off-record message. For example, imagine a real estate agent saying to potential buyers of a different racial background than the local majority: 'perhaps you would feel more comfortable located in a more... transitional neighbourhood, like Ashwood?' (Camp, 2018, 43). When buyers perceive the explicit suggestion to look for a house elsewhere (on-record content), the implicit insinuation is that they should feel uncomfortable in this neighbourhood because they do not belong to the majority group (off-record content). The AICA headline can be described in a similar way. While the headline reports that Pañuelos verdes attacked a private clinic that refused to perform abortions (on-record content), one can read the off-record message that Pañuelos verdes usually provoke these actions and that they are therefore violent/delinquent.

Unlike Camp's (2018) example, the implicit content of the headline is triggered by the choice of the predicate used. Domínguez-Armas & Soria-Ruiz (2021) describe insinuations of this kind which rely on the pragmatic mechanism of 'conversational eliciture' (Cohen & Kehler, 2021). A conversational eliciture arises when, by choosing a particular predicate (among others), a speaker elicits inferences on the part of the audience that would not otherwise be drawn. [11] For instance, by choosing the predicate 'Una clínica privada de Neuquén que se opone a realizar esta prácitica [abortar] repudió escrache de los "pañuelos verdes"' one is eliciting the inference that Pañuelos verdes typically attack private clinics,

---

[11] Their example can illustrate this mechanism 'the drug-addled undergrad fell off the Torrey Pines cliffs', where the sentence elicits on the audience that drugs caused the undergrad to fall off the cliffs (Cohen and Kehler, 2021, 3).

thereby licensing the inference that the pañuelazo pose a social risk because they oblige health centres to perform abortions.

How does the distortion of Pañuelos verdes protests affect the abortion debate in Argentina? Standard theories of argumentation (i.e., pragma-dialectics) describe argumentation as requiring certain conditions to be met (van Eemeren & Grootendorst, 1984). Argumentation, they maintain, requires that the listener recognises the arguer's utterances as counting as an attempt to justify their opinion. Van Eemeren and Grootendorst call them 'recognition conditions' (1984, 42). If the recognition conditions are not met, the act of argumentation is unsuccessful.

In §2, we reconstructed the pañuelazo as a contribution to the abortion debate in Argentina. The creation of the hashtag #AbortoLegalYa, the blockading of streets, the occupation of public spaces, and the organisation of rallies are understood by political theorists as the way in which Pañuelos verdes presented their views in support of the legalisation of abortion (Dulbecco & et al., 2021). By using hate speech, social actors in positions of power were refusing to meet the recognition conditions for Pañuelos verdes protests; thus reducing their significance to the abortion debate in Argentina. This effect was produced by referring to the pañuelazo with hate speech. This idea is supported by the testimonies of representatives of Pañuelos verdes who claimed that their voices 'were not heard on equal terms' in the abortion debate due to the hate speech used against them (Amnesty International, 2018).

4. Argumentative moves to confront hate speech

The question is: What should social movements do when social actors in positions of power use hate speech to discredit protests? The answer to this normative question is not widely explored in the literature among scholars of argumentation. To fill this gap, this section examines two hypothetical argumentative strategies that argumentation theory can provide: (a) arguing with hate speakers, and (b) advocating for a dialogue with restrictions. I will interpret as an argumentative strategy the communicative practices that contribute to the management of disagreement and the production and exchanging of reasons (§2). After considering the limitations of (a) and (b), I will describe a third alternative argumentative move: (c) opting for 'communicative activism'.

(a) To argue with hate speakers

The first argumentative strategy is to argue with social actors in positions of power by exposing the flaws and undesirability of hate speech with the aim of eventually making them reject their views.[12] One way to develop this argumentative strategy is, for example, to report the slur and its offensive character to the social actors in positions of power in the hope that it will be enough to make them reconsider their positions (see Cepollaro et al., 2019).

Perspectives advocating this argumentative strategy assume that arguers evaluate and consider arguments independently of the roles of the participants in the conversation. These types of arguers are willing to engage in arguments and evaluate, or even retract, their reasons if necessary. This ideal model of arguers is represented in Habermas' (1984) definition of the public sphere as governed by communicative rationality. According to him, public deliberation is guided by 'the force of the better argument', which depends on the formation of competent arguers capable of engaging in communicative exchanges and critically evaluating reasons.

Adopting this vision about arguers has the advantage of explaining the use of hate speech by social actors in positions of power in two ways:

(a) Social actors in positions of power are incompetent at recognising the salient facts of the abortion debate and use hate speech as a reprehensible strategy to win the argument (Bowell, 2021).
(b) Social actors in positions of power are unaware of the consequences of hate speech (Lepoutre, 2019).

In both cases, arguing with social actors in positions of power and exposing the flaws and undesirability of hate speech becomes an (almost) pedagogical activity with two sought-for outcomes: (i) it makes social actors in positions of power aware of the undesirability of using hate speech and (ii) it may end up convincing them to retract their words.

However, this strategy faces significant difficulties. To begin with, Allen (2012) objects to Habermas (1984) in that he leaves the interests of the ideal model of competent arguers unproblematised. Habermas (1984) presupposes that the interests of the arguer are genuine and cannot be a function of unjust social power relations (Allen, 2012). Therefore, the use of hate speech cannot be explained by (a). Hate speech would merely be an action carried out by speakers who are unaware of its consequences (b).

Even if social actors in positions of power are unaware of the consequences of hate speech and arguing with them leads them to retract their words, the effects of hate speech cannot be retracted (Howard, 2021; Kukla & Steinberg, 2021; Zurita & Pérez-Navarro, 2021). Hate speech

---

[12] Langton (2018a) examines a similar strategy by examining Lous Brandeis's idea that, by continuing speaking in contexts affected by hate speech, 'the truth will come out'. She raises similar objections to the ones pointed out in this section.

sticks its harm to society because it constitutes (and not merely causes) such damage (McGowan, 2019). Hate speech enacts hateful social norms that legitimise discriminatory behaviour towards the victims in their context. Once produced these norms cannot be undone by means such as retractation (McGowan, 2019).[13]

Moreover, responses to surreptitious forms of hate speech, such as provocative insinuations, are ineffective at dealing with its hateful effects (Domínguez-Armas and Soria-Ruiz, 2021). Consider again the AICA (2018) headline 'Una clínica privada de Neuquén que se opone a realizar esta práctica [abortar] repudió escrache de los "pañuelos verdes" ['A private clinic in Neuquén that opposes the practice [abortion] repudiated the "Pañuelos verdes" attacks]. The speaker could be accused of insinuating that 'Pañuelos verdes are violent/delinquent' by saying, for example: "Are you saying that Pañuelos verdes are delinquents?". However, the speaker can cancel the insinuation by saying: "No, Pañuelos verdes are not delinquents" or denying ever having intended to convey that content with: "I didn't say/mean that!".[14] This type of response can make the off-record content the topic of conversation. Participants would leave the flow of the abortion debate in Argentina to discuss whether or not Pañuelos verdes are violent/delinquents and whether or not the pañuelazo poses a social risk.

I consider that these theoretical problems make the argumentative strategy inapplicable for social movements. In practice, representatives of Pañuelos verdes tried to implement this strategy. For example, Natalia Mira (a young activist), participated in an interview with journalist Eduardo Feinmann for the TV channel 'A24!'. Mira explicitly stated that legislators, politicians and the media (which I call social actors in positions of power) often discredit their protests by using hate speech. However, Eduardo Feinmann continuously interrupted her because of the gender-neutral words Mira casually used. After three minutes of the interview, Mira refused to continue talking with Eduardo because 'he would not listen to reasons' (Schmidt, 2019). Subsequently, the interview gave rise to multiple reactions on the internet, none of them related to the harms of hate speech, but rather described Mira's intervention as 'typically made by a feminazi', 'ridiculous', etc. (Schmidt, 2019).

---

[13] McGowan (2019) says that retracting hate speech is analogous to un-ringing a bell. For instance, in contexts where sexist comments are uttered that tend to activate sexist associations between women and negative characteristics, saying to the speaker 'but women are not submissive' tends to reinforce this association, rather than making the association un-salient.

[14] Camp (2018) says that speakers can convey insinuations without being held accountable. She describes insinuations as at the same time intentionally recognizable while being communicated. Fricker (2012) says that insinuations are cancellable (she says 'deniable') and disavowable: the speaker can cancel their content and deny having had the intention of conveying that content.

(b) Advocating for a dialogue with restrictions

If arguing with social actors in positions of power is ineffective in dealing with the effects of hate speech, then one might advocate for establishing, as an appropriate strategy, a 'dialogue with restrictions' on what can be said. Waldron (2012) argues for the need for coercive action (i.e., imposing prohibitions on what can be said) to repress the harm of hate speech in society. The criminal effects of hate speech, he says, easily spread to individuals with similar characteristics to the targets because 'even the flare-up of a few particular incidents can have a disproportionate effect' (Waldron, 2012, 17).

Lepoutre (2017) notes that Waldron's (2012) perspective has multiple advantages. First, Waldron's (2012) account is methodologically appealing: he adopts a non-ideal theory by calling for an analysis of the harms produced by hate speech from the perspective of the victims. Second, Waldron (2012) conceives the harm of hate speech as intrinsically related to its use. Therefore, by imposing bans on hate speech, the harm should disappear. In my view, Waldron's (2012) perspective has a third advantage for social movements. By imposing bans on 'what can be said', social movements could feel safe when participating in political debates. Social movements would not have to worry about being discredited by hate speech since it would be forbidden. Therefore, establishing a dialogue with restrictions seems to be an effective argumentative strategy. Imposing bans can be formulated as an agreement upon the rules of the debate. Pragma-dialectics describe this agreement as belonging to one of the stages of an ideal argument (i.e., the opening stage) (Eemeren and Grootendorst, 1984, 163).

However, this optimistic view confronts several difficulties. Generally speaking, policies to ban hate speech are designed to target hate speakers and their behaviour. Gelber (2012) studies different laws against hate speech and concludes that they are premised on an attempt to change the speakers' behaviour. Their ability to ameliorate the harm of hate speech and/or reduce its incidence is limited. The remedy to the harm of hate speech comes from the hate speaker, leaving unanswered the question of how anti-hate speech policies could directly serve to assist the victims of hate speech themselves (Gelber, 2012, 206).

Moreover, advocating for the imposition of bans on what can be said is still couched in the zero-sum debate of free speech versus hate speech. I will not enter into the debate here due to lack of space, but it has been widely analysed in philosophy (Gelber, 2012) and political theory (Waldron, 2012). The idea is that 'free speech' defenders would see this strategy as reprehensible because it censors the free speech of arguers (McGowan, 2021). Additionally, the imposition of bans cannot cover surreptitious practices by which hate speech is promulgated (Tirrell,

2012). Moreover, it can be counterproductive since it prompts hate speakers to look for original ways to spread hate speech (Aikin & Talisse, 2020).

Let us analyse this second strategy in Argentina. Law 23592 of the Argentinian Penal Code establishes discriminatory acts as crimes, based on Article 16 of the National Constitution, describing as 'discriminatory acts or omissions those determined on grounds such as race, religion, nationality, ideology, political or trade union opinion, sex, economic position, social status or physical characteristics' (Sanc. 3/VIII/1988; prom. 23/VIII/1988; B.O., 5/IX/1988). Although the law collects the explicit derogatory terms described in §2, social actors in positions of power keep using them to describe the pañuelazo (France 24, 2018). Furthermore, restricting the surreptitious linguistic practices by which hate speech against Pañuelos verdes was promulgated was an arduous task. Provocative insinuations, as described above, do not fall under the definition of discriminatory acts given in Law 23592 because they do not explicitly discredit Pañuelos verdes. Therefore, a dialogue with restrictions seems to be insufficient to deal with hate speech targeted at social movements.

(c) Communicative activism

We seem to find ourselves in a paradoxical situation: social movements cannot resolve discussions affected by hate speech through 'standard' argumentative strategies. Non-coercive means, such as arguing with hate speakers, are theoretically and practically problematic. Coercive means, such as imposing bans, contribute to the zero-sum debate of free speech versus hate speech.

I believe that argumentation theory can benefit from contemporary advances in political theory and the social philosophy of language to solve this paradox. On this basis, I propose a third strategy: communicative activism. Communicative activism is a category of disruptive public responses (so-called counterspeech) of social movements to hate speech to counter its harm (Lepoutre, 2019). Communicative activism, in my view, is an argumentative strategy because it builds on the ameliorative project of ensuring the dialectical conditions for discussions by empowering social movements in arguments where their voices are not equally listened to (Tirrell, 2019). To this end, communicative activism raises the salience of the hate speech (Lepoutre, 2019) to third parties in order to bring them into the protests. The key role of third parties in communicative activism highlights the importance of the notion of polylogue (§1) for communicative activism. I see communicative activism as being based on three principles:
    (a) It is composed of communicative practices that make explicit and challenge hate speech (Langton, 2018a).

(b) It is used to make a correct association of concepts about the protests of social movements (Lepoutre, 2019).

(c) It aims to bring bystanders into the protests (Bohman, 1996).

There is only space to introduce two modes of communicative activism which address the explicit and implicit forms of hate speech presented in §2, namely: slurs and provocative insinuations. These examples are taken from protest banners carried by members of Pañuelos verdes:

(1) ¡Es feminista no feminazi! [It's feminist, not feminazi!][15]

(2) Quienes estamos a favor del aborto no te obligamos a abortar. Luchamos por el derecho que tienes a decidir si continúas con tu embarazo [Those of us who are pro-choice are not forcing you to have an abortion. We fight for your right to decide whether to continue with your pregnancy.][16]

(1) addresses the derogatory term 'feminazi'. (1) is an obvious form of communicative activism by rejecting the hateful term used to refer to social movements and posing a correct association as salient (Gray & Lennertz, 2020).[17] I think that (1) can be read according to the principles of communicative activism: (1) explicitly challenges the term feminazi (a) and it is used to highlight the association of 'feminazi' to their protests as inappropriate (b) for those reading the sign (c). However, it is less clear that (1) aims to draw bystanders to the debate (c). One can think that (1) is aimed exclusively at those who refer Pañuelos verdes as 'feminazis'. Even if this criticism was correct, (1) can be described as a form of communicative activism by virtue of (a) and (b).

Describing (2) is a more difficult task. An insightful description would describe (2) as 'blocking' (Langton, 2018a) implicit hate speech used to discredit the pañuelazo. To block is to 'hinder the passage, progress, or accomplishment of something by, as if by, interposing an obstruction' (Langton, 2018a, 145). Blocking occurs when speakers interfere by hindering the accommodation of implicit information taken for granted by other participants in a conversation. For example, a situation in which a fan of a football team shouts to a sluggish player 'Get on with it, Laurie, you great girl' and an alerted speaker replies 'Hey, what's wrong with a girl?' (Langton, 2018a, 145, italics in the original). The speaker is blocking the hateful presupposition of the fan that women are gentle and obliging by making explicit and challenging the implicit content (Langton, 2018a, 147). But blocking can also be done by other means (e.g., rephrasing what is said, raising an eyebrow, a joke, etc.).

---

15 https://twitter.com/profetaargen/status/972277617582895104?lang=eu
16 https://efeminista.com/lucha-local-despenalizacion-aborto- Argentina/
17 Gray and Lennertz (2020) name the act of explicitly refusing to take up the hateful associations that hate speech aims to make salient as 'linguistic disobedience' because the speaker disobeys the social norm that hate speech enacts in the context where is uttered (McGowan, 2019).

(2) can be described in a similar vein. (2) makes explicit and challenges the hate association that surreptitious hate speech like the AICA headline is conveying (i.e., Pañuelos verdes forces abortion). However, an important aspect of (2) is its target Instead of the hate speaker, as in Langton's (2018a) blocking example, (2) addresses bystanders of the abortion debate. By saying 'we fight for your right to decide', Pañuelos verdes are making explicit who the audience is (i.e., women). Therefore, (2) seems to be slightly different from the standard definition of blocking.

I suggest analysing (2) as 'reflective' blocking.[18] Like Langton's (2018a) definition, reflective blocking challenges implicit hate speech that otherwise would be accommodated. However, reflective blocking 'reflects' hate speech by directing the challenge to bystanders. Reflective blocking is particularly interesting in the case of social movements, as one of their goals is to recruit bystanders to join their protests (§2) (Barker & Kennedy, 1996). Therefore, reflective blocking is a manner of drawing unwitting bystanders to the political debate (Bohman, 1996).

In short, communicative activism is a category of confrontational communicative practice that challenges hate speech and aims to bring bystanders to the protests. But why is communicative activism preferable for social movements compared with the previous argumentative strategies? First, without rejecting the importance of competent arguers in the public sphere, communicative activism is not necessarily based on an ideal conceptualisation of arguers. Rather than trying to convince the hate speaker, communicative activism aims to bring third parties to the discussion. Secondly, opting for communicative activism does not entail entering the free speech debate. Communicative activism is a non-coercive strategy that does not impose prohibitions on what can be said. Moreover, communicative activism does not contribute to the use of surreptitious practices, as it is based on making the implicit content of these forms of hate speech both explicit and explicitly challenged.

However, the characterisation of communicative activism as an argumentative strategy might raise objections in argumentation theory. For example, pragma-dialectics would describe communicative activism as *fallacious*, as it hinders the accomplishment of critical discussions, namely, the resolution (on the merits) of a difference of opinion. Pragma-dialectics identify several rules for an ideal model of a critical discussion that serves as a code of conduct for rational discussants (van Eemeren & Grootendorst, 1984). Each rule violation is a potential threat to the successful conclusion of the discussion. Rule violations are what they call 'fallacies'.

Pragma-dialectics would describe communicative activism as a violation of the 'Rule IV: a standpoint may be defended only by advancing

---

[18] Cepollaro (ms) describes a taxonomy of the different varieties of blocking to which 'reflecting' blocking could be added.

argumentation relating to that standpoint' (van Eemeren and Grootendorst, 1992). For example, (1) 'It's feminist, not feminazi!' can be described as unrelated to the point of view of Pañuelos verdes towards the legalisation of abortion. In this case, communicative activism is a form of 'non-argumentative means of persuasion' that aims to gain audience approval by other means than persuasion (van Eemeren & Grootendorst, 1992).[19]

However, recall that pragma-dialectics describe the rules of critical discussions as an ideal model of conduct that presumes ideal conditions. Van Eemeren et al. (1993) describe the model by assuming that three 'higher-order' conditions must be met for the resolution of critical discussions:

1. 'First-order' conditions: the ideal model of critical discussion should be considered as a 'code of conduct' for discussants.
2. 'Second-order' conditions: discussants should want to resolve the dispute and have the skill and competence in the topic under discussion to engage in an argument.
3. 'Third-order' conditions: every discussant must have the right to present their point of view. Participants must be surrounded by a socio-political context of equality where none of them is dependent, subordinate, or inferior.

The 'third-order' conditions are particularly relevant for communicative activism. As van Eemeren et al. (1993) argue, the resolution of conflicts (on the merits) is 'incompatible with situations in which one standpoint or another may enjoy a privileged position by virtue of representing the status quo' (van Eemeren et al., 1993, 33). Therefore, they say, the third-order conditions underline the importance of political principles such as nonviolence, freedom of speech, and intellectual pluralism.

In my view, communicative activism can be defended as a broadly rational argumentative strategy that social movements use while they aim to guarantee the third higher-order conditions for critical discussions. Although communicative activism violates the first-order conditions, it should be considered an argumentative movement worth taking. I second

---

[19] Another objection may be that communicative activism is unnecessary because, rather than trying to reach a resolution to the conflict over abortion in Argentina, Pañuelos verdes are engaged in a different kind of dialogue (Walton, 1998). For example, Pañuelos verdes could be 'negotiating' or engaging in an 'erstic' dialogue. However, these reconstructions leave unexplored the harm of hate speech for the arguers and they undermine the argumentative activity to which Pañuelos verdes aims to contribute. According to Walton, political debates are based on persuasive or deliberative types of dialogue (Walton, 1998, 172) in which perspectives on possible courses of action are discussed (see also fn. 1). The pañuelazo were described as a contribution to the political debate on abortion (§2) in order to defend the legalisation of abortion.

that the fallaciousness of violating the first-order conditions requires a contextual evaluation (Henkemans & Wagemans, 2015; Jacobs, 2009; Mohammed, 2006). I suggest that in situations where the unfairness of the argumentative context is caused by representatives of institutions (i.e. through the use of hate speech) whose purpose is the representation of democratic systems, the first-order conditions can (and should) be sacrificed in order to bring about a 'good' deliberation procedure (Jacobs, 2009). For example, Pañuelos verdes based their actions on achieving a 'horizontal' process of decision making (Sitrin, 2012). A horizontal organisation relies on a flat plane of communication where there is open participation and no hierarchies between the participants. The concept 'horizontalism' played a central role in Pañuelos verdes protests, including communicative activism as described here (Sitrin, 2012). Notice that 'horizontalism' is based on equal participation and freedom of speech, in line with the third-order conditions of van Eemeren et al. (1993) describe. Horizontalism was conceived as a sine qua non condition for achieving successful political decision-making for Pañuelos verdes (Sitrin, 2012). Therefore, communicative activism should be described as an alternative argumentative strategy for social movements as it favours the third-order conditions for argumentation.

5. Conclusion

This paper defended communicative activism as an argumentative strategy that social movements can use in arguments affected by hate speech. Communicative activism is a category of communicative practice that challenges hate speech and aims to attract bystanders to the protests. However, there are certain problems with this strategy. First, communicative activism might be inapplicable in countries with high levels of citizen repression. Second, communicative activism may not be sufficient to remedy the harm of hate speech. However, communicative activism can be accompanied by the development of more refined hate speech laws which address surreptitious hate speech that communicative activism makes explicit. Third, measuring the success of communicative activism is problematic. My intuition is that the success of communicative activism is related to the resolution of conflict contaminated by hate speech. In the case of Pañuelos verdes, for example, on the 29th December 2020, the Argentinian Senate legalised abortion. Political theorists defended the pañuelazo as a determinant for such an achievement (Dulbecco & et al., 2021). Pañuelos verdes gained significant support in the public sphere between 2018 to 2020. The 'pañuelazo' became the 'Marea verde' [green tide] with rallies composed of one million Argentinian citizens (Dulbecco & et al., 2021). Was the gaining of support, however, due to communicative activism? The question about the success of communicative activism will guide my future research.

References

Aakhus, M., & Lewiński, M. (2017). Advancing Polylogical Analysis of Large-Scale Argumentation: Disagreement Management in the Fracking Controversy. *Argumentation, 31(1),* 179–207.
Aikin, S. F., & Talisse, R. B. (2020). Political Argument in a Polarized Age Reason and Democratic Life. John Wiley & Sons.
Allen, A. (2012). The Unforced Force of the Better Argument: Reason and Power in Habermas' Political Theory. *Constellations, 19(3),* 353–368.
Argentinas a favor del aborto denuncian incremento de ataques (2018, July 26). *France 24.* https://www.france24.com/es/20180726-argentina-aborto-mujeres-victimas-ataques
Atacadas por usar pañuelos verdes: casos de violencia en el contexto del debate por el aborto legal (2018, December 18). *Amnesty International.* https://amnistia.org.ar/atacadas-por-usar-panuelos-verdes-casos-de-violencia-en-el-contexto-del-debate-por-el-aborto-legal/
Austin, J. L. (1962). *How to Do Things with Words.* Oxford University Press.
Barker, C., & Kennedy, P. (1996). To Make Another World. Studies in protest and collective action. Routledge.
Bessone, P. G. (2020). Debates about the legalisation of abortion in Argentina: The catholic church and its relationships with the presidential governments under democracy (1983-2018). *Apuntes, 47(87),* 87–117.
Bettcher, T. M. (2014). Trapped in the wrong theory: Rethinking trans oppression and resistance. *Signs, 39(2),* 383–406.
Bianchi, C. (2020). Discursive Injustice: The Role of Uptake. *Topoi, 40(1),* 181–190.
Blay, E. (2013). El control policial de las protestas en España. *InDret: Revista Para El Análisis Del Derecho, 4,* 1–32.
Bohman, James. (1996). Public deliberation: pluralism, complexity, and democracy. MIT Press.
Bondy, P. (2010). Argumentative injustice. *Informal Logic, 30(3),* 263–278.
Bowell, T. (2021). Some Limits to Arguing Virtuously. *Informal Logic, 41(1),* 81–106.
Camp, E. (2018). Insinuation, Common Ground, and the Conversational Record. In D. Fogal, D. W. Harris, & M. Moss (Eds.), *New Work on Speech Acts* (pp. 40–66). Oxford University Press.
Cepollaro, B. (ms). Blocking Toxic Speech Online-A Qualitative Study on Social Media.
Cepollaro, B., Sulpizio, S., & Bianchi, C. (2019). How bad is it to report a slur? An empirical investigation. *Journal of Pragmatics, 146,* 32–42.
Chrisman, M., & Hubbs, G. (2021). Protest and speech act theory. In R. K. Sterken & J. Khoo (Eds.), *The Routledge Handbook of Social and Political Philosophy of Language* (pp. 179–193). Routledge.
Cohen, J., & Kehler, A. (2021). Conversational Elicuture. *Philosophers' Imprint, 21(12),* 1–26.

Cortina, A. (2017). Aporofobia, el rechazo al pobre: un desafío para la democracia. Espasa Libros.
Curran, J. (2002). *Media and Power*. Routledge.
Dahl, R. A. (1961). *Who Governs?*. Yale University Press.
Díez, B. (2020, December 31). Aborto en Argentina: "Te condenan y te estigmatizan, es el prejuicio constante al que se ve sometida una mujer que decide no maternar". BBC News. https://www.bbc.com/mundo/noticias-america-latina-55492942
Domínguez-Armas, Á., & Soria-Ruiz, A. (2021). Provocative insinuations. *Daimon*, 83, 63–80.
Dulbecco, P., Cunial, S., Jones, D., Calvo, E., Aruguete, N., Paola, I., Gómez, C., Pérez, S., Aymá, A., Moragas, F., & Kejner, E. (2021). *El aborto en el Congreso: Argentina 2018-2020*. Centro de Estudios de Estado y Sociedad-CEDES.
Eemeren, F. H. van, & Grootendorst, R. (1984). Speech Acts in Argumentative Discussions Studies of Argumentation in Pragmatics and Discourse Analysis. De Gruyter Mouton.
Eemeren, F. H. van, & Grootendorst, R. (1992). Argumentation, Communication, and Fallacies. A Pragma-dialectical Perspective. Routledge.
Eemeren, F. H. van, Grootendorst, R., Jackson, S., & Jacobs, S. (1993). *Reconstructing Argumentative Discourse*. University of Alabama Press.
Fairclough, N., & Fairclough, I. (2012). Political discourse analysis: A method for advanced students. Routledge.
Fricker, E. (2012). Stating and Insinuating. Proceedings of the Aristotelian Society Supplementary Volume, 86(1), 61–94.
Fricker, M. (2007). *Epistemic Injustice: Power and Ethics*. Oxford University Press.
Gelber, K. (2012). "Speaking Back": The Likely Fate of Hate Speech Policy in the United States and Australia. In I. Maitra & M. K. McGowan (Eds.), *Speech and Harm. Controversies over Free Speech* (pp. 50–71). Oxford University Press
Gray, D. M., & Lennertz, B. (2020). Linguistic Disobedience. *Philosopher's Imprint*, 20(21), 1–16.
Habermas, J. (1984). *The theory of communicative action*. Beacon Press.
Henkemans, A. F., & Wagemans, J. H. M. (2015). Reasonableness In Context: Taking Into Account Institutional Conventions in The Pragma-Dialectical Evaluation Of Argumentative Discourse. In F., van Eemeren, & B. Garssen (Eds.). *Reflections on Theoretical Issues in Argumentation Theory* (pp. 217–226). Springer, Cham.
Howard, J. W. (2021). Terror, Hate and the Demands of Counter-Speech. *British Journal of Political Science*, 51(3), 924–939.
Innocenti, B. (2022). Demanding a halt to metadiscussions. *Argumentation, 36(3)*, 345–364.
Jacobs, S. (1989). Speech acts and arguments. *Argumentation*, 3(4), 345–365.
Jacobs, S. (2009). Nonfallacious Rhetorical Design in Argumentation. In F. H. van Eemeren & B. Garseen (Eds.), *Pondering on problems of Argumentation* (pp. 55–78). Springer.
Jeshion, R. (2013). Slurs and Stereotypes. *Analytic Philosophy, 54(3)*, 314–329.
Kukla, Q., & Steinberg, D. (2021). "I Really Didn't Say Everything I Said" The Pragmatics of Retraction. In L. Townsend, P. Stovall, & H. Bernhard Schmid (Eds.), *The Social Institution of Discursive Norms* (pp. 223–247). Routledge.
Kukla, R. (2014). Performative force, convention, and discursive injustice. *Hypatia, 29(2)*, 440–457.

Kukla, R. (2018). Slurs, Interpellation, and Ideology. *Southern Journal of Philosophy, 56(S1)*, 7–32.
Langton, R. (1993). Speech Acts and Unspeakable Acts. *Philosophy and Public Affairs, 22(4)*, 293–330.
Langton, R. (2015). How to Get a Norm from a Speech Act. *The Amherst Lecture in Philosophy*, 10, 1–33. http://www.amherstlecture.org/langton2015
Langton, R. (2018a). Blocking as Counter-Speech. In D. Fogal, D. W. Harris, & M. Moss (Eds.), *New Work on Speech Acts* (pp. 144–165). Oxford University Press.
Langton, R. (2018b). The Authority of Hate Speech. *Oxford Studies in Philosophy of Law, 3*, 123–152.
Lepoutre, M. (2017). Hate Speech in Public Discourse: A Pessimistic Defense of Counterspeech. *Social Theory and Practice, 43(4)*, 851–883.
Lepoutre, M. C. (2019). Can "More Speech" Counter Ignorant Speech? *Journal of Ethics and Social Philosophy, 16(3)*, 155–191.
Lewiński, M. (2017). Practical argumentation as reasoned advocacy. *Informal Logic, 37(2)*, 85–113.
Lewiński, M. (2019). Argumentative Discussion: The Rationality of What? *Topoi, 38(4)*, 645–658.
Lewiński, M. (2021). Conclusions of Practical Argument: A Speech Act Analysis. *Organon, 28(2)*, 420–457.
Lewiński, M., & Mohammed, D. (2016). Argumentation Theory. In K. B. Jensen, R. Craig, J. Pooley, & E. Rothenbuhler (Eds.), *The International Encyclopedia of Communication Theory and Philosophy* (pp. 1–15). John Wiley & Sons.
Limbaugh, R. (1992). *The Way Things Ought to be*. Pocket books.
McGowan, M. K. (2019). *Just words. On speech and hidden harm*. Oxford University Press.
Meyer, D. S., Whittier, N., & Robnett, B. (2002). *Social Movements: Identity, Culture, and the State*. Oxford University Press.
Mohammed, D. (2006). Comments on "nonfallacious rhetorical strategies: Lyndon Johnson's Daisy Ad". *Argumentation, 20(4)*, 443–445.
Muzi, C. (2018, August 5). La historia del pañuelo verde: cómo surgió el emblema del nuevo feminismo en Argentina. *Infobae*. https://www.infobae.com/cultura/2018/08/05/la-historia-del-panuelo-verde-como-surgio-el-emblema-del-nuevo-feminismo-en-argentina/
Pope thanks Argentinian women's network for pro-life commitment (2020, November 26). *Vatican News*. https://www.vaticannews.va/en/pope/news/2020-11/pope-francis-thanks-argentinian-women-s-network-for-their-commit.html
Rosende, L. (2018, July 31). Quiénes llevan los pañuelos celestes. *Revista Anfibia*. https://www.revistaanfibia.com/quienes-llevan-los-panuelos-celestes/
Schiappa, E. (2003). *Definitions and the Politics of Meaning*. Southern Illinois University Press.
Schmidt, S. (2019, December 5). Teens in Argentina are leading the charge to eliminate gender in language. *The Washington Post*. https://www.washingtonpost.com/dc-md-va/2019/12/05/teens-argentina-are-leading-charge-gender-neutral-language/
Searle, J. R. (1975). A taxonomy of illocutionary acts. In K. Günderson (Ed.), *Language, mind, and knowledge* (Vol. 7, pp. 344–369). University of Minnesota Press.
Sitrin, M. A. (2012). *Everyday Revolutions. Horizontalism and autonomy in Argentina*. Zed Books.

Tirrell, L. (2012). Genocidal language games. In I. Maitra & M. K. McGowan (Eds.), *Speech and Harm. Controversies over Free Speech* (pp. 174–222). Oxford University Press.

Tirrell, L. (2019). Toxic Misogyny and the Limits of CounterSpeech. *Fordham Law Review, 87,* 2433–2352.

Una clínica privada de Neuquén repudió escrache de los "pañuelos verdes" (2018, June 28). *AICA*. https://aica.org/noticia-una-clnica-privada-de-neuqun-repudi-escrache-los-pauelos-verdes

Waldron, Jeremy. (2012). *The Harm in Hate Speech*. Harvard University Press.

Walton, D. (1998). The New Dialectic: Conversational Contexts of Argument. University of Toronto Press.

Wenzel, J. W. (1990). Three Perspectives on Argument, Rhetoric, Dialectic, Logic. In R. Trapp & J. Schuetz (Eds.), *Perspectives on argumentation: essays in honor of Wayne Brockriede* (pp. 9–26). Idebate Press.

Williams, Z. (2015, September 15). Feminazi: the go-to term for trolls out to silence women. *The Guardian*. https://www.theguardian.com/world/2015/sep/15/feminazi-go-to-term-for-trolls-out-to-silence-women-charlotte-proudman

Witek, M. (2013). How to Establish Authority with Words: Imperative Utterances and Presupposition Accommodation. In A. Brożek (Ed.), *Logic, Methodology and Philosophy of Science at Warsaw University* (pp. 145–157).

Zheng, R. (2022). Theorizing social change. *Philosophy Compass, 17(4),* 1–14.

Zurita, A. M., & Pérez-Navarro, E. (2021). The resistant effect of slurs: A nonpropositional, presuppositional account. *Daimon, 84,* 31–46.

Commentary On "Communicative Activism As An 'Alternative' Argumentative Strategy To Confront Contexts Affected By Hate Speech", By Álvaro Dominiguez Armas

SARA GRECO
USI-Università della Svizzera italiana
sara.greco@usi.ch

1. Introduction

Álvaro Dominiguez Armas' paper deals with practices of communicative activism, which are defined as "those confrontative argumentative practices that aim to hinder the passage of hate speech and pull third parties into the protests" (p.1). Communicative activism, thus, is an argumentative means to counter those moves that are based on hate speech or, as I will propose to define them (as based on existing literature), soft hate speech or discriminatory discourse. Dominiguez Armas' work mainly contributes to a philosophical investigation of argumentation, but it is also grounded empirically in a dataset of public discourse surrounding the abortion debate in Argentina.

I believe that the scientific posture the author assumes is particularly important for argumentation studies and defines an ambitious goal for his paper. In fact, Dominiguez Armas goes beyond analyzing and evaluating discriminatory discourse, attempting to reflect about how we can improve argumentation in society by creating healthier discussion spaces. Although the concept of *design* is not mentioned explicitly, this paper may legitimately count as a reflection on *argumentation design* in the sense proposed by Aakhus (2007) and Jackson (2015a, 2015b). Design thinking draws attention "to what can be done to make individuals and societies more or less reasonable" (Jackson 2015b: 242); by so doing, design "does not replace, but complements, theory built to aid in *argument* appraisal—evaluation of any actual product of argumentation. Both are extremely important" (Jackson 2015a, p. 244, emphasis in the original).

Design necessarily involves taking the risk of making practical proposals or "innovations" (Jackson 2015a). These innovations will then require appraisal to verify whether they fulfil their aim to build better spaces for argumentation. This necessarily entails some risk, because analyses grounded in philosophical and empirical considerations on existing practices need to make a step forward and culminate in design proposals, which are tentative by definition. However, I believe it is a risk to be taken for argumentation studies to be relevant in our society. This is important in all domains but particularly urgent in cases of hate speech and, more in general, discrimination and violence. By way of example, I would like to mention a virtuous case of a recent book in linguistics (Scarpa, 2021) that, after having identified linguistic and rhetorical markers of domestic violence, reflects on how this type of linguistic and stylistic analysis may be useful to detect and stop or prevent violence at an early stage. Álvaro Dominiguez Armas's paper shares a similar ambition, namely to reflect on existing argumentative practice, improve them and counter violent or fallacious behavior; and this is a reason for praise.

Moving beyond the potential of this paper for argumentation design, there are several other interesting aspects in the contents of this contribution. I chose to comment on a central point and two minor questions that I will mention cursorily.

2. On the concept of *hate speech*

A very important point that needs some further clarification concerns the definition of *hate speech*, which is at the core of this paper. Discourse analytical accounts of hate speech have observed a number of conditions that restrict the scope of properly defined hate speech and distinguish it from other forms of discriminatory discourse. Firstly, hate speech is a public communication act (Baider, Assimakopolous and Millar, 2017, p. 3), which means that this phenomenon does not cover private forms of discrimination and violent verbal behavior. Secondly, according to Baider, Assimakopolous and Millar (2017), who reflect on the definition emerging from the Council Framework Decision 2008/913/JHA, "one could isolate the criteria qualifying speech as hate speech in the EU as follows:
(1) A call motivated by racial/ethnic/national bias;
(2) A call for violence;
(3) A call punishable by the criminal law of the country where it occurs"
    (Baider, Assimakopolous and Millar, 2017, p. 4).

Conditions (1) and (3) imply that hate speech is directed against a minority group that is recognized as vulnerable and is protected by law. If these conditions are met (i.e., for example, if the call for violence against a protected minority is explicit), according to these authors, we can talk of *hard* hate speech, which "comprises prosecutable forms that are prohibited by law" (ibid). Notably, the definition of hate speech depends on what groups are recognized as protected minorities in each country. Outside the boundaries of hard hate speech, however, "there could still be cases of inflammatory, offensive comments or comments characterized by prejudice and intolerance that would not meet the threshold provided in the description above" (ibid.), but that can still be considered as *soft* hate speech. Soft hate speech is more indirect and less punishable because, for example, the incitement to violence is not explicit or it is more subtle (e.g. uses insinuations or other indirect forms), such as it happens in the cases mentioned in Dominiguez Armas' paper. Besides, the notion of *soft* hate speech, or (more generally) exclusionary or discriminatory discourse, is relevant to this paper because the considered targeted group is composed by activists (in this case, *Pañuelos verdes*, who supported abortion in the campaign in Argentina). Depending on the contexts, political activists may not be considered a minority or a protected group, thus condition (1) of hard hate speech may not be present.

Is this terminological and legal distinction relevant at all? I believe it is, given the general aim of this paper. In fact, if one reflects on counter-strategies against discrimination, it is very important to know what forms of discrimination we are targeting exactly. Explicit forms of hard hate speech are punishable by law and, in such cases, the means to design a better argumentative discussion may include formal appeals to criminal law procedures. The strategy of communicative activism proposed in this paper may be more appropriate in cases in which the opponents are using subtle discriminatory forms of exclusion, such as soft hate speech. It may also apply to cases in which participants to a discussion violate the "freedom rule" of argumentation (van Eemeren and Grootendorst, 2004) by abusing their power to depict activists as non-legitimate participants to an argumentative discussion.

At the level of empirical argumentation analysis, as this work proceeds, more examples from the examined corpus will be needed to understand the forms of discrimination that are present in the selected controversy.

3. Local definitional points

I will now turn to two local definitional points.

Firstly, in Footnote 1, Dominiguez Armas notes that the pragmatic debate about the legalization of abortion in Argentina is a practical reasoning question. He notes that other authors (notably, Schiappa, 2003) observed that evaluative or conceptual issues are involved in this otherwise pragmatic debate; in particular, the following evaluative issue: "what is X?" (i.e. a question referred to X being or not being a human life) is necessarily included. I would add that these two types of issues (pragmatic and evaluative) and the correlated standpoints in the debate are not just co-existing, but they are profoundly related, as the one presupposes the other (in fact, pragmatic standpoints often presuppose an evaluative component). While I am not familiar with the specific controversy analyzed in this paper and its cultural context, my suggestion is to consider both issues as interrelated to have a clear picture of the complexity of the arguments.

Secondly, an important concept to be refined in future research is "communicative activism". In particular, the noun "activism" requires further reflection. In the literature and in everyday language, activism tends to be related to somewhat regular activities of campaigning, e.g. by social movements that advocate for a certain cause; in this paper, communicative activism is clearly related to social movements campaigns. But aren't *all* forms of activism communicative, as campaigns (demonstrations as well as social media posts) are communicative activities? What distinguishes communicative activism from other communicative activities carried out by social movements? Is "communicative activism" a good term to identify the phenomenon that the author studies?

These final questions are meant to encourage further research in this domain, both at the level of analysis and at the level of design.

References

Aakhus, M. (2007). Communication as design. Communication Monographs, 74(1), 112-117.

Baider, F. H., Assimakopolous, S., & Millar, S. (2017). Introduction and background. In S. Assimakopolous, F. H. Baider, & S. Millar (Eds.), Online hate speech in the European Union. A discourse-analytic perspective (pp. 1-16). Cham: Springer.

Eemeren, F. H., van, & Grootendorst, R. (2004). A systematic theory of argumentation: The pragma-dialectical approach. Cambridge: Cambridge University Press.

Jackson, S. (2015a). Design thinking in argumentation theory and practice. Argumentation, 29(3), 243-263.

Jackson, S. (2015b). Deference, distrust and delegation: Three design hypotheses. In F. H. van Eemeren, & B. Garssen (Eds.), Reflections on theoretical issues in argumentation theory (pp. 227-243). Cham: Springer.

Scarpa, R. (2021). Lo stile dell'abuso. Violenza domestica e linguaggio. Roma: Treccani.

Schiappa, E. (2003). Defining reality: Definitions and the politics of meaning. Carbondale: Southern Illinois University Press.

USING ARGUMENTATION SCHEMES TO MODEL LEGAL REASONING

TREVOR BENCH-CAPON
University of Liverpool
tbc@liverpool.ac.uk

KATIE ATKINSON
University of Liverpool
katie@liverpool.ac.uk

Abstract
We present argumentation schemes to model reasoning with legal cases. We provide schemes for each of the three stages that take place after the facts are established: factor ascription, issue resolution and outcome determination. The schemes are illustrated with examples from a specific legal domain, US Trade Secrets law, and the wider applicability of these schemes is discussed.

1. Introduction

Reasoning with legal cases, especially as conducted in common law jurisdictions such as the UK and USA, is a form of argumentation much studied in Artificial Intelligence and in computational argumentation. The formal procedure within which it is conducted and the extensive documentation which records the argument presented for each side and an assessment of these arguments, make it a fruitful area for study. As described in [35], there may be several types of reasoning involved, including the use of rules, the balancing of factors, analogy and the use of policies to achieve particular purposes. All of these have been modelled in AI and Law, and this work suggests that reasoning with legal cases can been seen as going through a series of stages at which different reasoning styles are appropriate. This view will be elaborated in Section 2.
  One way of modelling a reasoning task [24] is to present it as a set of argumentation schemes [38]. In this paper we will use this method to articulate the reasoning required at each of the stages.
  Although legal reasoning is worthy of study in itself, we believe that the insights are also applicable to other, less formal, domains where it is necessary to balance reasons for and against particular options to come to

a decision. While there are some similarities with practical reasoning tasks, such as choosing which restaurant to go to (e.g. [5]) and which car to buy, (e.g. [17]), we suggest that the methods are more applicable to classification tasks, because the use of binding precedents to determine preferences is not applicable in many domains, whereas the examples used to train a classification system can act like a set of precedents in a settled domain of law where preferences are resolved. Legal reasoning is indeed used as the model for explaining the output of machine learning systems in
[26] for several diverse classification tasks: customer churn, poisonous mushrooms and university admissions.

2. Stages in Reasoning with Legal Cases

A legal case will be governed by laws which state what the plaintiff or prosecution must show to establish their claim. The law can be seen as a set of definitions and represented as rules [32]. At some point, however, the terms in the rules will be undefined [33], and it must be determined whether or not these terms apply in the particular case. This gives rise to a set of what are termed issues in AI and Law [1]. These issues must be resolved in favour of either the plaintiff or the defendant, and the reasons for each side are termed factors in AI and Law [1]. Which reasons apply in a given case will be determined by the facts. The facts themselves will be decided on the basis of the evidence presented.

Thus deciding a legal case involves
- accepting facts on the basis of the evidence
- ascribing factors on the basis of the facts
- resolving the issues on the basis of the factors
- deciding the outcome on the basis of the issues

We will not say more about reasoning with evidence here, since it is not especially legal. In common law jurisdictions, deciding which facts to accept is the responsibility of lay jurors, not legal professionals, and the reasoning as to what to believe is no different from that used in everyday life. A set of argument schemes for reasoning with evidence, based on several common schemes, was presented in [13]. We will now say some more about the reasoning involved at each of the other stages.

2.1. From Issues to Outcome

Because they are the base level terms in a set of definitions, issues generally can be viewed as supplying necessary and sufficient conditions. We illustrate this point with reference to US Trade Secrets Law, which has been widely studied in AI and Law (e.g. [30], [2], [15], [16], [27], [7], [25]), and which we will use for our examples in this paper. Cases falling within this domain cover scenarios where there is a claim that a trade secret has been misappropriated: thus the information being considered in a case must be both a *Trade Secret* and have been *misappropriated*. These terms may be further defined: for information to be a Trade Secret it must be both *valuable* and its *secrecy maintained*. For information to be misappropriated it must have been *used* despite a *confidential relationship*, or obtained through *improper means*. Thus the issues can be expressed as what is termed a "logical model" in [15], as shown in Figure 1.

Once each of the leaf issues is resolved, the outcome can be deter- mined.

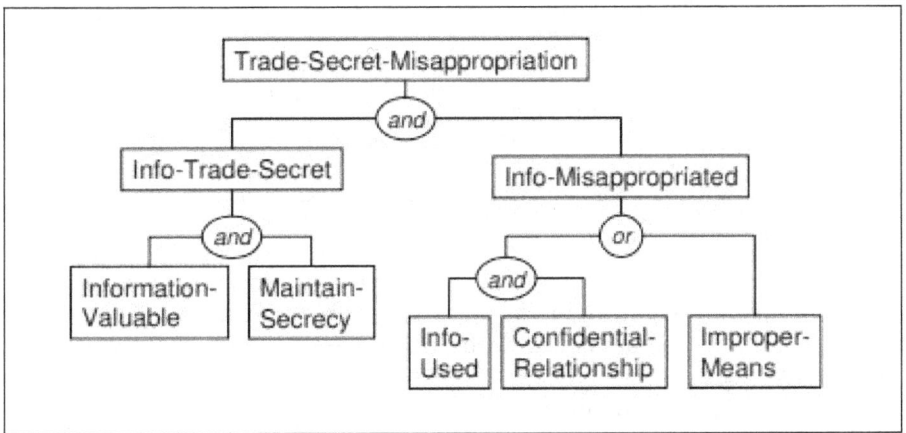

Figure 1. Issues in US Trade Secrets Law [15]

2.2. From Factors to Issues

The resolution of the issues, however, tends to be less clear cut. Typically there will be reasons both why the issue should be resolved one way and why it should be resolved the other way. For example, with respect to the existence of a *confidential relationship*, a person may have disclosed the information in negotiations, which is a reason to find for the defendant, but the defendant may have been made aware that the disclosure was in confidence, which is a reason to find for the plaintiff, even if there was no formal agreement. Each issue will thus be associated with a set of factors, providing reasons to resolve the issue either for the plaintiff or the defendant. The factors associated with the issues in US Trade Secrets in [3] are shown in Table 1

Issue	Plaintiff Factors	Defendant Factors
InfoValuable	F8p Competitive Advantage F15p Unique Product	F16d Info Reverse Engineerable F20d Info Known to Competitors F24d Info Obtainable Elsewhere F27d Disclosure In Public Forum
Secrecy Maintained	F4p Agreed Not To Disclose F6p Security Measures F12p Outsider Disclosures Restricted	F10d Secrets Disclosed Outsiders F19d No Security Measures
ImproperMeans	F2p Bribe Employee F7p Brought Tools F14p Restricted Materials Used F22p Invasive Techniques F26p Deception	F17d Info Independently Generated F25d Info Reverse Engineered
InfoUsed	F7p Brought Tools F8p Competitive Advantage F14p Restricted Materials Used F18p Identical Products	F17d Info Independently Generated F25d Info Reverse Engineered
Confidential Relationship	F4p Agreed Not To Disclose F13p Noncompetition Agreement F21p Knew Info Confidential	F1d Disclosure In Negotiations F23d Waiver of Confidentiality

Table 1. CATO factors grouped by Issues [3]

These *factors* must be balanced against one another. Which set of reasons is considered stronger will depend on the preferences of the person making the choice: some will require a formal agreement, while others will not. In law these preferences are revealed in the past decisions, and courts are bound by these precedents[1]. Where no precedent exists, the judges must express a preference, creating a precedent that will govern future cases. Many questions arise concerning the construction of preferences in general [22], but in AI and Law the ideas of [12], [11] and [18] have generally been followed and it is taken that the preference will be according the purpose or social value promoted by the decision. This may involve argument as to which value should be preferred: this is modelled in [9]. We will not, however, discuss the question further in this paper: our discussion will be in terms of settled law, where precedents are available to resolve such preference questions.

The reasoning at this stage thus involves, for each issue, identifying the relevant factors for and against, and then, in a case with factors for both sides, choosing the stronger set, in accordance with precedents. Formal accounts of this style of reasoning with precedents are given in [19] and [25], and a set of argument schemes modelling it is given in [27]. These formal accounts are at the whole case level, but can readily be adapted to consider issues rather than complete cases [7].

---

[1] For example, National Instrument Labs, Inc. v. Hycel, Inc., 478 F.Supp. 1179 (D.Del.1979) provides a precedent in favour of the existence of a confidential relationship where there had been *DisclosureInNegotiations* (F1d), but the defendant

2.3. From Facts To Factors

What factors are present in a given situation will depend on the facts. But the presence of a factor is not straightforwardly determined by the facts. For example, the security measures taken by the plaintiff will be relevant, but whether they are considered *adequate* or *inadequate* will depend on what is considered reasonable, and the strictness of the standard applied. If considered *adequate*, then there is a reason to find that secrecy had been maintained and that factor F6p applies. If considered *inadequate*, then there is a reason to decide that secrecy had not been maintained, and so F19d applies. Thus the various relevant facts must be assessed for their significance for the decision maker, and this significance recorded by ascribing the appropriate factor to the case. Where the line is drawn will require a judgement. In common law, precedents will determine these "switching points" [28].

The ascription of factors has received rather less attention in AI and Law than the other stages. In [8] four different ways of ascribing factors were described: these will be discussed below in Section 6

2.4. Summary

Different AI and Law approaches have covered different stages in this process, as shown in Table 2. Thus [32] models only the legislation (of the British Nationality Act), and does not represent the interpretation of terms in the legislation through case law. In HYPO the dimensions are neutral and the user must decided which party they favour. In CATO the party favoured on a dimension is identified through the use of factors, but the user must decide which side is favoured on the balance of these factors. IBP can predict an outcome by using a logical model of issues once the issues have been resolved on the balance of factors. The focus of Bex *et al.*'s hybrid approach [14] is on the move from evidence to facts.

In this paper we will model the three stages after the facts have been agreed, using argumentation schemes and their critical questions [38].

Stage	BNA [32]	HYPO [30]	CATO [1]	IBP [15]	Bex [14]
Outcome	X			X	
Issues	X		X	X	
Factors		X	X	X	
Facts		X			X
Evidence					X

Table 2. Stages in a Legal Decision and Some Example Systems

3. Argument Schemes for Modelling Reasoning

Argument schemes are typically seen as a form of defeasible inference rules, but as argued in [24] they can also be used to model a reasoning process, by articulating the arguments that can be used and the objections that can be made at each stage of the process. Such sets of argument schemes can be used as the specification of dialogue system to realise this process (e.g. [4]). Here we will use argument schemes to describe the process of reasoning with legal cases.

To present reasoning with legal cases as an argument, a form of three ply argumentation was introduced in HYPO [30]. First a claim is made by a proponent, then the opponent challenges the claim, and finally the proponent tries to rebut these challenges. A repertoire of moves for each of these stages was developed [1]. This structure fits well with the notion of argumentation schemes as proposed by Wal- ton [36]. Here an instantiated scheme is put forward, challenges made in the form of critical questions, and then the proponent attempts to provide answers to these questions. This correspondence allows the argument moves of the original systems to be seen in terms of argumentation schemes and their associated critical questions. Argumentation schemes provide an excellent way of making a reasoning procedure more precise, and indicating the ways in which assertions within that procedure may be challenged.

In this paper we will describe the three stages of the reasoning with legal cases identified above as argumentation schemes.

4. Determining the Outcome

Once resolved, issues can form the leaves of a logical model as depicted in Figure 1. This can then be seen as a standard example of logical reasoning, using the law represented as a set of rules. We can express the schemes thus:

Issue to Outcome Scheme (IO):
Warrant Premise: The law yields a rule $R: I_1, I_2...I_n \rightarrow O$
Issue Premise: $I_1, I_2...I_n$ are (not) satisfied
Conclusion: Outcome is (not) O

Examples in US Trade Secrets are:
Example 1
The law yields a rule that if the information is Trade Secret and Misappropriated the plaintiff should win

The information is Trade Secret and Misappropriated Therefore, The plaintiff should win
Example2
The law yields a rule that if the information was Used and there is a Confidential Relationship or the information was Misappropriated the plaintiff should win
The information was not Misappropriated Therefore, The defendant should win[2]

Because these are strict rules, critical questions must take the form of questioning the premises.
IOCQ1: *Exception*: Is there an issue I_{n+1} which is satisfied and which provides an exception to R? (Where the argument is for O)
IOCQ2: Unneeded Premise: Is one of the issues $I1$ ... In not required for O? (Where the argument is for not O)
IOCQ3: Issue Incorrectly Resolved: Is one of the issues $I1$ ... In in fact (not) satisfied?

We could thus pose the following critical questions in Example1:
IOCQ1Ex1: That the employee was the sole developer of the information provides an exception to the rule.
IOCQ3Ex1: The information was not, in fact, Misappropriated.

And we can object to Example2 with IOCQ2:

IOCQ2Ex2: It is not necessary that the information be Misappropriated where the information was Used and there is a Confidential Relationship.

5. Determining the Outcome

As described in [21], after an initial period of flux, sufficient precedents are established for an area of law to be considered settled, until some events bring about a period of reinterpretation. Thus for much of the time, precedents will be available to decide questions of preference between factors: except for newly enacted legislation, landmark cases setting new precedents are relatively rare.

---

2 As we shall see as the example develops, this argument will be defeated by a critical question.

5.1. Resolving Issues With Precedents

A set of schemes to balance factors within a case in accordance with a set of precedents was proposed in [27]. These, however, ignored the intermediate stage of issues, and moved straight from factors to decisions. The schemes can, however, be easily adapted to resolve issues by considering not the complete set of factors but only those pertaining to the issue under consideration. It is generally straightforward to allocate factors to issues, as shown in Table 1. Thus in the following F_p and F_d should be taken to comprise only factors relevant to the issue I.

Citation Scheme (C):
Factor Premise: Case C has factors F_i ... F_n in common with precedent P
Precedent Premise: Issue I was resolved for the plaintiff (defendant) in P
Conclusion: Issue I should be resolved for the plaintiff (defendant) in C
For an example from US Trade Secrets we will consider resolution of the issue of *Secerecy Maintained*.

Example3
Factors associated with Secrecy Maintained include the plaintiff factors *SecurityMeasures* (F6p) and *OutsiderDisclosuresRestricted* (F12p) and the defendant factor *DisclosuresToOutsiders* (F10d). Sup- pose we have a case, *Restricted*, in which all three of these factors are present. We have a precedent, *Bryce*[3], in which *SecurityMeasures* was present and the other two factors were absent, and the issue was resolved for the plaintiff. On the basis of *Bryce* we can argue:
Restricted has factor *SecurityMeasures* in common with *Bryce*. The issue *Secrecy Maintained* was resolved for the plaintiff in
Bryce.
Therefore, the issue *Secrecy Maintained* was resolved for the plain- tiff in *Restricted*.
Two critical questions may be posed against this scheme, either pointing to a counterexample, a precedent in which the issue was resolved differently, or by pointing to another factor present only in the case or the precedent which weakens the argument.

---

[3] M. Bryce & Associates, Inc. v. Gladstone, 107 Wis.2d 241, 319 N.W.2d 907 (Wis.App.1982)

CCQ1: *Counterexample*: Is there another precedent P' with fac- tors in common with C in which the issue was resolved for the defen- dant (plaintiff)?

CCQ2: *Distinction*: Is there a factor F_{n+1} present in only the case or the precedent which weakens the case for the plaintiff (defendant)?

These two critical questions are found in [27]. But that paper takes the factors present in a case as given. If the presence of a factor can be disputed we get two more critical questions:

CCQ3: *Factor Not Present*: Is one of the factors $F_1 ... F_n$ not in fact present in the case?

If the conclusion of the critiqued argument is to resolve the issue for the plaintiff, the factor favours the plaintiff, and if to resolve for the defendant, the factor favours the defendant.

CCQ4: *Additional Factor* : Is there an additional weakening fac- tor F_{n+1} present in the new case?

If the conclusion of the critiqued argument is to resolve the issue for the plaintiff, F_{n+1} favours the defendant, and if to resolve for the defendant, F_{n+1} favours the plaintiff.

Counterexamples to pose CCQ1 are put forward using the Citation Scheme. Distinctions to pose CCQ2 are put forward using their own argument scheme.

Distinction Scheme (D):

Factor Premise A: There is a factor F present in only one of the current case C and Precedent case P

Polarity Premise: F weakens the case for the plaintiff (defendant)

Conclusion: Do not resolve Issue I for the plaintiff (defendant)

The polarity premise is needed because it is possible that the dif- ference will strengthen rather than weaken a case. What is required (when attacking an argument for the plaintiff) is a pro-defendant fac- tor in the current case but not the precedent, or a pro-plaintiff factor in the precedent but not the current case.

Example3 continued

Restricted has an additional defendant factor, *DisclosuresToOut- siders*, and this can be used to distinguish *Bryce:*

DisclosuresToOutsiders is present in *Restricted* but not in *Bryce*
DisclosuresToOutsiders weakens the case for the plaintiff
Therefore, Do not resolve Issue I for the plaintiff.

Critical questions can, of course, now be posed against the dis- tinction scheme. These are based on the notion of downplaying a distinction as developed in [1]. The idea is that there may be some other factor present in only one of the cases which strengthens the case for the original party. Such a factor may be used to substitute for a missing factor (if the party favoured is the same), or cancel out an additional factor (if the party favoured is different).

DCQ1: *Substitution*: Is there a Factor F' which can substitute for F?
DCQ2; *Cancellation*: Is there a Factor F' which can be used to cancel F?

Example3 continued
In the case of *Restricted* we do have a factor which can cancel out the additional factor *DisclosuresToOutsiders*, namely *OutsiderDisclosuresRestricted*, so we can pose DCQ2.
Whether the factor is considered sufficiently strong to cancel out the distinguishing factor is something that must be decided by the court. If it were considered that it was strong enough, it would still be possible to pose CCQ3 by asking whether the security measures taken in *Restricted* were indeed adequate.

5.2. Where Precedents are not Needed

The above schemes do rely on the existence of precedents to determine the preference for the factors associated with one side rather than the other. Because, as can be seen from Table 1, there is only a relatively small number of factors associated with each issue, only a few leading cases will be needed to supply the necessary preferences. It should, however, be noted that in some cases the preference will be obvious from the nature of the factors; for example if the plaintiff has given a *WaiverOfConfindentiality* (F23d), then the issue will be resolved for the defendant, even if there had been an agreement not to disclose (F4p), or even a formal non-disclosure agreement (F13p). Such factors were termed *knockout* factors in [15].

For such factors we can have another scheme:

Knockout Factor Scheme (KO)
Factor Premise: Factor F, relating to issue I, is present in Case C
Knockout Premise: Factor F favours plaintiff (defendant) and is, by its nature, preferred to any factors favouring the defendant (plain- tiff)
Conclusion: Issue I should be found for the plaintiff (defendant)
Critical questions can concern either whether the factor is present (to be resolved using the schemes for factor ascription presented in the next section), or whether it is indeed decisive (by citing a counterexample).

KOCQ1: Is factor F really present in C?

KOCQ2: Is there a precedent case *P* containing factor *F* in which issue *I* was resolved for the defendant (plaintiff)?

5.3. Cases which Require a Preference

If we have an issue which cannot be resolved using either a precedent or a knockout factor, then the judges must themselves express a preference between the pro-plaintiff and pro-defendant factors. As discussed in [12] and [11], this will involve expressing a preference for the values promoted by finding for the plaintiff or those promoted by finding for the defendant. Such arguments are likely to be very varied, including such things as an appeal to established values in society ("life is more important than property"), feasibility ("where this to be decided, the floodgates for acrimonious litigation would be opened"), analogy with a different area of law ("this preference is established in contract law and should apply here also") or the constitutional remit of judges ("such a decision can only be made by the legislature"). The computational deployment of such arguments was discussed in [9]. The schemes used for such arguments are, however, not specifically legal, and may use a variety of the established schemes to be found in [36]. We will therefore not discuss them further here, but limit our discussion to areas of law sufficiently settled that the required precedents are available.

6. Ascribing the Factors

6.1. Dimensions and Factors

We now come to the ascription of factors. Factors were a develop- ment from dimensions in HYPO. In HYPO dimensions were aspects of cases which were potentially relevant to the outcome. In general, dimensions were a range of values which increasingly favoured one of the parties. Thus the number of disclosures was one such dimension, and the more disclosures, the more favourable the dimension was to the defendant. A small number of disclosures might be considered not to constitute a reason to find for the defendant, but at some point the number of disclosures would be sufficient. If a particular case falls in the range where the defendant is favoured, this means that the pro-defendant factor *DisclosuresToOutsiders* applies, but fewer dis- closures means that no factor applies on this dimension. The point at which the factor starts to apply was termed the switching point in [29]. Most dimensions are either neutral or favour a particular side. Some, however, such as *SecurityMeasures*, favour the plaintiff at one end and the defendant at the other, giving rise to both a pro-plaintiff and a pro-defendant factor, possibly with a neutral range between them. Other dimensions, such as disclosures give rise to two factor for the same side: if the disclosure is in the public domain, the stronger pro-defendant factor *DisclosureInPublicForum* applies rather than the normal *DisclosuresToOutsiders*.

However, in HYPO many of the dimensions (10 out of the 13) were, in fact Boolean. Here one of the values would give rise to a factor: thus it is either true or false that the information was disclosed to the defendant in the course of negotiations, and if true then the pro-defendant factor *DisclosureInNegotiations* applies.

Finally we can have a factors which arise from two dimensions: use of a trade secret may save the defendant time, money or both. If these savings are significant, then the factor pro-plaintiff *CompetiveAdvantage* (F8p) will apply. The significance will require consideration of both time and money: the more time saved, the less money need be saved and vice versa. Thus we get a trade off of the sort shown in Figure 2.

These different relationships between factors and their dimensions mean that there are several ways in which factors are ascribed:

- For Boolean factors, factors may be ascribed on the basis of facts supplying necessary and sufficient conditions (i.e. the factor can be interpreted using its ordinary meaning);
- Also Boolean factors which may be ascribed according to anal- ogy (for example employing a former employee of a competitor at well over the market rate in order to obtain information ac- quired in the former employment may be considered analogous to bribery);
- Factors ascribed according to whether some threshold (switching point) is passed (as in the case of disclosures);
- Factors which involve a trade off between two facts (such as time saved and money saved).

Each of these will have their own associated argument scheme.

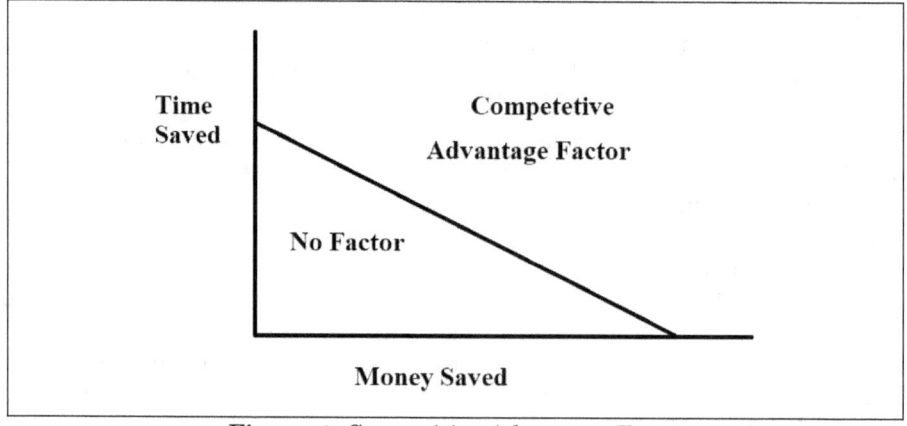

Figure 1. Competitive Advantage Factor

6.2. Ordinary Meaning Scheme

The most straightforward scheme is where the facts of the case justify the ascription of a factor, on a ordinary interpretation of the terms involved.

Ordinary Meaning Scheme
Facts Premise: Facts $a_1...a_n$ are true in Case C_1
Usage Premise: As F is ordinarily understood, $a_1...a_n$ are sufficient for Factor F to be considered present in C_1
Conclusion: F is present in C_1

The following is a set of critical questions to enable the scheme's components to be questioned:

MCQ1: Do $a_1...a_n$ really justify the ascription of F? There might be some additional fact which is needed. For example we might require specific mention of the allegedly misappropriated information in a non-disclosure agreement.

MCQ2: Does some other fact, b, provide an exception which prevents the ascription of F? There might be some unusual feature in the situation which should prevent ascription. For example, although the information was disclosed in negotiations, the defendant entered the negotiations under false pretences

MCQ3: Do other facts $b_1...b_n$ justify the ascription of factor F_2, which is incompatible with F? For example if the defendant had used restricted materials when developing his project, that should not be considered an example of reverse engineering, and so *RestrictedMaterialsUsed* (F14p) should apply and *ReverseEngineered* (F25d) should not.

6.3. Analogy Scheme

There are a number of schemes for analogy in the literature. Different schemes are given in [38], [37] and [34]. Here we give one tailored to our need to analogise between aspects of cases rather than cases as a whole.

Analogy Scheme
Base Premise: A situation S_1 is described in precedent P_1.
Derived Premise: Factor F is plausibly ascribed to P_1 on the basis of S_1.
Case Premise: Case C_1 contains situation S_2.
Similarity Premise: As it relates to F, situation S_2 is similar to situation S_1.
Conclusion: Factor F is plausibly ascribed to C_1.

Example 4

In the case *Space Aero*[4], the defendant had acquired the information while an employee of the plaintiff and the issue of Confidential Relationship was resolved for the plaintiff, on the basis of the factor *KnewInfoConfdential* (F21p). Suppose in a new case, *Subcontract*, the defendant had acquired the information while an employee of a subcontractor working for the plaintiff.

We can now suggest an analogy between the two cases:

In *Space Aero*, the information was acquired while an employee of the plaintiff

KnewInfoConfdential was ascribed to Space Aero on this basis

As it relates to *KnewInfoConfdential*, being an employee of a sub-contractor of the plaintiff is similar to being an employee of the plain- tiff

KnewInfoConfdential is plausibly ascribed to Subcontract

The following set of critical questions is based on the account given in [38] for the basic scheme for argument from analogy.

ACQ1: Are there respects in which P_1 and C_1 are different that would tend to undermine the force of the similarity with respect to F? For example, a sub-contractor has a transient relationship, whereas an employee is in a more stable relationship.

ACQ2: Is the similarity sufficient for F to be ascribed? Employees of a sub-contractor have no direct relationship to the owner of the information.

ACQ3: Is there some other precedent P_2 that is also similar to C_1, but in which F was not ascribed? Suppose there was a precedent with a fixed term employee, where the relationship was not considered sufficient to ascribe *KnewInfoConfidential*.

6.4. Switching Point Scheme

The next scheme is based on Rigoni's notion of a *switching point* [29]. If we consider a dimension with a factor favouring the plaintiff at one end and a factor favouring the defendant at the other, there will be points (possibly the same) at which one factor ceases to apply and the other factor begins to apply. These are the *switching points*. Thus given a precedent more favourable on the dimension than the new case, we can say that the factor applies to the new case. Similarly, if the new case is less favourable, we can argue that the factor does not apply. We can use this notion as the basis of an argument scheme:

---

[4] Space Aero Products Co. v. R.E. Darling Co., 238 Md. 93, 208 A.2d 74 (1965).

Switching Point Scheme
Precedent Premise: P_1 is a precedent with location L_1 on dimension D at which factor F is present.
Case Premise: C_1 is a case with L_2 on dimension D
Party Premise: F favours the plaintiff (defendant)

Value Premise: L_2 is more (less) favourable to the plaintiff (defendant) than L_1
Conclusion: F applies (does not apply) to C_1

Example 5
In the case *National Rejectors*[5] some engineering drawings were sent to customers and prospective bidders without limitations on their use. There were perhaps 100 such recipients and it was held that the factor *DisclosuresToOutsiders* applied. Suppose we have a new case, *Leaky*, in which there had been a similar practice with 150 recipients. We can say for the argument:
National Rejectors is a case with 100 disclosures and DisclosuresToOutsiders is present
Leaky is a case with 150 disclosures DisclosuresToOutsiders favours the defendant 150 is more favourable to the defendant than 100 DisclosuresToOutsiders applies to Leaky

We can question an instantiation of this scheme with the following critical questions:
SCQ1: *Is L_2 so much more favorable that a different factor ap- plies?* For example in the US Trade Secret domain of [1] there are two pro-defendant factors on the disclosures dimension, *DisclosedToOut- siders* and the stronger *DisclosedInPublicForum*.
SCQ2: When arguing that the factor does not apply because L_2 is less favourable: *Is L_2 sufficiently close to L_1 that the same factor applies?* It is possible that P_1 does not precisely identify the switching point, and that C_1 may become a new precedent for the factor, giving a more generous switching point. For example, had there been only 90 disclosures in a new case *LessLeaky*, it could still be that *DisclosedToOutsiders* applied.
SCQ3: *Is there another precedent, P2, which can ground an instantiation of the switching point scheme to give an argument that the factor does not (does) apply?* It may be that some additional information is needed to say which precedent should apply. Suppose there were a case with 200 disclosures where DisclosedToOutsiders was held not to apply. Further examination of this case and National

---

5 National Rejectors, Inc., v. Trieman, 409 S.W.2d 1 (Mo.1966).

Rejectors would be needed to explain this different treatment, and the explanation applied to Leaky.

6.5. Trade Off Scheme

The next scheme concerns trade-offs between two dimensions, as described in [6]. For example in the US Fourth Amendment domain there is a trade-off between being able to enforce the law and respect for privacy [10]. The factor involves balancing these two concerns and is something like "Sufficient respect for privacy while enabling enforcement". The idea in [6] is that a line, e.g $a.D_1 + b.D_2 + c = 0$, can be fitted to the precedents[6], separating the pro-plaintiff and pro- defendant regions of the case space. In the equation a and b are the coefficients of the variables D_1 and D_2, representing the values on the two dimensions. This determines the gradient of the line which indicates how much more D_1 is needed to compensate for less D_2, and c is a constant showing where the line crosses the axes given these coefficients.

Trade Off Scheme
Precedents Premise: $P_1...P_n$ are precedent cases in which factor F is present
Locations Premise: Precedent $P_i \in \{P_1.., P_n\}$ has locations D_{1_i} and D_{2_i} for dimensions $D1$ and $D2$,
Case Premise: C_1 is a case with L_1 on dimension D_1 and L_2 on dimension D_2
Line Premise: All $a.D_{1_i} + b.D_{2_i} + c \geq 0$ *Point Premise*: $a.L_1 + b.L_2 + c \geq (<) 0$ *Conclusion*: F applies (does not apply) to C_1

Example6 Suppose we have two precedents to which CompetitiveAdvantage applied. In one (Slow) 8 months and $500,000 was saved, and in the other (Fast) 24 months were saved, but only $100,000 was saved. A line passing through these two points is 4(money) + 1(time) − 28 = 0. Suppose we have a new case, Useful, in which $300,000 and 20 months were saved. We can form the argument:

Slow and *Fast* are precedent cases in which *CompetitiveAdvantage* is present
Slow and *Fast* have locations (5,8) and (1,24) on the money and time dimensions
Useful has location 3 on the money dimension and 18 on the time dimension

For both *Slow* and *Fast* 4(*money*) + 1(*time*) − 28 = 0 12 + 18 - 28 = 2, which is greater than 0

---

[6] Of course, more complicated curves can be used, but a straight line is the simplest.

Therefore, *CompetitiveAdvantage* is present in *Useful*.

Now suppose we have a precedent, *Useless*, in which $300,000 was saved but only 14 months. Now 12 + 14 - 28 = -2, and so we can argue that *CompetitiveAdvantage* is not present in *Useless*.

For this scheme we have the following key critical questions;
TCQ1: Is there a counter example, a precedent, P_{n+1}, for which $a.D_{1_{n+1}} + b.D_{2_{n+1}} + c < (\geq) 0$? There might be a precedent which does not fit the line. For example if we had a precedent which saved $400,000 and 13 months in which *CompetitiveAdvantage* was held to be absent. This would suggest that we need a more complicated curve than the simple line used in the argument, and it is possible *Useful* would fall on the wrong side.
TCQ2: Can the line be drawn less (more) tightly? The constant c can be adjusted to raise (lower) the line to allow (disallow) more cases to qualify unless this creates a counter example. For example, if we lower the line by reducing the constant c to 26, then *Useless* will fit and *CompetitiveAdvantage* can also be ascribed to this case.

7. Discussion

The schemes presented here are derived from reasoning with legal cases in which precedents are available to resolve the issues that arise and the ascription of factors. Some of the procedures can be adapted to practical reasoning in less formal circumstances. For example, when choosing a restaurant [5], we can identify issues such as value for money, convenience and quality of experience. Each of these will have associated factors: value will depend on cost and quality, convenience on distance and speed of service, and the quality of experience on the nature of the cuisine and the noise level and so on. Whether these dimensions will give reasons to choose or reject a venue will be a matter of judgement: a restaurant may not be expensive enough that this is a reason to avoid it, nor cheap enough that it is a reason to select it. Thus far, the problem is similar to the legal situation, in that it can be decomposed into the same elements. Some schemes may also be applicable; once the issues have been resolved, we can probably apply a rule, for example that the venue is at least satisfactory on each issue. Identifying reasons for and against also involves the move from points on relevant dimensions to reasons for or against, and this too could make use of the schemes in Section 6. This big difference lies in the way competing arguments are resolved. In law it is desirable that like cases are treated in a like manner, but this is not so with decisions like choosing a restaurant; in fact, variety may be a reason for acting differently from the last time. In such discussions it is important that the preferences accepted by the group be established[7], but this has to be done other than by appeal to precedent. This need to establish preferences other than by using precedents limits the applicability of these legal schemes to practical reasoning tasks.

Classification tasks are, however, a different matter. The style of explanation given by the above schemes has been used in [26] to ex- plain the outputs from machine learning systems for a variety of non legal domains: customer churn, poisonous mushrooms and university admissions. In such domains it is possible to treat the known cases as precedents, and to explain a classification in terms of similarity of features. If I know one red mushroom with white spots is poisonous, this may well explain why I think a new red mushroom with white spots is also poisonous. Note that in such a domain there are no preferences; the relationship between features and classification is assumed to be a matter of causal fact, not personal choice, as in the case of practical reasoning tasks. The direction of fit [31] is crucial: for classification we must fit our beliefs to the world, whereas in practical reasoning we attempt to make the world fit our desires.

Thus it may be possible to use these schemes in domains out- side law, but this will require that the reasoning does not require preferences, as with classification, or that these preferences can be established by past choices.

8. Concluding Remarks

In this paper we have modelled reasoning with legal cases as a set of argument schemes, with different schemes relevant to the different stages of the process. These schemes will, in particular, support the presentation of justifications for, and explanations of, the reasoning in such cases. By presenting instantiations of the schemes and al- lowing the user to pose critical questions, it is possible to tailor the explanations to the needs of particular users, and provide contrastive explanations [23] by explaining why things do not hold as well as why they do.

References

1. Vincent Aleven. Teaching case-based argumentation through a model and examples. PhD thesis, University of Pittsburgh, 1997.
2. Kevin Ashley. Modeling legal arguments: Reasoning with cases and hypotheticals. MIT press, Cambridge, Mass., 1990.
3. Kevin Ashley and Stefanie Brüninghaus. Automatically classifying case texts and predicting outcomes. AI and Law, 17(2):125–165, 2009.

---

[7] This point applies more generally within deliberation dialogues, as demonstrated through the model provided in [20].

4. Katie Atkinson, Trevor Bench-Capon, and Peter McBurney. A dialogue game protocol for multi-agent argument over proposals for action. *Autonomous Agents and Multi-Agent Systems*, 11(2):153–171, 2005.
5. Katie Atkinson, Trevor Bench-Capon, and Douglas Walton. Distinctive features of persuasion and deliberation dialogues. *Argument and Computation*, 4(2):105–127, 2013.
6. Trevor Bench-Capon. Using issues to explain legal decisions. In *Proceedings of XAILA 2021*. arXiv preprint arXiv:2106.14688, 2021.
7. Trevor Bench-Capon and Katie Atkinson. Precedential constraint: The role of issues. In *Proceedings of the 18th ICAIL*, pages 12–21, 2021.
8. Trevor Bench-Capon and Katie Atkinson. Argument schemes for factor ascription. In *Proceedings of COMMA 2022*, pages 68–79, 2022.
9. Trevor Bench-Capon and Sanjay Modgil. Case law in extended argumentation frameworks. In *Proceedings of the 12th ICAIL*, pages 118–127, 2009.
10. Trevor Bench-Capon and Henry Prakken. Using argument schemes for hypothetical reasoning in law. *AI and Law*, 18(2):153–174, 2010.
11. Trevor Bench-Capon and Giovanni Sartor. A model of legal reasoning with cases incorporating theories and values. *Artificial Intelligence*, 150(1-2):97–143, 2003.
12. Donald Berman and Carole Hafner. Representing teleological structure in case-based legal reasoning: The missing link. In *Proceedings of the 4th ICAIL*, pages 50–59, 1993.
13. Floris Bex, Henry Prakken, Chris Reed, and Douglas Walton. Towards a formal account of reasoning about evidence: argumentation schemes and generalisations. AI and Law, 11(2):125–165, 2003.
14. Floris Bex, Peter Van Koppen, Henry Prakken, and Bart Verheij. A hybrid formal theory of arguments, stories and criminal evidence. AI and Law, 18(2):123–152, 2010.
15. Stephanie Brüninghaus and Kevin Ashley. Predicting outcomes of case based legal arguments. In Proceedings of the 9th ICAIL, pages 233–242. ACM, 2003.
16. Alison Chorley and Trevor Bench-Capon. An empirical investigation of reasoning with legal cases through theory construction and application. AI and Law, 13(3):323–371, 2005.
17. Thomas Gordon and Nikos Karacapilidis. The Zeno argumentation framework. In Proceedings of the 6th ICAIL, pages 10–18, 1997.
18. Matthias Grabmair. Modeling Purposive Legal Argumentation and Case Outcome Prediction using Argument Schemes in the Value Judgment Formalism. PhD thesis, University of Pittsburgh, 2016.
19. John Horty and Trevor Bench-Capon. A factor-based definition of precedential constraint. AI and Law, 20(2):181–214, 2012.
20. Yanko Kirchev, Katie Atkinson, and Trevor Bench-Capon. Demonstrating the distinctions between persuasion and deliberation dialogues. In Proceedings of the 39th SGAI International Conference on AI, pages 93–106, 2019.
21. Edward Levi. An introduction to legal reasoning. *The University of Chicago Law Review*, 15(3):501–574, 1948.
22. Sarah Lichtenstein and Paul Slovic. *The construction of preference*. Cambridge University Press, 2006.

23. Tim Miller. Explanation in Artificial Intelligence: Insights from the social sciences. Artificial intelligence, 267:1–38, 2019.
24. Henry Prakken. On the nature of argument schemes. Dialectics, dia- logue and argumentation. An examination of Douglas Walton's theories of reasoning and argument, pages 167–185, 2010.
25. Henry Prakken. A formal analysis of some factor-and precedent-based accounts of precedential constraint. AI and Law, 29(4):559–585, 2021.
26. Henry Prakken and Rosa Ratsma. A top-level model of case-based argumentation for explanation: formalisation and experiments. Argument and Computation, 13(2):159–194, 2022.
27. Henry Prakken, Adam Wyner, Trevor Bench-Capon, and Katie Atkin- son. A formalization of argumentation schemes for legal case-based reasoning in ASPIC+. Journal of Logic and Computation, 25(5):1141–1166, 2015.
28. Adam Rigoni. An improved factor based approach to precedential con- straint. AI and Law, 23(2):133–160, 2015.
29. Adam Rigoni. Representing dimensions within the reason model of precedent. AI and Law, 26(1):1–22, 2018.
30. Edwina Rissland and Kevin Ashley. A case-based system for Trade Secrets law. In Proceedings of the 1st ICAIL, pages 60–66. ACM, 1987.
31. John Searle. Rationality in action. MIT press, 2003.
32. Marek Sergot, Fariba Sadri, Robert Kowalski, Frank Kriwaczek, Peter Hammond, and Terese Cory. The British Nationality Act as a logic program. Communications of the ACM, 29(5):370–386, 1986.
33. Edwina Skalak, Davidand Rissland. Arguments and cases: An in evitable intertwining. *AI and Law*, 1(1):3–44, 1992.
34. Katharina Stevens. Reasoning by precedent—between rules and analo- gies. *Legal Theory*, pages 1–39, 2018.
35. Nina Varsava. How to realize the value of stare decisis: Options for following precedent. *Yale Journal of Law and the Humanities*, 30:62–120, 2018.
36. Douglas Walton. *Argumentation schemes for presumptive reasoning*. Lawrence Erlbaum Associates, 1996.
37. Douglas Walton. Similarity, precedent and argument from analogy. *AI and Law*, 18(3):217–246, 2010.
38. Douglas Walton, Christopher Reed, and Fabrizio Macagno. *Argumentation schemes*. Cambridge University Press, 2008.

CONTEXTUAL AND EMOTIONAL MODULATION OF SOURCE-CASE SELECTION IN ANALOGICAL ARGUMENTS

MARCELLO GUARINI
Department of Philosophy, and the Centre for Research in Reasoning, Argumentation, & Rhetoric, University of Windsor, Canada
mguarini@uwindsor.ca

Abstract

In making analogical arguments about actions, is more similarity between the source case and the target case always better? The argument presented herein is that more similarity is not always better. The reason is that the context of the argument, including emotional considerations, modulates the selection of the source case to service the goals of the argument. Sometimes, the goals of the argument are such that very high levels of similarity between source and target would be ineffective.

1. Introduction

This paper examines analogical arguments about action. In this domain, is more similarity between the source case (the one from which we argue) and the target or disputed case always better? The argument presented herein is that more similarity between the source and target cases is not always better. The reason is that the context of the argumentation, including emotional considerations, modulates the selection of the source case to service the goals of the argument. Sometimes, the goal of the argument is such that very high levels of similarity between source and target would be ineffective. In the process of making this point, the importance of emotion in some analogical arguments will be discussed, as will the significance of the preceding points for argument construction, interpretation, and assessment. Hafner *et. al.* (2002) and Walton and Hyra (2018) have argued for the importance of contextual considerations in analogical arguments. This paper adds to that kind of work by focusing on how emotion can impact source-case selection.

2. A Case Study: Reconstructing Nathan's Argument to David

The story[1] of David and Bathsheba is well known, though what might be less well known is Nathan's argumentative intervention with David. Let us begin with a review of the story (found in *2 Samuel* of Alter, vol. 2, 2019).

Notwithstanding the fact that King David has several wives and multiple concubines, he is very much taken with Bathsheba, who was married to Uriah (a loyal and devoted solider in David's army). David has Bathsheba brought to him while Uriah is away, and she becomes pregnant by him. David attempts to cover up the fact that this is not Uriah's child by having him sleep with Bathsheba, but in accordance with custom Uriah will not do so while some of his comrades are in battle. To hide the affair with Bathsheba, and to have Bathsheba for himself, David instructs Joab to send Uriah to the front lines and engineer a scenario where he is sure to be killed, and events unfold as planned. With Uriah's death, David is free to marry Bathsheba, which is exactly what he does.

This was not David's finest hour.

Nathan (a prophet) is one of the people who knows what David has done. He approaches David and tells this story:

> Two men there were in a single town, one was rich and the other poor. The rich man had sheep and cattle, in great abundance. And the poor man had nothing save one little ewe that he had bought. And he nurtured her and raised her with him together with his sons. From his crust she would eat and from his cup she would drink and in his lap she would lie, and she was to him like a daughter. And a wayfarer came to the rich man, and it seemed a pity to him to take from his own sheep and cattle to prepare for the traveler who had come to him, and he took the poor man's ewe and prepared it for the man who had come to him. (*2 Samuel* 12: 1-4, in Alter, vol. 2, 2019.)

---

[1] Some (though not I) take this story as a literal portrayal of what happened, i.e., as a completely accurate historical account. Others see it as a work of complete fiction. Still others see it as something like Shakespear's Richard III, i.e. based on historical figures with many fictional details added to help the author(s) achieve their purposes. For ease of exposition, this paper is written as if the events transpired exactly as presented in the relevant texts, but there is no commitment to that view. Notwithstanding considerations of historicity or lack thereof, the story is instructive and worthy of examination.

David flies into a rage. Notice how he cites a lack of pity as his reason for wanting to see the rich man severely punished:

> And David's anger flared hot against the man, and he said to Nathan, "As the Lord Lives, doomed is the man who has done this! And the poor man's ewe he shall pay back fourfold, inasmuch as he has done this thing, and because he had no pity!" And Nathan said to David, "You are the man! (2 Samuel 12: 5-7).

Alter (2019, vol. 2) describes Nathan's response as an "accusatory explosion" (p. 351), a snapping shut of a "rhetorical trap" (p. 352), and a "knife thrust" (p. 352). Crucially, David is *angry* at the rich man's lack of *pity*. Nathan is looking to transfer these emotions to the situations at hand, namely, the adulterous affair and the killing of Uriah. I interpret this as analogical arguing. There is much more going on in the relevant texts, but the focus in this paper will be on the arguments by analogy and their analysis.

Several comparisons are being made between the David-Uriah-and-Bathsheba (DUB) situation and the rich-man-poor-man-and-sheep (RPS) case. In fact, DUB consists of at least two different actions or sets of actions that can be called into question.

A_1: *David sleeps with Bathsheba while she is still married to Uriah.*

A_2: *David conspires to have Uriah killed so that he will be free to marry Bathsheba and cover up the affair and pregnancy; Uriah is killed; David marries Bathsheba.*

Let us say that

DUB_1 *is the portion of the story that concludes with A_1,* and

DUB_2 *is the entire story, including A_1 and A_2. DUB_1 is a substory of DUB_2.*

Let us call DUB_1 and DUB_2 the target cases – i.e., the cases about which Nathan wants to make his point. The RPS case is the source. The arrows indicate mappings or comparisons. An "X" indicates a consideration that is not mapped.

Source Case: RPS		Target Case: DUB_1
The rich man	→	David
The poor man	→	Uriah
The poor man's sheep (ewe)	→	Bathsheba
The poor man owns one sheep	→	Uriah has only one wife, Bathsheba
The poor man is kind to and cares for/protects his one sheep, an ewe	→	Uriah is kind to and cares for/protects Bathsheba
The rich man has many sheep and cattle	→	David has many wives and concubines

The rich man steals the poor man's only sheep	→	David sleeps with Uriah's only wife, Bathsheba
The rich man kills the sheep	X	
The rich man does not feel pity	→	David does not feel pity
The rich man's theft is unacceptable in virtue of his action being a grossly self-indulgent act of further benefiting himself at the expense of an innocent man who is less well off. The rich man should experience some pity for the poor man.	→	David sleeping with Bathsheba is unacceptable in virtue of his action being a grossly self-indulgent act of further benefiting himself at the expense of an innocent man who is less well off. David should experience some pity for Uriah.

In general, killing is something that is morally salient, so the killing of the sheep is included in the source case, but if the ewe is mapped to Bathsheba, then there is nothing in the target to which we can map the killing of the ewe, so the killing of the ewe has not been mapped to anything. With respect to sheep considered more generally, the comparison of humans to sheep was not uncommon in this culture. Think of the 23rd psalm – "The Lord is my shepherd, I shall not want…" (Alter 2019, vol. 3) – which implies that those who recite, sing, or pray the psalm –including David – are like sheep under the guidance and protection of a divine shepherd. In this literary, cognitive, affective, spiritual, and performative environment, the intended comparisons would have been transparent to David.

Let us consider a second set of mappings (below). Note that the rich man stealing the sheep needs to be mapped to a sequence of actions in DUB_2. On its own, David marrying someone would not be grossly self-indulgent in the way that rich man stealing the sheep is. To capture the idea that David has engaged in a kind of grossly self-indulgent taking, his marrying Bathsheba must be seen as part of a *sequence of events*.

A Third type of interpretation is logically possible. We could read Nathan as saying that the *combination* or *conjunction* of sleeping with Bathsheba (A_1) and having Uriah killed so he can marry Bathsheba (A_2) is problematic, but individually there is no problem. The gross self-indulgence could be seen as resulting from having done *both* of A_1 and A_2. The mappings would look different on that interpretation. The source case would only be deployed once, and the rich man stealing the sheep could be mapped to (i) David sleeping with Bathsheba *and* (ii) the sequence of events starting with the conspiracy to kill Uriah and ending in David marrying Bathsheba. Again, the idea would be that the *combination* of those actions is problematic. Given that *Exodus* 20: 13-17 (Alter 2019, vol. 1) lays down commandments against adultery and coveting as well as killing, it is reasonable to think that Nathan was objecting individually to (i) the adulterous affair and (ii) the conspiracy to commit murder and the actual murder, hence the reconstruction where the same source case, PRS, is used to map to two different targets.

Let us return to the mapping below. Nathan's argument is not fully articulated. He never claims that both A_1 and A_2 are forms of self-indulgent taking. However, by saying that David is like the rich man, he is saying that A_1 is like stealing the sheep, and he is saying that A_2 is like stealing the sheep as well. If stealing the sheep is mapped simply to conspiring to kill Nathan – the first of the three elements in the mapped sequence below – it is not clear how that amounts to any kind of taking (self-indulgent or otherwise) since the sheep is mapped to Bathsheba, and Nathan is the one being killed. If stealing the sheep is mapped to David's conspiracy and its successful execution by Joab – the first two elements in the three-element sequence below – we have the same problem. In what sense has David self-indulgently taken anything or anyone? If stealing the sheep is mapped one-to-one to the act of getting married, it is unclear how that is problematic because marriage is not generally self-indulgent taking. The one-to-three mapping captures a taking that is grossly self-indulgent.

With respect to an argument reconstruction, Nathan's case could be summarized as follows.

Source Case: RPS	Target Case: DUB₂
The rich man	→ David
The poor man	→ Uriah
The poor man's sheep (ewe)	→ Bathsheba
The poor man owns one sheep.	→ Uriah has only one wife, Bathsheba.
The poor man is kind to and cares for/protects his one sheep.	→ Uriah is kind to and cares for/protects Bathsheba.
The rich man has many sheep and cattle.	→ David has many wives and concubines.
The rich man steals the poor man's one sheep.	David sleeps with Uriah's only wife, Bathsheba. Bathsheba becomes pregnant with David's child. David wants to cover up his affair with Bathsheba. David wants to have Bathsheba for himself. David conspires with Joab to kill Uriah so David can marry Bathsheba and cover up his affair. Joab assigns Uriah to a battle group given a dangerous task, one engineered to lead to losses; Uriah and others in his group are killed in battle. Bathsheba is widowed. David marries Bathsheba.
The rich man kills the sheep.	X
The rich man does not feel pity for steeling the sheep.	→ David does not feel pity for having. Uriah killed and marrying Bathsheba.
The rich man's theft is unacceptable in virtue of his action being a grossly self-indulgent act of further benefiting himself at the expense of an innocent man who is less well off. The rich man should experience some pity the poor man.	→ David taking Uriah's wife by having him killed is unacceptable in virtue of his action being a grossly self-indulgent act of further benefiting himself at the expense of an innocent man who is less well-off. David should experience some pity for having Uriah killed to cover up the affair and marry Bathsheba.

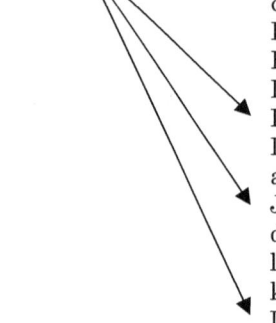

P1. The poor man took good care of and protected his one sheep, an ewe.
P2. The rich man has many sheep and steels the poor man's one sheep.
P3. What the rich man does is unacceptable in virtue of his action being a grossly self-indulgent act of further benefiting himself at the expense of an innocent man who is less well-off. The rich man should have pity for the poor man.

P4. Bathsheba is the only wife of Uriah, and he took good care of and protected her.

P5. David has many wives and concubines, yet he uses his power as king to have Bathsheba brought to him so he could sleep with her.

C1a. Just as the rich man steeling the sheep is unacceptable, David sleeping with Bathsheba is unacceptable in virtue of his action being a grossly self-indulgent act of further benefiting himself at the expense of an innocent man who is less well-off.

C1b. David should experience some pity for Uriah.

P6. Bathsheba becomes pregnant with David's child.

P7. David wants to cover up that he had an affair with Bathsheba.

P8. David wants to marry Bathsheba.

P9. David has many wives and concubines and conspires with Joab to kill Uriah so that he can take Uriah's wife as his own and coverup his affair.

P10. Joab assigns Uriah to a battle group given a dangerous task, one engineered to lead to losses; Uriah and others in his group are killed in battle. Bathsheba is widowed

C2a. Just as the rich man steeling the sheep is unacceptable, David conspiring to kill Uriah so he can marry Bathsheba is unacceptable in virtue of his action being a grossly self-indulgent act of further benefiting himself at the expense of an innocent man who is less well-off.

C2b. David should experience some pity for Uriah

It might be wondered why Nathan does not simply appeal to the relevant commandments against murder and adultery to persuade David that he has gone astray. Among other things, the next section will engage that question.

3. The Role of Emotions in Nathan's Argument

To understand what Nathan was doing, it helps to introduce some contemporary terminology. People are engaged in *motivated inference* when their desires or emotions inform the inferences they draw; i.e., the inferences are not based simply on the evidence and are self-serving (Kunda 1990; Kunda 1999, chapter 6.). (See also Dunning et. al. 1995.) It is not hard to imagine that David's desire for Bathsheba could have been motivating his reasoning about what counts as acceptable or unacceptable behaviour for a king. Let us see how this might work.

David was a king in command of military forces. It is not unusual for generals and kings to send soldiers to their death. Indeed, both *1 Samuel* and *2 Samuel* are filled with stories of King Saul and King David waging war and being involved in the killing of *many* people. Clearly, the commandment against killing was not being interpreted as absolute. It is possible that David might have thought that by sending Uriah to the front lines, he was acting within the powers of a king. The desire to cover up his affair and have Bathsheba for himself prevents him from seriously considering that there are limits to the power of a king; consequently, sending Uriah to his death may have seemed an acceptable use of his power. To be sure, such reasoning is self-serving, but that is how motived inference works. Nathan realized he was not going to get through to David by appealing to the relevant commandments, especially since the one against killing seemed to have known exceptions. Nathan needs to tread carefully. David sent an innocent and loyal man (Uriah) to his death to get his way. The PRS story is insightful (and careful) for at least two different reasons. Both have to do with emotion. Moreover, the argument itself appears to be an attempt not only to persuade David in some respects, but to get him to *feel* a certain way. Let us examine Nathan's emotionally informed approach to persuading David.

One reason Nathan's argument is insightful is that it does not contain content that might interfere with the desired response from David. RPS will allow Nathan to argue that what David did was wrong without mentioning adultery or the murder of a human being in the source case. Any source case that mentions adultery or murder is likely to get David's "back up." For reasons that will be explained below, Nathan needs to elicit a strong response against a certain kind of action in his source case; doing so would likely require steering clear of pecuniary interests, lest he make David suspicious that he has come to preach to him about adultery. Adultery and conspiracy to kill are nowhere mentioned in the source. While it is not flattering for Bathsheba to be compared to a sheep, using a story that makes use of a sheep, rather than a person, is part of why the analogy appears to work on David. He will not think the source case has anything to do with his pecuniary interests or with the killing of a human being, and that helps Nathan get the response he needs from David.

A second reason Nathan's argument is insightful is that the story of sheep theft is likely so speak to David in a powerful way. When he was younger, David was neither a king nor a prince. He lived a much humbler existence tending his father's sheep, protecting them from all comers, human or animal. David took great pride in being a capable shepherd. Nathan must have known that the story of sheep theft – especially from a man who had only one sheep – would elicit a powerful response from David. *And it did. Exodus* 21: 37 (Alter 2019, vol. 1) prescribes that the theft of one sheep is to be repaid with four sheep; David goes beyond this and says the thief is "doomed" (which is the biblical way of saying he should die for his offense, either by execution or other means). David was *livid*, and part of the reason for that was the thief's lack of pity. This was exactly the response Nathan needed. With it, he could compare the RPS case with what David had done.

While there is both ancient (Aristotle, 1984, book 2) and recent work (Gilbert, 1994, 2004, & 2014) on emotion in argument, much excellent, philosophical work on analogy has not discussed emotion. Juthe (2014) and Alvargonzález (2020) have surveyed much literature and have developed classification schemes for analogy, but emotion is not discussed. Even Ribeiro's (2014) excellent anthology does not contain words such as "emotion" or "affect" in the index. Thagard and Shelley (2006) are exceptions. They discuss three kinds of analogy that *transfer* emotions: those used in persuasion, empathy, and reverse empathy (Thagard and Shelley 2006, pp. 43-49). In empathy and reverse empathy, the primary function is to transfer emotion to achieve an *understanding* of someone's emotional state. In some analogical arguments, there can be an transfer of emotion from source to target. To the examination of this we now turn.

To start, let us draw a distinction between *being persuaded* that a particular emotional response is appropriate, on the one hand, and *having* that emotional response, on the other. For example, one might say, "I am persuaded that I *should* pity Uriah, but I don't feel any pity." In other words, there can be a kind of doxastic transfer from what one believes about emotion in the source to what one believes about emotion in the target. It does not follow that one will feel the way one thinks one should feel. Analogical transfer of emotion itself requires that one or more feelings about the source is (are) transferred to (i.e. *felt about*) the target. Nathan's argument is trying to achieve not only a doxastic transfer about what David believes should be felt, but also the transfer of emotion itself.

Arguably, Nathan does not want simply to persuade David to believe something; he also wants to change how he feels about what he did. With respect to DUB_1, culminating with David sleeping with Bathsheba, it is possible that David was emotionally motivated in a way that prevented him from seeing that he did was wrong. It is also possible that he may have seen that what he did was wrong – the text does present him as wanting to cover up Bathsheba's pregnancy – but he did not experience any regret, concern, or pity. Being emotionally motivated might prevent someone from believing something they should; it can also prevent someone from feeling a certain way. Perhaps, *if* David did think he did something wrong by sleeping with Bathsheba while she was married, he was still not having any of the feelings he might normally have when he does something he believes to be wrong. It is possible Nathan not only wanted David to see that sleeping with Bathsheba was wrong, but that he should also be more emotionally worked up about what he did. If he feels some compassion or pity for the people he mistreated and is at least a little upset about what he did, he might be less likely to make such a mistake in the future. The analogical transfer of emotion from the source case, RPS, to the target, DUB_1, helps achieve the desired goal of changing how David feels. He is angry at what the rich man did and the fact that he experienced no pity; if the analogical transfer of emotion is successful, he would be upset at himself and experience pity for those he mistreated. In other words, Nathan may be trying to help David cultivate a motivational and affective make-up that will improve his behaviour as well as changing his beliefs. With respect to David marrying Bathsheba after having Uriah killed, there is no evidence that David thought he did anything wrong. Here, Nathan really has his work cut out for him. He wants (i) to persuade David that what he did was wrong and that he should feel a certain way about what he did, and (ii) he wants to change how David feels about what he did. What is striking is that Nathan's source case does double duty: it can be applied both to the adultery and the killing, serving both to persuade David to think differently and to change how he feels.

Clearly, Nathan was not lacking in rhetorical savvy. He knew his audience well, with a firm grasp of what to avoid saying in the presentation of his source case, and an equally firm grasp of what would move David.

4. Other Contextual Considerations

There is much we can learn from this example about analogical arguments. To start, the selection of the source is impacted by the context of persuasion. That should be clear from the previous remarks, but to reinforce the point, let us consider a different kind of source, an example of what I will call near-identity. For brevity, I will focus on the issue of conspiring to kill someone (DUB2). Imagine that Nathan argued in this way.

Q1. A previous king that David has never heard of, having many wives and concubines, wanted to take another woman for his wife, but she was married.
Q2. That king had the husband of that woman sent to the front lines and had his generals engineer a scenario that would guarantee the death of the woman's husband so he could marry the woman.
Q3. What that king did was unacceptable in virtue of it being a self-indulgent use of his power to further benefit himself at the expoense of an innocent man who is less well off.
Q4. David has many wives and concubines, and he wanted to take Bathsheba for his wife, even though she was married to Uriah.
Q5. David has Uriah sent to the front lines and had his generals engineer a scenario that would guarantee the death of Uriah so David would be free to marry Bathsheba.
D1. Just as the previous king acted in an unacceptable manner, so too did David.

It is often said that for analogical arguments to work, the source and target cases need to be appropriately similar. In the context we are considering, selecting a near-identical source would likely not work. It would be an example of too much similarity. If David thinks that, as king, he has the authority to send a solider to his death in a war, then simply using the argument from near identity will not persuade him that he is in the wrong. In David's time, it would have been recognized that kings do have the authority to send soldiers into combat. Nathan needs to persuade David that there are limits to that authority. Sending someone into combat for the good of the nation is one thing. It is something quite different for the king to conspire to engineer a combat scenario that is guaranteed to get a soldier killed because the king wants the soldier's wife. This latter scenario is an abuse of power in virtue of its shockingly self-indulgent nature. It has nothing to do with the good of the nation. Even though Nathan's sheep-theft case is not as similar to DUB_2 as the near-identity case, it is a better choice. *There is such a thing as too much similarity[2] in analogical argumentation.* That said, it is not difficult to imagine a context where a near-identical source might be effective in analogical argumentation. Let us consider an example.

While fanciful, this scenario helps to make a plausible point about the role of context. Imagine David was brought to trial[3] for sending Uriah to the front lines in the way he did. We will imagine the tribunal in question is one governed by *stare decisis*, and that there is a precedent case that is near-identical to what David did, and there are no countervailing precedents. David's attorney insists that he plead guilty because of the precedent case. Being a difficult client, David insists that his plea will be innocent, and that his lawyer is being paid handsomely to defend him, so that is what he needs to do. The case goes to trial. The prosecutor appeals to the near-identical precedent, and the defense has nothing that can overcome the force of the near-identical precedent. The prosecution wins.

---

2 This paper draws on a pre-theoretic understanding of similarity. Whether it is structure mapping (Gentner and Markman, 1997) or factors or dimensions (Ashley, 1990) or other approaches discussed in this paper, there are different ways of understanding and quantifying similarity between source and target. As it should be uncontroversial on any approach to similarity that the near-identity case is more similar to what David did than Nathan's sheep story, this paper does not need to select a specific quantitative metric for similarity. Moreover, straying into that territory in any detail would require a separate paper.

3 This example assumes we are dealing with a criminal trial. Some jurisdictions ban the use of analogical argument in criminal trials. If this is concerning to the reader, we could imagine the example is civil in nature, where David is being sued for damages due to his causing a wrongful death, and we could imagine the precedent case was civil in nature. We could even imagine a variant on the example that is strictly ethical in nature. Consider a near-identical case where the king in the precedent is highly regarded by David, and that king eventually agreed that what he did was unethical. In such a situation, the near-identical case might carry some weight with David.

In this context, it is not David who needs convincing; it is the tribunal hearing the case. Dispute contexts where there are near-identical precedents are unlikely to end up in court very often since competent and honest lawyers will often be able to convince their clients that they have no chance of winning if there is a near-identical precedent that could be used against them. However, some clients may be insistent, and some such cases may end up in court. In such a context, arguments from near-identical source would be very convincing.

How much similarity is needed in analogical arguments depends on the context, including the goals of the argument and the state and composition of the audience. When trying to convince a tribunal that is (at least for the most part) committed to deciding cases as they have been decided in the past, then selecting a case that was decided in the past and is near-identical to the disputed or target case may be a very good idea (though see Hafner et. al. 2002 for further contextual considerations). When Nathan was trying to make his point to David, it would have been useless to appeal to a near-identical case because David would not have conceded that the near-identical source was an example of someone having done something wrong. He had no antecedent commitment in favour of treating the near-identical case as an example of problematic behaviour. In the fanciful scenario where David is tried for his actions, he might remain completely unconvinced, but that does not matter from the perspective of the prosecution, because they are trying to convince the tribunal. The prosecution might use multiple cases, such as the near-identical source and Nathan's RPS source, but such is not required given the way the example has been set up.

When it comes to source-case selection in analogical arguing, context modulates how much similarity is advisable between the source and the target cases. To be sure, context can also modulate which things are seen as similar. This is a point that has been discussed in the literature (Holyoak and Thagard 1995; Walton 2013; Macagno et. al. 2017). Let us call this *modulation of similarity*. For example, in many contexts it would seem bizarre to say that steeling a sheep is like committing adultery or conspiring to commit murder, but context can modulate our sense of similarity so that what does not seem similar in some contexts is similar in other contexts. That context modulates which things count as similar in the first place (and even the magnitude of the similarity) is an important point, *but this paper is making another point*. Not only can context (i) modulate what counts as similar, it can (ii) modulate how much similarity and what kind of similarity we are looking for between source and target in deciding which source cases to select. This is why the title of the paper focuses attention on the *modulation of source-case selection*. While related, modulation of similarity and modulation of source-case selection should not be conflated.

5. Argument Construction, Interpretation, and Assessment

There are dialectical, rhetorical, and narrative considerations to Nathan's use of analogical arguments. When Nathan presents his narrative of sheep theft, he does so to elicit a response from David, which consists in both the statements David makes and the emotions he experiences and expresses about the rich man. Nathan's response – "You are the man!" – uses David's response. Nathan develops his argument through the dialectical exchange. Regarding the narrative of sheep theft, it is selected to elicit the emotional response Nathan wants David to have. In his supportive discussion of narrative arguments, Christopher Tindale (2021, p.116) asks what value is added to argument or the analysis of argument by invoking the category of narrative argument. One possible response that is suggested by Nathan's narrative is that they can elicit an emotional response. It does not follow that this is their only role, but it is certainly one role. In constructing his argument, Nathan makes use of a narrative analogy in a dialectical exchange for the purpose of eliciting a response from David that he will then use (i) to persuade David that he has erred and (ii) to modify his affective and motivational states. The selection of the narrative for this dialectical exchange is governed by rhetorical considerations regarding what may or may not be effective in eliciting a dialectical response that will help Nathan to achieve his argumentative goals. Interpreting and assessing Nathan's argument requires an understanding of its rhetorical, narrative, and dialectical dimensions.

The critical questions that are often proposed for evaluating analogical arguments generally deal with its logical or structural dimensions, such as whether the relevant similarities between source and target are outweighed or defeated by the relevant differences (Walton et. al. 2008, chapter 2; van Eemeren et. al. 2009, chapter 7; never mind the myriad of textbook treatments that take a similar approach). There is no doubt that the logical/structural dimension is essential to assessing analogical arguments. A common (and legitimate) concern in assessing analogical argument is that the differences may outweigh the similarities, making it a bad argument. In other words, there is concern over whether the cases are *too different*. The possibility that, in some contexts, the cases can be *too similar* – and that the differences can be helpful – goes almost completely unnoticed. To be sure, analogical arguments from near-identical sources have the logical virtue of one or more weighty, relevant similarities and little or no relevant differences. However, in the case of persuading David, an argument from a near-identical source is unlikely to succeed. As discussed above, the source not being directly about adultery or murder – *important differences indeed* – is what allows Nathan to get a better hearing than he otherwise might have received. While David might be an extreme case, he is not a special case. Arguments about actions often deal with subject matters that are emotionally charged. Thinking critically about analogical arguments needs to take that into consideration. Some might find it tempting to think that interpretation and evaluation of analogical arguments need not be especially concerned with whether the audience would concede or agree with the premises attributed in the argument reconstruction. I join with Juthe (2020, p. 29) in rejecting this approach to interpreting analogical arguments.

Regarding the structure of analogical arguments, Walton has suggested that narrative analogical arguments can be understood using story scripts. The example he uses makes use of very neat one-to-one mappings between source and target. Juthe (2014, not focusing on narrative analogical arguments) treats analogical arguments in general as involving one-to-one mappings from source to target. One of the features of Nathan's argument is its use of a one-to-many mapping of relations. As Bartha (2010, chapter 3) has pointed out, it was also a limit of earlier models of cognitive science models of analogies, but more recent work makes use of one-to-many mappings. Bartha also points out that more is needed than structure mapping. Above, the rich man steeling the sheep was mapped one-to-many to a sequence of actions and events in DUB_2. Some one-to-many mappings may be acceptable, and some may not. Structure alone will not determine this. A model that includes semantic constraints (Holyoak and Thagard, 1995) may help in some cases, but it unclear that it would help in the case under consideration. Holyoak and Thagard's approach even includes the goals of the analogy as constraints on how to carry out the mapping between source and target. While that marked an important addition to structure mapping, it does not capture why the one-to-many mapping from RUB to DUB_2 is appropriate (and why other one to many mappings may not be). We are being asked to interpret the stealing of a sheep as similar to *the sequence* of conspiring to have a man killed, having him killed, and marrying his wife. There is a *normative valancing* of the stealing of the sheep that Nathan wants to transfer to the sequence of actions and events in DUB_2, and that matters with respect to what is mapped to what in normative domains. (If "semantic" constraints in the Holyoak Thagard approach were expanded to include normative valancing, it would be a more powerful model.) Case-based reasoning models in AI, especially in legal domains, do have a kind of normative valancing in the factors or dimensions that are used to compare cases, but those models tend to have dimensions or factors that work within a domain (eg, think of Ashely's 1990 work in trade secret law). Even story scripts, another AI modeling technique (Walton, 2013), tend to focus on within-domain comparisons. One of Walton's examples is of a radio-active substance found outside its proper container being compared to an animal getting out of a cage. While the objects are different, both cases are clearly about strict liability. Comparing the theft of a sheep to a sequence of actions involving (i) conspiracy to kill, (ii) killing, and (iii) marriage, involves the comparison of very different actions. An action such as getting married is not something that would usually be valanced in a normatively negative manner, but when placed in the sequence of actions and events that take place in DUB_2, it makes sense for the *sequence* of 3 relations to receive the transfer of negative valence. If David had Uriah killed for completely different reasons and had no interest in Bathsheba (and did not sleep with or marry her) the mapping of sheep theft to David's killing of Uriah would fail because there is no sense in which David self-indulgently "took" Bathsheba. Conspiracy, killing, and marriage each *have their own normative regulations*, and Nathan reframes them so that when combined in a particular way, they can be seen as a kind of wrongful (grossly self-indulgent) taking, which is normatively governed as well. As Macagno et. al. (2017) remind us, what counts as similar often emerges in the context under consideration. Similarity metrics based on structure alone, domain specific valancing of factors, or weighing similarities and differences (independent of contextual considerations) does not provide the needed flexibility. Bartha's own modal model of parity between source and target is (mostly) focused on achieving a particular threshold of similarity and is not sensitive to degrees of similarity and the sort of contextual considerations (emotion, constitution of the audience, *et cetera*) considered herein. Assessing levels of similarity in a manner that is sensitive to context, and then determining the extent or type of similarity it would be effective to use in a given context are areas where more work is needed, whether in philosophy, cognitive science, or AI.

6. Conclusion

The role emotion plays in the story of David helps us to see both the challenges posed by emotion and the opportunities they present. That emotion and desire led David astray is not hard to see. Nathan's response was to use an analogy that spoke to David in a deep way, one with sufficient similarity to what he had done and a powerful emotional valence that it could shift David's doxastic, emotional, and motivational states. Crucially, the analogy could not be too similar to what he had done for the reasons articulated above. While attempting to change someone's motivational or emotional states may seem manipulative or otherwise problematic, it need not be more problematic than changing someone's doxastic states. Just as there are ways of changing what someone believes or accepts that are logically, epistemically, or ethically reprehensible, there are ways of changing how someone feels that would be problematic for various reasons. It does not follow that there are no acceptable ways of achieving the same goals. The story of Nathan's intervention with David is compelling for many reasons, not the least of which is that David did some terrible things, did not seem to believe he had erred, had no remorse for what he did, and no compassion for those he harmed, and yet Nathan found a way to reach David. Nathan's use of analogical argument was a reasonable response to David's transgressions.

References

Alter, Robert. 2019. *The Hebrew Bible: A Translation with Commentary*, three volumes. New York and London: W. W. Norton and Company.

Alvargonzález, David. 2020. "Proposal of a Classification of Analogies." *Informal Logic*, 40(1): 109-137.

Aristotle, 1984. *Rhetoric*. Translated by Rhys Robers. In Johnathan Barnes, ed., *The Complete Works of* Aristotle, the revised Oxford translation, vol. 2. Princeton, New Jersey: Princeton University Press.

Ashley, Kevin. 1990. *Modeling Legal Argument: Reasoning with Cases and Hypotheticals*. Cambridge, Mass. and London, UK: MIT Press.

Bartha. Paul. 2010. By Parallel Reasoning: the Construction and Evaluation of Analogical Arguments. Oxford: Oxford University Press.

Braman, Eileen, and Thomas E. Nelson. 1994. "Mechanism of Motivated Reasoning? Analogical Perception in Discrimination Disputes." *American Journal of Political Science*, 51(4): 940-956.

Dunning, David, Ann Leuenberger, and David A. Sherman. 1995. "A New Look at Motivated Inference: Are Self-Serving Theories of Success a Product of Motivational Forces?" *Journal of Personality and Social Psychology*, 69(1): 58-68.

Eemeren, Frans van, Bart Garssen, and Bert Meuffels. 2009. *Fallacies and Judgments of Reasonableness: Empirical Research Concerning the Pragma-Dialectical Discussion Rules*. Volume 16 of Argumentation Library. Dordrecht, Heidelberg, London, New York: Springer.

Gentner, Dedre, and Markman, A.B. 1997. "Structure mapping in analogy and similarity." *American Psychologist*, 52(1): 45-56.

Gilbert, Michael A. 1994. "Multi-Modal Argumentation." *Philosophy of the Social Sciences*, 24(2): 159-177.

Gilbert, Michael A. 2004. "Emotion, Argumentation, and Informal Logic." *Informal Logic*, 24(3): 245-264.

Gilbert, M.A. 2014. *Arguing with People*. Peterborough, ON: Broadview Press.

Hafner, Carole, Donald H. Berman. 2002. "The Role of Context in Case-Based Legal Reasoning: Teleological, Temporal and Procedural." *Artificial Intelligence and Law* 10(1-3): 19-64.

Holyoak, Keith J., & Thagard, Paul (1995). *Mental Leaps. Analogy in Creative Thought*. MIT Press.

Juthe, André. 2014. "A Systematic Review of Classifications of Arguments from Analogy," chapter 7 of Ribeiro, Henrique Jales, editor, 2014. *Systematic Approaches to Argument by Analogy*. Amsterdam: Springer.

Juthe, André. 2020. "A Defense of Analogy Inference as Sui Generis," *Logic and Logical Philosophy*, 29: 259-309.

Kunda, Ziva. 1990. "The Case for Motivated Inference," *Psychological Bulletin*, 108: 480-498.
Kunda, Ziva. 1999. *Social Cognition*. Cambridge, Mass: MIT Press.
Macagno, Fabrizio, Douglas Walton, and Christopher Tindale. 2017. "Analogical Arguments: Inferential Structures and Defeasibility Conditions." *Argumentation*, 31(2): 221-243.
Ribeiro, Henrique Jales, editor. 2014. *Systematic Approaches to Argument by Analogy*. Amsterdam: Springer.
Thagard, Paul, and Cameron Shelley. 2006. "Emotional Analogies and Analogical Inference," chapter 3 of Paul Thagard, *Hot Thought: Mechanisms and Applications of Emotional Cognition*. MIT Press.
Tindale, Christopher. 2021. *The Anthropology of Argument: Cultural Foundations of Rhetoric and Reason*. Routledge.
Walton, Douglas. 2010. "Similarity, precedent and argument from analogy." *Artificial Intelligence and Law*, 18(3), 217-246.
Walton, Douglas. 2013. Argument from Analogy in Legal Rhetoric. *Artificial Intelligence and Law, 21(3): 279-302.*

DISTANT SIMILARITY AS AN ADVANTAGE OF (SOME) ARGUMENTS FROM ANALOGY? COMMENTARY ON MARCELLO GUARINI'S 'CONTEXTUAL AND EMOTIONAL MODULATION OF SOURCE CASE SELECTION IN ANALOGICAL ARGUMENTS'

MARCIN KOSZOWY
Laboratory of The New Ethos, Warsaw University of Technology
marcin.koszowy@pw.edu.pl

1. Introduction

In his contribution, Marcello Guarini delves into the contextual features and emotional components of analogical arguments. In order to attempt at systematically capturing these two phenomena, the author employs the concepts of contextual and emotional modulation in the process of selecting a source case of an argument from analogy (i.e. an original case to set up the argument; see e.g. Walton, 2014, pp. 24-25). As an illustrative material, Guarini uses the biblical story of king David, Batsheba, and Uriah, and focuses on the argument performed by Nathan that made David admit to the conspiracy to kill Uriah which ended in David marrying Batsheba. As I find the proposed theoretical framework worth employing in contemporary argument studies, in this commentary I will focus on two topics which I find particularly worth exploring in this respect, namely: Guarini's emphasis put on some key dialogical aspects of analogical arguments (Section 2), and the applications of the proposed tools to study some contextual features of analogy in argumentation (Section 3). As one of the potentially most impactful aims of Guarini's contribution is an attempt at modifying a typical argumentation theory perspective on analogical arguments ("the more similarity the better") towards the view according to which distant similarity works better for argument's persuasiveness, in conclusion Section I will point to possible ways of developing this idea.

2. Dialogical dimension of arguments by analogy

One of the key issues related to the dialogical dimension of analogical arguments Guarini puts forward, is an interpretation proposed by Alter (2019, vol. 2, p. 352) according to which Nathan's argument is a kind of a "rhetorical trap", as it helps, in my view, emphasising the role of studying dialogical factors in the process of assessing analogical arguments. Guarini's analyses help to observe that in some cases, not revealing by a speaker that a current dialogue move is part of a process of building an analogical argument is crucial for successful persuasion, otherwise the 'rhetorical trap' would not work. As Guarini states (see Section 3 of Guarini's paper): "one reason Nathan's argument is insightful is that it does not contain content that might interfere with the desired response from David [...] Any source case that mentions adultery or murder is likely to get David's 'back up'".

If Nathan's story of a sheep was interpreted in terms of *anticipatory dialogue moves,* then explaining analogical arguments with a framework of dialogue games could be one of possibly useful research perspectives. In particular, the following passage from the section devoted to "The Role of Emotions in Nathan's argument", seems to provide some reasons for employing this approach:

> *Nathan needs to elicit a strong response against a certain kind of action in his source case; doing so would likely require steering clear of pecuniary interests, lest he make David suspicious that he has come to preach to him about adultery. Adultery and conspiracy to kill are nowhere mentioned in the source.*

This explanation shows that argument by analogy, when used as an *anticipatory dialogue move(s)* even should in some cases have a weaker connection between the source and the target, simply because of the fact, that when performing his series of speech acts, Nathan does not reveal that the construction he starts performing is a part of an analogical argument. In other words, Guarini's analysis may point us to the *anticipatory context* of dialogue moves. Hence, it may seem to be a quite interesting to establish, as a posible line of inquiry, dialogue protocols that help achieving a persuasive goal of such arguments by analogy which should not be noticed by the recipient. A posible direction could be to employ the profiles of dialogue method (e.g. Krabbe 1992) to modelling legitimate and illicit dialogue moves in analogical arguments. This kind of study would be in line with e.g. the study of arguments from ignorance showing how presumptions that arise from the context along with the dialogical setting create expectations on the subsequent moves of a dialogue (Walton 1999; Macagno 2022, p. 90). Given the potential of dialogue profiling tools for the mapping of anticipatory 'rhetorical traps' in arguments from analogy, I would encourage Guarini to consider possible advantages of this take as a part of future inquiry.

3. Contextual aspects of analogical arguments

Guarini's focus on the contextual character of arguments by analogy boils down to the claim that in some cases, under certain circumstances, more distant analogy can be 'better' than a close similarity between a source and a target case. While explaining the goal of the paper in the Introduction section, Guarini poses a question: "In this domain [i.e. arguing about action – MK], is more similarity between the source case and the target or disputed case always better?" Guarini's argument in favor of the claim that more similarity between the source case and target cases is not always better seems to take into account just one criterion of the goodness of an argument, namely its persuasiveness.

The very claim that arguing by indicating a close similarity between a source and a target case can be in some cases less persuasively successful than analogical arguments which point to a more distant similarity seems to be itself far from being a novel observation. Quite typically, indications of close similarities can in some cases be not as much persuasively effective as indirect analogies with an apparently small relevance to the discussed topic. However, this observation made by Guarini constitutes a convincing motivation to employ the notion of modulation and some related concepts to be able to model some contextual features of arguments by analogy. Guarini explains modulation in terms of the source selection process that consists of: (i) modulating what counts as similar, and (ii) modulating "how much similarity and what kind of similarity we are looking for between source and target in deciding which source cases to select" (p. 10).

This program of studying modulation is illustrated in the paper by an imagined alternative argument that Nathan could have presented to David (see a near-identical argument from analogy on p. 8). Guarini, by selecting for his comparison two very distant kinds of argument: a very metaphorical one and a direct one shows that a distant similarity between a source and a target case can constitute an advantage of an argument by analogy, contrary to the rather more common intuition that the more similarity there is between the two, the better.

The comparison between two argumentative strategies (the actual one made by Nathan, and the imagined one that could have been made) points us to the need for a *systematic inquiry into what are the types of contextual factors that actually make arguments by analogy with a looser similarity between the source and the target case stronger*. But it might be also claimed that we do not really know if the second strategy would not work.

It might be less likely to happen, but it is not an impossible scenario that David, once having heard an imagined argument by analogy or even a direct accusation such as "you did a horrible thing" did not admit to his actions. So we could be more careful in claiming that alternative argument strategies would not work without an empirical evidence for that (which does not undermine the claim that it is likely to be the case that Nathan's argument is persuasively better than the constructed one).

4. Concluding remarks

Guarini's paper brings in an interesting insight into arguments by analogy that has a potential of modifying argumentation scholars' approach to similarity between the source and the target cases. Given that close similarity between the two does not have to be an advantage of an argument, the next possible step would be to discuss not only extreme cases of close and distant similarity but also those which are in between the extremes.

What also seems to be important about Guarini's contribution is that the paper outlines a rich potential of some conceptual frameworks to grasp both the structural complexity and persuasive potential of analogical arguments. To make this potential more explicit, the next natural step would be to analyse (i) a richer illustrative material to collect empirical evidence about the types of arguments by analogy, which would show how the initial structural approach to analogical arguments introduced by Guarini in fact helps grasping differences between various arguments by analogy (e.g. one-to-one mapping vs. one-to-many mapping).

One of the possible next research steps that can be inspired by Guarini's detailed analysis of Nathan's argument being part of a brief dialogue with David, would be to undertake an empirical study of the effectiveness of arguments by analogy. Basing on Guarini's ideas, a crowdsourcing experiment could be designed in which the participants where to assess for instance thee arguments by analogy to be used under the same circumstances, which would have, respectively, (i) near-identity between a source and a target, (ii) medium-identity, and (iii) quite distant (metaphorical) similarity between a source and a target.

In footnote 2 (see Section 4 of Guarini's paper), the author makes an important remark about the lack of need for employing any of the theoretical accounts of similarity, despite the fact that the very notion of similarity has been employed whenever the target and a source in analogical argument is mentioned. Guarini claims that "as it should be uncontroversial on any approach to similarity that the near-identity case is more similar to what David did than Nathan's sheep story, this paper does not need to select a specific quantitative metric for similarity". Despite the particular goal of this paper, it might be interesting to learn if

Guarini sees any possible applications of the works on similarity (e.g. those mentioned in the paper, by Gentner and Markman (1997) and Ashley, (1990)) in the future study on how source cases for analogical arguments should be selected.

Acknowledgement

The work reported in this paper has been supported by the Polish National Science Centre under Grant 2020/39/I/HS1/02861.

References

Alter, R. (2019). The Hebrew Bible: A Translation with Commentary, vol. 2. New York and London: W.W. Norton & Company.
Ashley, K. (1990). Modeling legal argument: Reasoning with cases and hypotheticals. Cambridge, Mass. & London, UK: MIT Press.
Gentner, D, & Markman, A.B. (1997). Structure mapping in analogy and similarity. American Psychologist, 52, 45-56.
Krabbe, E.C.W. (1992). So what? Profiles for relevance criticism in persuasion dialogues. Argumentation 6, 271-283.
Macagno, F. (2022). Argumentation Profiles: A Tool for Analyzing Argumentative Strategies. Informal Logic, 42(1), 83–138.
Walton, D. (1999). Profiles of Dialogue for Evaluating Arguments from Ignorance. Argumentation 13, 53-71.
Walton, D. (2014). Argumentation Schemes for Argument from Analogy. In: H. Ribeiro (Ed.), Systematic approaches to argument by analogy (pp. 23–40). Argumentation Library, vol 25. Cham: Springer.

Koszowy

Developing Requirements for Reconstructing Visual Arguments Using Meta-visual Disputes

BITA HESHMATI
University of Groningen
b.heshmati@rug.nl

Abstract

Reconstruction is a method for analyzing visual arguments in which visual messages are (verbally) expressed in terms of conclusion and premises. In this paper, I aim to develop requirements for reconstructing multimodal visual arguments using a notion called meta-visual disputes. Meta-visual disputes are disputes about the use of pictures and potentially other visual elements. As such, they help us understand how visual elements should be incorporated into the conclusion and premises. More specifically, meta-visual disputes show that reconstruction should be multimodal and non-reductionist in order to make a comprehensive analysis of visual arguments possible.

1. Introduction

In this paper, I aim to develop requirements for reconstructing visual arguments. My method for identifying and establishing these requirements is aided by, first, arguments about the use of words and expressions or meta-linguistic disputes; second, arguments about the use of pictures (and other visual signs) or meta-visual disputes. Both meta-linguistic and meta-visual disputes fall under the broader notion of *argumentation about meaning* where the overall meaning can be understood as pragmatically emerging through communicative devices such as pictures and words.

The structure of the paper will be as follows: first, I discuss the prevalence of reconstructions in visual argumentation literature and highlight the importance of developing requirements for properly reconstructing them (section 2). Then, I introduce the notion of meta-linguistic disputes in which the disputants argue over the use of terms and expressions. I advance Poppy Mankowitz's propositional account for explaining this phenomenon (section 3). Based on the notion of meta-linguistic disputes, I introduce the analogous notion of 'meta-visual disputes' in which the disputants argue over the appropriate use of pictures and how they function (argumentatively) in their contexts-of-use (section 4). Finally, I develop requirements for reconstructing visual arguments by drawing on the defining characteristics of meta-visual disputes. I argue that meta-visual disputes show that reconstruction should be non-reductive and multimodal (section 5). Finally, I conclude the paper by identifying future research for further exploring the notion of meta-visual disputes (section 6).

2. A problem in visual argumentation literature

The distinctive quality of multimodal visual arguments is that they contain both visual and verbal elements. Examples of visual arguments can be found in online socio-political discourse (e.g., news articles, social media posts, etc.) where pictures and photographs are commonly used. Given that pictures and other visual elements appear to be qualitatively distinct from words in how they communicate information a salient question that arises is how and whether pictures can function argumentatively in particular cases. Subsequently, we can ask if visual arguments can be evaluated similarly to standard verbal arguments (Blair, 2015). The question of how to analyze visual elements has been an essential component of visual argumentation debates.

A method of analysis for visual arguments is verbal reconstruction, a method pragma-dialecticians introduce for better understanding and analyzing disputes in a critical discussion. Frans van Eemeren and Rob Grootendorst introduce the notion of "normative reconstruction" as a "framework for analysis" for understanding argumentative discourse (1989, p. 374). Broadly understood, reconstruction allows the implicit and explicit elements of the argument to be identified and their argumentative function is made explicit through the conclusion and premises. In visual argumentation literature, pictures and other visual elements are typically treated as implicit or hidden premises that are made explicit in the verbal reconstruction by, for instance, translating visually-expressed propositions into verbal claims in the conclusion and premises. An idea behind verbal reconstruction is that translating the content of pictures into words would allow them to be analyzed using the normative-evaluative that are standardly applied to verbal arguments (e.g., standards of acceptability in informal logic or pragma-dialectical rules for critical discussion).

While scholars use reconstructions to analyze various kinds of visual arguments, they do not directly discuss how visual arguments should be reconstructed and the specific requirements for reconstructing them.

However, specifying such requirements is invaluable and likely necessary because they would guide us in translating visual elements to verbal ones in a manner that preserves their argumentative meaning and relevance. Furthermore, considering that reconstructions are commonly used for evaluating visual arguments, it is crucial that pictures be correctly interpreted and incorporated into the reconstruction. This issue is especially relevant if reconstructions are considered accurate representations of visual arguments in their original context. In this case, establishing norms and requirements for reconstruction would be indispensable for the proper analysis of visual arguments.

The requirements for reconstruction are also necessary to compare various methods of reconstruction in their application. How scholars reconstruct visual arguments, broadly speaking, follows two approaches. The first approach treats pictures and other visual elements as implicit premises (or enthymemes) that should be made explicit through words. The underlying assumption here is that pictures contain visually-expressed propositions and/or visual messages that can be translated into verbal counterparts. In this sense, pictures are analogous to words or verbal claims. This approach is *reductionist* as the visual element is reduced to and replaced by verbal claims and descriptions. The second method of reconstruction also treats pictures and other visual elements as implicit premises; however, it does not reduce or replace them with words. In this case, visual content is not transformed into verbal ones. That said, the argumentative function of the picture is made explicit by being directly placed or referred to in the conclusion and premises. In this sense, pictures remain an integral part of the reconstruction. So, the latter is a *non-reductionist* approach. Specifying requirements for reconstruction would be invaluable for understanding whether and how to use reductionist and non-reductionist methods for analyzing visual arguments.

The question of how to reconstruct an argument comes logically before the question of how an argument should be evaluated. By implication, identifying the requirements for reconstructing visual arguments is the more fundamental step in the study of visual argumentation. The reason is that the extent to which we can apply normative-evaluative rules depends on the overall plausibility of the reconstruction. For instance, in a reductionist approach to reconstruction, the pictorial content of the picture can be evaluated using informal logic because it is translated into standard verbal claims and descriptions. However, the full persuasive force of, for instance, a picture cannot be evaluated if the picture has a strong emotional appeal and can evoke emotions in the viewer when words cannot achieve the same level of persuasiveness. Conversely, on the non-reductionist approach, both the function and the pictorial content of the picture can be evaluated as the visual element is not replaced by words and so does not lose, for instance, evidentiary and emotional force.

Studying the issue of reconstruction is aided by disputes about the meaning of words, pictures and other communicative devices. A classic example is the case of meta-linguistic disputes where the disputants argue over the meaning of linguistic features, namely, words and expressions. According to David Plunkett (2015) who introduces the notion, a (normative) meta-linguistic dispute "is a dispute in which speakers each use (rather than mention) a term to advocate for a normative view about how that term should be used" (p. 832, emphasis added). Following this notion, an example of a meta-linguistic would be the following: the proponent utters the claim, 'Barack Obama is tall' and the opponent utters the claim, 'Barack Obama is not tall' while both disputants know Obama's (approximate) height. Given the latter, the disputants cannot be arguing about Obama's height. That said, they engage in a *meta-linguistic* dispute about the term 'tall' and whether it should be used to describe Obama.

An analogous concept to meta-linguistic disputes exists for multimodal visual arguments where pictures and other visual elements are commonly used. The analogous concept is of meta-visual disputes where the disputants argue over the particular use of generally pictures and their (potential) argumentative meaning in the context of use. Consider the following example discussed by Cara Finnegan which revolves around a photograph of a cow's skull (Fig. 1) taken by Arthur Rothstein in the mid-1930s. The picture was used in multiple newspapers and magazines to represent the drought in the Dakotas at the time. As a result, a dispute shaped around whether the photograph of Cow's Skull should have been used in the discussions about this drought. The dispute was about whether the cow's skull, in fact, symbolizes the drought in Dakotas or can show the drought getting worse. As such, the dispute can be called a *meta-visual* dispute. Similar to the case of meta-linguistic disputes, the disputants had the same baseline understanding of what is directly communicated. Specifically in this case, the viewers agree that the picture shows the skull of a cow against dry land. However, they disagree on *whether the picture can be used to show* the cow died as a result of the drought and the general drought conditions in the Dakotas.

Figure 1. Arthur Rothstein, Cow's Skull

There are multiple reasons why studying meta-visual disputes aids the process of reconstruction. Broadly speaking, meta-visual disputes show that the disputants cab directly argue about the particular use of visual elements. Considering that, meta-visual disputes strongly suggest that pictures and other visual elements cannot be reduced to verbal claims and descriptions. In this sense, they demonstrate the argumentative relevance of visual elements in their context of use. Furthermore, the particular points of discussion in meta-visual disputes show what considerations are relevant for the analysis of pictures.

3. Meta-linguistic Disputes

3.1. Linguistic versus non-linguistic disputes

To understand the notions of meta-linguistic and meta-visual disputes, a broader distinction can be made between non-linguistic and linguistic disputes. In the former, the object of dispute concerns an empirical fact or some aspect of reality, independently of the language used to express it. In the latter, the object of dispute concerns an *a priori* linguistic fact or aspect of language as used to describe reality. Put simply, in non-linguistic disputes, arguments concern worldly phenomena whereas, in linguistic disputes, arguments concern the language used to describe worldly phenomena. In this section, I distinguish meta-linguistic disputes from non-linguistic disputes. Then, I advance Mankowitz's propositional account for explaining how the disputants communicate their views about language use in meta-linguistic disputes. The propositional account applies to classic cases of meta-linguistic disputes in which the disputants pragmatically convey their meta-linguistic views through their stated utterances. This section makes clear how the intuitions behind meta-linguistic disputes can be applied to multimodal visual arguments or argumentative discourses containing visual elements.

The first question to consider is how meta-linguistic disputes, in general, can be delineated from other kinds of disputes. Disputes can be divided into non-linguistic and linguistic disputes. The former are disputes in which two or more disputants disagree over a non-linguistic issue, e.g., whether the proposition, the temperature of the earth has risen by 1 degree Celsius, is true. Evaluating and resolving a dispute about the temperature rise of the earth depends on settling an empirical fact; namely, how many degrees the earth has become warmer over the past five decades. Considering that, settling this issue depends on knowledge about the earth independently of the language used to convey that knowledge. In this case, the disputants take the dispute as hinging on and settling an empirical fact about the earth, as opposed to, *apriori* linguistic descriptions.

In the case of linguistics disputes, arguments center around linguistic issues, namely, terms used to describe particular facts and worldly phenomena. If the disputants agree about empirical facts expressed through the statement, then their arguments can only concern linguistic matters. The underlying assumption in these cases is that the disputants share the same knowledge; however, they have differing views about the terms of discussion (whether and how linguistic terms should be used to express the same facts) (Plunkett and Sundell, 2013, p.7). Consider the following example: a recent graduate, Sheila, is interviewing for a teaching position. The interviewer, Mark, asks whether Sheila has written any books, in response to which she claims she has written a book. She explains she has completed her manuscript which is going to be published soon. Mark then claims that Sheila has not written any books. The dispute can be summarized as follows:

Mark: Have you written a book?
   Sheila: Yes, I have written a book. My manuscript will be published soon.
Mark: You have not written a book.

In this scenario, Mark and Sheila are both aware of the process of writing a book, namely, writing the manuscript and having it published. Yet, they disagree on whether the term, 'book', can be used to describe Sheila's achievement. The dispute takes place as a result of Mark and Sheila using these terms in contrary ways. Sheila uses the term 'book' to mean, roughly, 'unpublished manuscript (of appropriate length)' and Mark uses 'book' to mean 'published manuscript'.

A distinguishing feature of meta-linguistic disputes is that the disputants use terms and expressions to pragmatically convey how they should be used. Drawing on Barker, Plunkett (2015) define the latter as "meta-linguistic usage" where the disputants "use (rather than mentions) a term to express a view about how that term should be used" (p. 1). Consider this widely-discussed example: a dispute takes place on a radio show about horse races about a racehorse called Secretariat. The dispute is as follows:

(1) Secretariat is an athlete.
(2) Secretariat is not an athlete.

Similar to the previous example, both disputants are aware of the relevant facts. They are both aware of Secretariat's participation in races and its athletic competency and attributes. In this sense, the disputants are not using the term 'athlete' to make a factual statement about Secretariat. Instead, they use this term to make a claim about how and whether the term, 'athlete' should be used in various cases. Their disagreement directly concerns the term, 'athlete' and why it can or cannot be used for racehorses, or more generally, non-human animals. In this sense, they use the term, 'athlete' *meta-linguistically* to convey when this term should be used.

The concept of meta-linguistic usage can provide insights into the potential way in which the meta-linguistic dispute can be resolved. In a dispute, understanding that the other disputant is using terms or utterances meta-linguistically helps address the issue. As in meta-linguistic disputes, the immediate object of the dispute is a term(s) and the convention of use of that term. Therefore, settling the dispute amounts to resolving the meta-linguistic dispute. Notably, awareness plays an important role: if the disputants are unaware of their meta-linguistic use of terms, they risk talking past each other and engaging in a verbal dispute (a pseudo-disagreement). In this case, the meta-linguistic dispute would be reduced to a verbal dispute in which the real argument, which is about the use of words and expressions, remains unaddressed and hence, unresolved.

3.2. The propositional account

To engage in a meta-linguistic dispute, the disputants should be able to communicate effectively. However, given that the disputants do not directly express their meta-linguistic views, the question arises, how can the disputants effectively understand each other? For instance, how can Mark understand that Sheila's mentioning of the terms, 'book' or 'written' imply something about their appropriate use in the context of the discussion? Following Grice's framework, Mankowitz (2021) argues the disputants communicate meta-linguistic propositions, or propositions about language use, by relying on shared conversational maxims and commitments.

Adopting the propositional account, we could discover and formulate the meta-linguistic views conveyed in the discussion. Notably, the disputants may not necessarily intend to advocate for a meta-linguistic view; nevertheless, they are committed to the meta-linguistic propositions that are pragmatically conveyed through their utterances. In the Obama example, the proponent claims, 'Obama is tall'. Assuming that the proponent is aware of the general meaning of the terms used and intends to convey the idea that Obama is tall, she commits to a set of propositions such as the following:

(1) Obama is of maximal height.
(2) The term 'tall' should be used to describe Obama.
(3) The proposition, 'Obama is tall' expresses the relevant standard for tallness.

The proponent commits to the above propositions where the first and second statements directly follow from the utterance 'Obama is tall' as they are semantically implied. The third statement indirectly follows from the utterance, 'Obama is tall' as they are meta-linguistically or pragmatically implied. The first proposition is a statement of fact and one that is agreed upon in the context of the meta-linguistic dispute (e.g., it is included in the common ground between the disputants). The last two statements convey meta-linguistic views about what it means to use the term 'tall'. Identifying these meta-linguistic propositions (and others) is necessary for understanding the disputants' meta-linguistic views and points of disagreement.

In summary, meta-linguistic disputes are disputes about whether and how terms should be used in various cases. Meta-linguistic disputes are different from other non-linguistic disputes. As will be clear in the following section, an analog of meta-linguistic dispute applies to visual arguments where the disputants disagree about the meaning of pictures or other visual elements. These disputes can be called meta-visual disputes as they mirror meta-linguistic disputes with the only difference that the latter concern a distinct communicative device, namely, visual elements.

4. Meta-visual Disputes

The notion of meta-visual disputes can be understood analogously to meta-linguistic disputes. While meta-visual disputes also rely on a verbal context, they are essentially about the use of pictures or other visual elements. Meta-visual disputes show how different individuals perceive or interpret pictures in conflicting ways and adopt conflicting views about the use of a picture in a particular case. In this section, I introduce the notion of meta-visual disputes and discuss examples that would show the relevance of this notion to reconstruction in the following section.

Before discussing how meta-visual disputes relate to reconstructing visual arguments, I discuss how they can be understood in relation to meta-linguistic disputes by revising the Obama example. In the meta-linguistic dispute surrounding Obama discussed earlier, the proponent utters 'Obama is tall' whereas the opponent utters the opposing claim, 'Obama is not tall'. In this manner, the disputants implicitly argue over the issue of whether the term 'tall' can be used to describe Obama. In the meta-visual case, the disputants would show a picture of Obama instead of making a verbal claim (see Fig. 2). In the latter case, the dispute can be formulated as follows:

P: The picture of Obama next to Clinton (Fig. 2) shows that Obama is tall.
O: The picture of Obama next to Clinton (Fig. 2) does not show that Obama is tall.

The same assumptions hold in both the meta-linguistic and meta-visual disputes about Obama. Both disputants know Obama's height. Importantly, they are also aware of the general meaning of the term tall as it applies across various contexts, e.g., taken to mean, 'a person of maximal height relative to some determined threshold'. Therefore, the latter dispute is not about Obama's height. That said, it is about *whether the picture of Obama/Clinton can be used to establish* that Obama is tall.

We can understand meta-visual disputes from a similar perspective as meta-linguistic disputes because the propositional account can explain both phenomena. In this case, the disputants pragmatically or implicitly convey views about how a picture should be used, i.e., meta-visual views. In the latter dispute, the proponent commits to the *meta-visual proposition* that the picture of Obama can be used to establish the claim that Obama is tall. The opponent commits to the contrary proposition that the picture of Obama cannot be used in establishing the stated claims. Accordingly, this is a meta-visual dispute about the use of the Obama/Clinton picture in this particular context.

Figure 2. Picture of Obama next to Clinton

In the above example, the disputants argue about the use of the Obama/Clinton picture verbally. However, it is important to note the disputants could also argue *solely by visual means or by using the pictures*. For instance, consider the picture of Obama standing next to LeBron James (see Fig. 3). In this case, the disputants argue by each *showing* a different picture. Namely, the proponent shows the picture of Obama standing next to Hilary Clinton and the opponent shows the picture of Obama standing next to LeBron James. By doing so, the opponent effectively communicates that the picture of Obama/Clinton (Fig 2.) should not be used to establish Obama is tall. In this case, the dispute can be summarized as follows:

P: *The picture of Obama next to Hilary Clinton (Fig. 2)* shows that Obama is tall. (conveyed visually)
O: *The picture of Obama next to LeBron James (Fig. 3)* shows that Obama is not tall. (conveyed visually)

The proponent (pragmatically) conveys that the standard for tallness for the use of the term 'tall' is represented by the picture of Obama standing next to Clinton. Conversely, the opponent (pragmatically) conveys that the relevant standard for tallness is represented by the picture of Obama standing to LeBron James. In this way, they engage in meta-visual dispute soley by means of pictures.

In sum, both meta-visual disputes revolve around issues about the use of pictures. However, the mode(s) of communication involved are different in each case where, for instance, in the latter dispute, communication is solely visual, whereas, in the former, it is multimodal as the dispute is carried out both through verbal and visual means of communicating. Arguably, many cases of meta-visual disputes are multimodal as they are embedded in broader online debates (and so have a verbal context).

Figure 3. Picture of Obama next to LeBron James

There are cases of meta-visual disputes where the visual elements relate to a substantive socio-political issue that underlies the dispute. Consider the following example: a Greek news outlet (Kathimerini) used a particular picture of refugees in an article about the refugee crisis. Some discourse analysts argue that the use of this particular picture of refugees (Fig. 4) misdirects the audience from the reality of the refugee crisis. They argue that the use of this picture in the article is problematic because it promotes actions against refugee migration and simultaneously advocates anti-immigration sentiments and policies (Serafis et al., 2019). While some are critical of the particular use of this picture in this news article, others find no issue with the use of the picture (the opponent treats the picture as non-argumentative and neutral towards the issue of refugee migration to Greece). The dispute can be summarized as follows:

P: The news outlet Kathimerini promotes action against the 'migratory phenomenon'.
The refugee migration to Greece is misrepresented by Figure 4.
(meta-visual proposition)
O: The news outlet does not promote action against the 'migratory phenomenon'.
The refugee migration is represented accurately by Figure 4.
(meta-visual proposition)

Figure 4. Serafis et al. (2020) Kathimerini, 10.07.2015.

Both disputants have substantive reasons for why they agree or disagree with the use of the picture of refugees (Fig. 4) by the Kathimerini news outlet. The proponent argues that the picture misrepresents refugees and the refugee crisis for the following reasons: first, the picture only includes men, so it creates the false perception that all refugees are men. Second, the fact that the refugees are not facing the camera suggests refugees do not belong to Greek society (Serafis et al., 2019). Third, the portrayal of refugees as sitting idly implies that refugees are disengaged with society and are/would be a burden to society. Their idea is that these subtle messages are visually conveyed through the picture to promote anti-immigration actions or sentiments. The opponent would argue that the refugee picture is neutral and is not used maliciously or even argumentatively by the news outlet. That said, they could consider the picture as presenting the reality of (a group) of refugees and argue against the claim that the picture is used fallaciously in promoting anti-immigration sentiments about the refugee crisis in Greece.

As mentioned, the disputants' conflicting views about the use of this picture and its potentially argumentative function, lead to a dispute about whether the news outlet promotes anti-migration ideas and policies. This dispute has potentially significant epistemic and practical implications: epistemically, it raises the question of whether and how pictures should be used in various contexts. Relatedly, what the conventions are for the use of pictures in potentially argumentative contexts (e.g., news outlets or social media platforms). From a practical perspective, meta-visual disputes raise the question of how pictures impact public perception and promote favor specific attitudes and policies over others. Both the theoretical and practical questions are directly relevant to visual argumentation in social-political public debates.

In summary, meta-visual disputes are about the particular use of pictures and/or other visual elements in potentially argumentative contexts. It will become clear in the following section that meta-visual disputes have a significant epistemic advantage for our understanding of visual argumentation. That is; they facilitate developing requirements for the analysis and evaluation of visual arguments.

5. Reconstruction and the role of meta-visual disputes

5.1. Methods of reconstruction

Considering how meta-visual disputes revolve around the use of pictures, they can provide insight into how we should analyze visual arguments. The study of arguments, as a whole, involve reconstructing them in terms of conclusion and premises. In the case of visual arguments, this process would also include visual elements. So, a salient question arises as to how and to what extent the visually-conveyed messages should be turned into verbally-stated claims. As I will show in this section, meta-visual disputes guide us in the process of reconstruction. More specifically, they show reconstruction should be multimodal: namely, pictures and other visual elements should be directly incorporated into the conclusion and premises.

Reconstructing arguments is a method through which various elements of a given argument can be analyzed and evaluated. In the case of multimodal visual arguments, various methods of reconstruction are possible and each would have an impact on the manner in which these arguments are evaluated. For instance, some scholars use *verbal* reconstruction to allow normative-evaluative rules that are conventionally employed for verbal arguments to be applied to visual arguments as well. Verbally reconstructing visual arguments is beneficial because it leads to a better understanding of the extent to which we can apply normative-evaluative models such as informal logic (e.g., the RSA criteria) and pragma dialectics. As Blair argues, this process would raise the question of whether visual arguments, in general, can be evaluated similarly to verbal arguments; relatedly, reconstruction allows us to explore whether special or further evaluative norms are needed for the evaluating visual arguments (Blair, 2015). Accordingly, questions of evaluation are particularly relevant when we think about reconstruction.

Some scholars use a method of reconstruction in which visual elements are directly incorporated into the reconstruction without any transformation of the visual elements to verbal ones. Groarke (2015) offers a model of reconstruction in which the visual element is directly included in the conclusion and premises. Groarke advances a 'key component table' which specify the structure of the argument as well as the communicative mode of arguing. The former states the conclusion and premises, i.e., the key components of the argument, and the latter states how the argument is communicated, e.g., the visual mode of communication. In Groarke's model, the argumentative function of the picture is made explicit, mainly without verbalizing the pictorial content and/or the visually-expressed propositions in the reconstruction. As a result, the argumentative function of the picture is made explicit without verbalizing the visual content in the reconstruction.

Groarke's method can be contrasted with the method of verbal reconstruction in which the picture is transformed into verbally-stated claims in the conclusion and premises. For instance, Blair (2015) reconstructs many arguments in a manner that translates visually-expressed propositions into verbal. Consider his example of a traffic sign warning drivers of the possibility of having a runaway car (Figure 5).

Figure 5. Blair. (2015). *Lock the Park Break.*

In the reconstruction of the argument presented through the traffic sign, Blair signifies the drawing of the stickman being hit by a car with the verbal statement, "a runaway car is at risk of hitting someone" (2015, p. 222). In his reconstruction, the drawing is reduced to a verbal description of it. The motivation behind using this way of reconstruction is to allow multimodal visual arguments to be evaluated similarly to standard verbal arguments.

Whether and when a method of reconstruction is used depends on contextual features. In the example above, the verbal claim corresponding to the drawing works well in the reconstruction as it clearly conveys the visual message and captures the persuasive function of the drawing in the argument. However, there are cases in which turning the visual message into a verbal claim would pose a challenge, especially if the picture serves more than one specific function. An example of the latter is the picture of Alan Kurdi (Figure 6) as used in online debates since 2015. One argument that has formed around this picture is that the condition of refugees has become direr over the years and that entails European countries should provide further help by letting more refugees into their countries. To follow the same reconstruction method as the one employed above, we would reconstruct the visual argument in the following manner:

1 We should help refugees.
   1.1 If we do not help refugees, they will die.
      1.1.1 Alan Kurdi died tragically in fleeing his country.

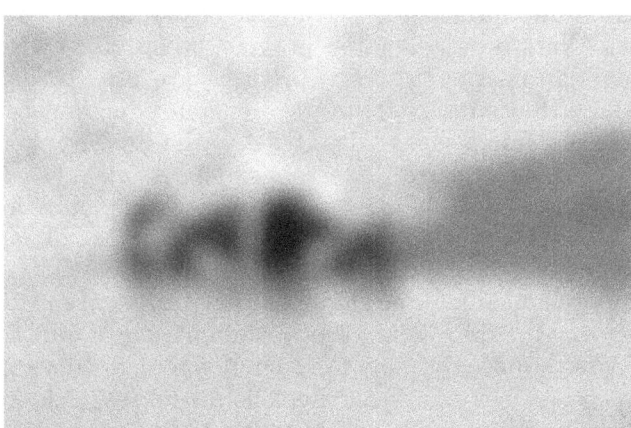

Figure 6. Alan Kurdi, the Syrian refugee (Photographer: Nilüfer Demir).

Similar to the traffic sign example, the latter reconstruction of the Kurdi argument does not contain a reference to the picture. Instead, it includes a verbal *description* of what is visually depicted or conveyed by the picture of Kurdi. The fact that there is no mention of the picture in the reconstruction entails that the visual argument is reduced to a verbal counterpart. In this sense, it is a *reductionist* reconstruction of the visual argument.

The Alan Kurdi argument can also be reconstructed following a different method of reconstruction. In this case, the reconstruction follows Groarke's method where the multimodal visual element is directly included in the reconstruction without any verbal transformation of the visual content. Consider the following reconstruction of the argument:

> 1 The European countries should let refugees in their countries.
> 1.1 If they do not help refugees, they will die.
> 1.1.1 *The picture of Alan Kurdi* shows that a refugee kid died tragically in fleeing his country.

The crucial difference between the latter reconstruction and the previous one is evident in how sub- premise (1.1.1) is formulated. The reductionist method excludes any references to the picture in the premise (1.1.1) whereas the non-reductionist method directly includes a reference to the picture and, in this sense, constitute a *multimodal* reconstruction rather a verbal one

5.2. Meta-visual dispute and the non-reductionist approach

As shown, meta-visual disputes are disputes in which the disputants seriously analyze and argue over the use of pictures in particular cases. Considering that, these disputes provide potentially valuable insights about visual arguments in which pictures are often used. Meta-visual disputes show that first, disputants are non-reductionists about visual elements. Second, meta-visual disputes are not (directly) about the content of pictures or visually-convey messages. That said, these disputes form around the disputants' subjective perception of the picture and conflicting perceptions of how the picture is related to the broader socio-political issues discussed (e.g., refugee crisis in Greece).

First, meta-visual disputes show that disputants are non-reductionists about the use of pictures. As in the Kathimerini example, the disputants take seriously the issue of how (and why) the picture of refugees sitting idly in the sun is used in the article. The disputants explore ideas about how the picture might have been used for specific argumentative purposes, namely, to promote anti-immigration actions and policies over pro-immigration ones. The existence of meta-visual disputes suggests a non-reductionist approach to visual arguments is more plausible. This is true because meta-visual disputes show pictures can become the direct object of dispute and so are indispensable to a plausible reconstruction of a multimodal visual argument.

Relatedly, meta-visual disputes point to the limitations of a reductionist method of reconstruction. Following a wholly reductionist reconstruction, the disputants cannot argue about the use of a specific picture and its persuasive function in a given context. For instance, on the reductionist reconstruction of the Kurdi argument, the disputants cannot ask whether it is acceptable to use the picture of a deceased refugee child to persuade an audience of the severity of the refugee situation. Relatedly, they cannot argue over the intentions behind using such a picture and its function in a particular context. In more technical terms, meta-visual propositions or arguers' commitments become inevaluable on the reductionist approach.

Secondly, meta-visual disputes shed light on the complexity of reconstructing visual arguments, surpassing a merely verbal reconstruction in which the pictures are in some manner reduced to words. They show that the particular perceptions around the use of a picture partly depend also on a verbal context. Importantly, disputants use terms to form their meta-visual views. Consider Kathimerini's picture of idle refugees again. The proponent argued against the legitimate use of the picture partly because of the word 'inflow' in the article. The idea is that the term, 'inflow' has a particularly negative connotation, namely, that it suggests that the 'wave' of refugees arriving in Europe is analogous to a natural disaster (Serafis et al., 2019). Similarly, the opponent disagrees with the proponent's critical view partly on the ground that the term, 'inflow' does not have such a negative connotation and is not used to promote anti-immigration views. This dispute shows pictures gain argumentative meaning and are interpreted in conjunction with particular terms and verbally-stated claims in the relevant context. Accordingly, an accurate reconstruction of a visual argument takes both verbal and visual elements into account.

Finally, meta-visual disputes show that the meaning of pictures emerges pragmatically in context instead of the level of visual representation. That is, the disputants interpret and analyze pictures not merely based on what is directly shown; but rather, through other (verbal) elements and potentially broader contextual features. As a result, the disputants can disagree about the argumentative function of a picture and its legitimate use while understanding and agreeing on what is visually shown. Arguably, the reductionist method is insufficient in many cases because it cannot account for the pragmatic meaning of pictures, namely, the meta-visual propositions that are communicated around the use of pictures. Instead, verbal reconstruction accounts for the representational meaning or the propositional content of pictures. As such, it fails to capture the full argumentative function of pictures in visual arguments.

In summary, studying meta-visual disputes is epistemically advantageous because these disputes offer insight into how visual arguments can and should be reconstructed. They show a non-reductionist verbal reconstruction is more plausible than a reductionist one because the former allows the picture and its persuasive function to be evaluated; whereas, the latter cannot allow for evaluating the function of the picture as a whole. Hence, a full evaluation of the visual argument is not possible on a reductionist account. Finally, meta-visual disputes show the function of pictures is determined on a case-by-case basis. Considering that, the following requirements can be established: first, pictures and other visual elements should be directly incorporated into the reconstruction of multimodal visual arguments. Second, the verbal context in which pictures appear should be taken into account when reconstructing visual arguments.

6. Conclusion

In this paper, I developed the notion of meta-visual disputes and how they revolve around the particular function of pictures in the context of use. I argued that meta-visual disputes are an invaluable tool for understanding how to reconstruct visual arguments and specify requirements for a plausible reconstruction. Since reconstruction is one of the primary steps in the analysis of visual arguments, the requirements for reconstruction facilitate the proper evaluation of visual arguments. As discussed, meta-visual disputes show that pictures can serve indispensable argumentative roles, such as in online discourse concerning the refugee crisis. Arguably, the disputants attribute argumentative value to pictures and closely consider their use in online discourse. In discussing pictures, the disputants endorse a non-reductionist method for reconstruction, showing that a non-reductionist method is most plausible as it includes pictures or other visual elements directly in the conclusion and premises. Accordingly, the first requirement is that pictures should be directly included in the reconstruction. The second criterion is that reconstruction should closely follow the visual argument in the manner presented in its original context. Consequently, reconstruction should include the particular terms used in the original context. The non-reductionist method promotes the idea that argumentative meaning emerges contextually through the use of various multimodal elements. So, changing or eliminating these elements would preclude proper reconstruction and evaluation of multimodal visual arguments.

Finally, the propositional account highlights a distinction that can be made between visual meaning and meta-visual meaning. The idea is that individuals can agree about the content of a picture (visual meaning) and, at the same time, disagree about its function and use in a given context (meta-visual meaning). In future research, I explore issues related to the connection between visual and meta-visual meaning; specifically, the issue of whether and how cognitive biases and the disputants' prior commitments can lead to the formation of various kinds of meta-visual disputes. Relatedly, I will explore whether and how studying meta-visual disputes can provide useful insights for the widely-discussed issues of vagueness and misinterpretation in visual argumentation.

References

Blair, J.A. (2015). Probative Norms for Multimodal Visual Arguments. Argumentation, 29, 217-233.
Champagne, M., & Pietarinen, A.-V. (2020). Why Images Cannot be Arguments, But Moving Ones Might. Argumentation, 34(2), 207–236.
Groarke, L. (1996). Logic, Art and Argument. Informal logic, 18 (2-3), 105-129.
Groarke, L. (2015). Going Multimodal: What is a Mode of Arguing and Why Does it Matter? Argumentation, 29 (2), 133-155.
Kjeldsen, J. E. (2015). The Rhetoric of Thick Representation: How Pictures Render the Importance and Strength of an Argument Salient. Argumentation, 29(2), 197–215.
Johnson, Ralph H. (2003). Why "visual arguments" aren't Arguments. Informal logic at 25, ed. H.V. Hansen. Windsor, ON: University of Windsor, CD Rom.
Mankowitz, P. (2021). How to have a metalinguistic dispute. Synthese, 199(3–4), 5603–5622.
Peach, H. (2021). Picturing a Thousand Unspoken Words: Visual Arguments and Controlling Force. Informal Logic, 41(1), 57-79.
Plunkett, D. (2015). Which concepts should we use?: Metalinguistic negotiations and the methodology of philosophy. Inquiry, 58(7-8), 828-874.
Serafis, D., Greco, S., Pollaroli, C., & Jermini-Martinez Soria, C. (2020). Towards an integrated argumentative approach to multimodal critical discourse analysis: evidence from the portrayal of refugees and immigrants in Greek newspapers. Critical Discourse Studies, 17(5), 545-565.
Van Eemeren, F. H., & Grootendorst, R. (1989). Speech Act Conditions as Tools for Reconstructing Argumentative Discourse. Argumentation, 3, 367-383.

MAPPING ARGUMENTS IN FAVOR OF THE PRESUMPTION OF INNOCENCE

MICHAEL J. HOPPMANN
Northeastern University, Boston
m.hoppmann@northeastern.edu

The presumption of innocence is an old and well-respected fundament of the rule of law in deliberative democracies. It is explicitly or implicitly recognized as an essential part of due process by almost all modern legal systems. At the same time, this status does not make the presumption of innocence sacrosanct. On the contrary, its justification is frequently questioned, and its applicability challenged. These challenges are rarely aimed at paradigmatic cases – say the respectable citizen standing in an ordinary court being accused of murder – but instead focus on chipping away around the corners: Is this presumption really required in minor cases? Or outside of formal courts? For foreigners? And terrorist? Or repeat offenders? Should those accused of non-legal offenses enjoy this presumption? Even if it sometimes delays or denies enforcement of important rules? The answer to all these questions is a resounding "yes!". But unfortunately, the reasons for that answer are not that easy to come by. Precisely because the presumption of innocence is such an old and revered concept, few see the need to explicitly argue for it anymore. This essay is an attempt to fill part of that gap. It is part of a larger project that aims at establishing a modern theory of stasis based on the presumption of innocence as a foundation. That modern stasis theory is a critical tool for citizens who are looking for guidance when speaking up for justice.

The presumption of innocence is first and foremost that: a presumption. As such is behaves like a default argument in a discourse but does not have some of the typical features of an argument. The scholarly discussion surrounding presumptions is both old [1] and vibrant [2], indicating the

---

[1] Richard Whately is widely credited with bringing the concept of presumption to prominence in rhetoric. His 1828 Elements of Rhetoric remains influential for contemporary discussions on presumptions in rhetoric and argumentation theory.
[2] Walton and Godden (2007) offer a careful analysis of the main lines of scholarship in contemporary presumption studies. Hansen et al. (2019) provides a selection of the most recent contributions. Among the most influential works for the purposes of this essay are Bodlović (2020a), Bodlović (2020b), Bodlović (2021), Godden (2017), Godden (2019), Godden & Walton (2007), Kauffeld (1998), Kauffeld (2003), Krabbe & van Laar (2011), Rescher (2006), Ullmann-Margalit (1983), van Eemeren & Houtlosser (2002), Walton (1988), Walton (2014), Whately (1963).

importance of this concept for modern rhetoric and argumentation theory. The resulting works in philosophy, argumentation theory, rhetoric, and related disciplines have led to a fairly detailed understanding of the functioning of presumptions in general. They have also created a variety of different theories of understanding their nature.[3]

Given that this essay is primarily concerned with the "why" of one particular presumption (the presumption of innocence), rather than the "what" and the "how" of presumptions in general, most of the details of presumption studies is not pertinent to this section. In the interest of a precise analysis of the foundations of the presumption of innocence I will nevertheless need to reference some important distinctions and clarify terminological choices. To put these theoretical aspects of presumption studies into their practical context, I want to start off with a description of the discursive effects of presumptions with the help of two well-known analogies: the weighing of arguments on a scale and discussions as a competition on a level playing field.

One of the established allegories of argumentation treats the reasonable resolution of a difference of opinion akin to the weighing of evidence on an ancient, two-armed apothecary scale. The most famous depiction of this analogy features (an often blind-folded) Justicia in the process of passing fair judgement about alleged crimes – but the allegory works just as well in reasonable debates at large. In the standard case of a (single mixed) difference of opinion, settled by reasonable discussion, two people encounter a crossroad and disagree about the best path ahead. Arguer 1 (the protagonist) champions path 1 and arguer 2 (the antagonist) prefers path 2. In a perfect procedure (and in an absence of a third option and a vast amount of other systematic complications), the protagonist will present arguments in favor of path 1 and critically test those for path 2, and the antagonist will do the opposite.

Allegorically speaking the two reasonable discussants agree to use the apothecary scale to determine the best solution. They encounter the scale in a neutral position and equally accessible to both discussants.[4] The arguers then take turn to add weights to the left side of the scale (arguments) or the right side (counterarguments). Additionally, they take off weights placed by the opponent that did not withstand scrutiny (undercutters, defeaters and other objections). If in the end of this process the scale tilts towards the left, the discussion is resolved in favor of the protagonist and if towards the right, then the antagonist succeeds.

Thus far this allegory assumed the pure, and extremely rare, case of a disagreement without the presence of a practical presumption. In the present allegory, these presumptions work like weights that have been screwed to the right arm of the scale, thus tilting it against the protagonist

---

3 See Walton and Godden (2007)
4 I.e., they assume dialectical egalitarianism (Bodlović 2020a, 1)

before the discussion even begins.⁵ By virtue of choosing this particular scale, both arguers engage in an unequal weighing of their respective evidence. Additionally, the presumption that has been screwed to the arm of the scale, is not subject to the same types of objections as ordinary arguments which can be added or removed from the scale, following the rules of critical questioning. ⁶ Instead, the main way of removing a practical presumption from a discussion, is by agreement of the arguers that this isn't the right scale. But why would arguers choose to use a rigged scale to settle their disagreement in the first place? To answer that question, a switch to the second analogy may be useful.

Metaphors of sports and war are a common way to explain the functioning of argumentation and individual moves of arguers.⁷ Some of these refer to parts of a fair set-up as "levelling the playing field". Presumptions also alter this proverbial playing field, but rather than appearing to level it, they create systemic advantages for agents that assume specific roles. To stay in the same picture, these practical presumptions add little bumps, larger hills, or sizeable plateaus to the playing field, thus encouraging certain communicative behaviors and facilitating the defense of select positions over others.

The main reasons for establishing these advantages on the playing field via the presumption hills and plateaus is to prevent harm or reduce cost to society. Put in other words, presumptions privilege positions that are (usually) socially desirable and thus encourage communicative agents to prefer them. One of Whately's most fundamental presumptions) – the one from which many others derive – is the presumption in favor of the status quo (2010, 114. In its essence this presumption instructs those that adopt it to always prefer of two similarly reasonable policies the one that is currently in already effect. If one where to argue for example in favor of introducing a new tax code that is similarly fair and effective as the present one, then the presumption in favor of the status quo weighs in and sways the overall argument in favor of the opponent. It is easy to see how this particular presumption improves the social economy and reduces cost, by preventing a community from spending its financial, social or

---

5 I am assuming the default case here in which the protagonist is the one facing a presumption against his position, for example because he accuses someone (presumption of innocence), proposes a new law or policy (presumption in favor of the status quo). It may in principle also be possible for a protagonist to move in favor of a presumption, and thus for the antagonist to encounter an additional weight against her side, but most of these cases would either be futile in the critical reasonableness aspect of the rhetorical performance (preaching to the choir) or could be reconstructed as answering an implicit protagonist, thus flipping the sides again.
6 These rules in turn are subject to extensive research that is beyond the scope of this essay. For an introduction and overview see Krabbe & van Laar (2011) and Hoppmann (2014).
7 For an overview of the pros and cons of their usage see Aikin (2011)

psychological capital on change without progress, thus freeing the same capital up for more pressing matters.

The presumption in favor of the status quo is just one (although an important) example of these types of tools that encourage desired communicative behavior, by altering the playing field. Many others, from small but significant one (such as the presumption of ownership from possession) to vast but very soft one (such as the presumption of veracity), create a discursive playing field that is far from flat and equal like a soccer stadium. Instead, it resembles a rich and diverse adventure garden, with a complicated topology that acknowledges equality as one, but not the only discursive value. Each feature of this communicative playing field is very well thought out, and each presumption needs to serve an important goal. Given that the presumption of innocence creates one of the most notable – and probably the single tallest – plateau in the entire proverbial playing field: What purpose does it serve and what justifications can it invoke? That is the central question that the arguments presented in this essay attempt to answer.

Arguing in favor of the presumption of innocence may at first glance appear like a stunningly futile endeavor. The presumption of innocence is one of the most recognized principles of procedural justice and the backbone of fair trial rules (Clooney & Webb 2020, 199). It is prominently listed in countless national constitutions and international agreements and confirmed by overwhelming precent in the courts. Treating the accused as innocent until proven guilty is an almost sacrosanct slogan, accepted by legal professional and lay people alike and so common that it made it into the Latin legal principles in multiple versions. Nicholas Rescher, one of the foremost scholars of presumption studies, goes as far as explaining the entire class of epistemic presumptions based on their similarity to the principle of innocent until proven guilty (Rescher 1988, 49) – methodologically speaking an extreme case of the tail wagging the dog and a powerful indicator of the conceptual stability of the presumption of innocence.

All that being said, there are three strong arguments in favor of the necessity to establish the presumption of innocence itself beyond reasonable doubt: a theoretical, a political and a practical.

Speaking about the theory of presumptions, Godden and Walton observe that the *force* of a presumption is directedly related to its *foundation* (2007, 314), meaning that we need to understand the strengths of the arguments for a presumption to fully comprehend its effect on the discourse. They explain: *"Presumptions that are based on very weak grounds should be easily overturned, while presumptions built on stronger foundations should stand even in the face of considerable evidence to the contrary (e.g., the presumption of innocence)"* (Ibid., 337f.).[8] Translated to the two allegories above, Godden and Walton's claim means that the strengths of a presumption's foundation (i.e., the sum of the arguments for establishing it for a particular circumstance) determines the precise amount of weight fixed to one arm of the scale and the correct altitude of the plateau on the playing field. Based on the close relationship between the strength of a presumption in favor of an antagonist and the corresponding elevated burden of proof for the protagonist, this correlation would also allow one to find a justified probative standard for the situation on a scale of simple preponderance of evidence all the way to a proof beyond reasonable doubt.

I am not certain that Godden and Walton's indicated correlation holds in all positive cases – I think a relatively weak presumption (such as the presumption in favor of the status quo) could be established by equality compelling arguments as a very strong presumption (such as the presumption of innocence). Yet, the Godden and Walton relationship is almost certainly true in the negative space: Compelling presumptions cannot rest on dubious justifications. This relationship alone is sufficient for the endeavor of this essay.

The political reason for defending the presumption of innocence is a very timely one. At this time in history, it is probably safe to say that no reasonable person in a secular rule of law country would oppose the presumption of innocence *in its core area of application.* There are a variety of aspects that help define this core area or standard case. Among the most important ones are a) personal status, b) personal history, c) type of proceeding and d) societal status. Accordingly, almost all would agree that a citizen or legal resident (a) without criminal record or previous conviction (b) accused of a crime in a state or federal court (c) in times of peace (d) should be presumed innocent until proven guilty (procedural right) and only be convicted if her guilt can be established beyond a reasonable doubt (substantive right).

Things get more complicated in the public perception when accusations happen beyond this standard case. Should the full presumption of innocence apply to foreigners captured in a war zone and classified as "enemy combatants" or people arrested near a border and accused of illegal immigration? How about previously rightfully convicted people accused of parole violations or current inmates charged with crimes in prison? Should people accused of violating rules that behave like laws, carry punishments similar to laws, but are decided by the para-courts of non-state actors have the right to be presumed innocent? Finally, how much of a reduction of the procedural or substantial presumption of innocence appears justified in

---

8 Note that Godden and Walton here offer yet another instance of taking the foundation of the presumption of innocence for grated to such a degree that it is the natural example of the most forceful presumption.

cases of war, natural disasters, or states of emergency?[9] In all these less-than-ideal cases there are significant instances in which popular support, public policy, or both lean against presuming the accused innocent.

A particularly interesting and highly dynamic example of accusations under less-than-ideal level c aspects (i.e., in para-courts), is currently provided by U.S. American Title IX proceedings. Title IX of the Education Amendments was originally established in 1972 as an instrument against sex-based discrimination in college, including and especially concerning access to collegiate athletics. Its application was noticeably extended over time by case law, starting with Alexander v. Yale (1980) and it is now one of the principal tools for prosecuting alleged sexual assault, harassment, and other transgressions in institutes of higher education. Title IX proceedings behave like criminal trials in numerous key regards: They judge people who stand accused of having committed offenses against rules that are substantially similar or even identical to criminal laws (sexual assault, sexual harassment, etc.) and impose punishments on those punishments on those they find guilty. These punishments are different from those available to state or federal courts (evidently, universities no longer have the ability to imprison their students), but similar in practical severity to mid-level criminal punishments. Since U.S. universities serve as gatekeepers to most high-income careers and social status, suspension, expulsion, public declarations of guilt, or other limitations to a person's education may well harm an individual more than criminal fines or even deferred prison sentences.

Yet, because Title IX proceedings are not classified as trials, most accused are denied the fundamental procedural rights that are taken for granted even in petty criminal trials: access to council, right to remain silent, right to be confronted with the accuser and right to a competent judge or jury. These core pillars of fair trials are seen as obstacles to the efficient enforcement of Title IX rules, and thus not only unnecessary but actively harmful for the establishment of a fair, equitable and non-discriminatory system of higher education. The fate of the most fundamental procedural right – the right to be presumed innocent until proven guilty beyond reasonable doubt – in this political development is particularly telling. In an attempt to accelerate the efficiency of the prosecution of alleged Title IX violation, the Department of Education's Office for Civil Rights of the Obama administration passed a rule in 2014 that punished all (partially) federally funded schools, who would not *abolish* a strong presumption of innocence and related burden of proof in favor of a weak "preponderance of evidence" standard. This weak

---

9 This question concerns the public support for these presumption under extraordinary circumstances, not the legal status of the presumption under these conditions. The legal answer to this last question is clear: None (Clooney & Webb 2020, 244ff.).

preponderance of evidence standard together with the absence of other basic procedural rights has led to countless reports of Title IX punishments imposed on evidently or presumably innocent students, often with a particularly adverse effect to members of minorities. [10] The Trump administration (who was not known to be a champion of other aspects of procedural rights, notably in the a and b aspect above), reversed the Obama administration rules and codified certain procedural rights in Title IX proceedings into law. The Biden administration is currently working and reversing these rules in turn and return to the Obama status quo ante.

The political developments surrounding U.S. American Title IX rules provides two takeaways that are highly relevant for the purposes of this essay: First, presumptions come at significant social costs. The main cost of the presumption of innocence is a reduced efficiency of prosecutions and punishments. Convincing stakeholders of the necessity of paying this cost can be an uphill battle especially in politically contentious topics and for offenses that disproportionally effect certain parts of a population (such as sexual assault and hate speech). Second, even the most well-acknowledged principles of fairness or procedural rights can quickly be eroded or outright opposed if they are being taken for granted for too long. This erosion usually manifests first in non-standard cases, but may ultimately also effect its core function.

The third argument in favor of dedicating a significant part of this project to a critical assessment of the foundation of the presumption of innocence is of a practical nature. This argument is related to the political one but takes a different spin. As indicated above, the ultimate aim of this project is to create a modern stasis model based on the presumption of innocence. The purpose of that model is to equip critical citizens with a tool to judge the fairness of judicial and para-judicial proceedings and to actively speak up for justice (in accusation or defense) themselves. Given the close dependency of the stasis axioms on a well-founded and forceful presumption of innocence defending an accused may lead to defending the vitality of a part of the stasis model, which in turn may lead to the necessity of defending the presumption of innocence itself.[11] An occasional third-level regress of this kind is especially likely in non-standard cases such as the Title IX proceedings mentioned above. Since these kinds of cases (non-legal accusations decided by para-courts, committees, representative bodies, or informal gatherings) are significantly more numerous yet equally influential for a just society, providing a detailed

---

10 Yoffe (2017a), Yoffe (2017b), Yoffe (2017c).
11 Godden and Walton make a closely related point when they write "To appreciate this point, consider the situation where a respondent does not recognize his obligation to accept some presumption in an argument. If the presumption is genuinely basic or primitive, then the dialogue simply becomes stuck. In order for the discussion to proceed, the nature and source of the respondent's obligation will have to be explained to him; ultimately he will have not only to understand, but to accept, the conditions which give rise to his obligation." (2007, 336)

synopsis of the fundament of the presumption of innocence seem pertinent. The arguments presented in this essay may be incomplete – both in total number and in individual execution – but they should offer a functional starting ground for any practical defense of a presumption of innocence.

Given the arguments and motivations above, the collection presented in this essay is a serious and open invitation to extension and to debate. I do not assume that there is much disagreement about the benefits of at least some level of presumption of innocence for at least some core cases. Accordingly, arguing for this concept proceeds from a position of strength. That means that the position can certainly afford a high level of critical scrutiny without endangering the overall social direction. To facilitate this scrutiny, I will clearly indicate any limitations, weaknesses, and rebuttal conditions of each argument wherever I can find it.

Furthermore, because this is a collection of arguments in favor of an already well-justified position and because at least some of their purpose is to offer a practical heuristic for critical citizens, I will not attempt to avoid systematic overlaps or redundancies. Many arguments cover ground that is alternatively defended by previous arguments but given the defeasibility of each and every one of the arguments in a specific case, I see this redundancy as a benefit rather than a weakness. Finally, because the ultimate goal of the larger project is to make a contribution to the critical toolset of speaking up for justice, any addition to the list of arguments below is very welcome. The collection of seven arguments in favor of the presumption of innocence in this essay is based a careful search of the relevant literature. But since explicit arguments in favor of this concept are actually surprisingly hard to find, and the scholarship in the fields in question (ethics, law, argumentation theory and related disciplines) is vast, I am certain that the list is incomplete. I hope that it will grow over time but also serve its current systematic function for now.

Two further preliminarily tasks remain before the synopsis of the actual arguments: A collection of terminological clarifications and theoretical disclaimers, and a brief discussion of taxonomic aspects of the set of arguments.

Armed with an extensive allegorical understanding of presumptions and their relationship to social effects, it is now time to specify some details. First, as briefly indicated above, the presumption of innocence is a practical presumption. That means, its purpose is to influence practical decision making in the cases of a difference of opinion under the conditions of limited time and imperfect information (Godden 2019, 214ff.). The general situational trigger for a presumption of innocence is an accusation against a person of having broken a norm to which she is subject combined with the request to a more powerful actor to punish this transgression. The evident paradigmatic case of this trigger is a criminal accusation in an ordinary court, but the situation is by no means limited to this standard case. What exactly triggers a presumption depends on its justification. As such, each of arguments below will specify its own triggering conditions that frequently overlap but are not perfectly coextensive.

Second, in a two-party disagreement (such as a standard accusation) each presumption for one party causes an equivalently elevated burden of proof for the other party. The apothecary scale allegory above illustrates this very clearly. There is some discussion to the exact nature of the burden caused by presumptions in general (Bodlović 2020a, 8ff.), but for the case of the presumption of innocence this question is rather straightforward: A weak presumption of innocence means that the evidence of a norm violation presented by the accuser must be a little more compelling than material invoked by the defense (e.g., fulfilling the "preponderance of evidence" standard), and a forceful presumption requires the accuser to produce significantly more compelling evidence than the defense to cause a justified conviction (thus fulfilling the "beyond reasonable doubt" standard). The strength of the presumption of innocence in a given case sits on a continuum between these two extremes and it is thus absolutely essential for any reasonable judge who wants to make a justified decision to understand which arguments apply to the situation at hand and which level of presumption is triggered.

Third, Godden (2019) rightfully observes that presumptions have very specific audiences. In his words: "*[...], the probative obligations generated by presumptive entitlements are only binding upon subjects of the presumption rule – i.e., those beholden to the non-alethic discursive goals backing the presumptive warrant.*" (215). For the discourse of accusation and defense this creates a set of interesting questions. The principal agent entitled to the presumption of innocence is the accused, and the principal agents bound by it are the decision makers. In the standard case of a criminal accusation these are judge and jury assisted by the presentation of the accuser. Things get a little murkier when the accused is not a living person anymore or when third-party audiences are involved. Can dead people carry presumptive entitlements when they are no longer potential parties to a dialogue? And if so, how does this shape or influence the presumptions?[12] And on the other end of the spectrum: Is an interested public bound by the same presumptive force as a jury? The answer to this problem depends at least in part on whether the public at large contributes to a potential punishment (by social ostracization or similar means) or if it is motivated by purely scientific inquiry. The questions created by the audience interactions of presumptions cannot be fully answered at this point, but they are important to keep in mind when assessing the set of arguments below.

Fourth and last, the presumption of innocence is a macro-level presumption that causes a macro-level burden of proof. This is an important and relatively straightforward clarification at first glance, but

---

[12] The famous rule of *De mortuis nil nisi bene* is probably an attempt to deal with this uncertainty.

it gets more complicated as one focuses on the detail. Distinctions between macro- and micro-level probative burdens are mostly scalar, but the basic idea is to differentiate between duties based on the overarching claim ("You murdered your father!") and those based on a discursive turn ("I went to sleep early that night"). Both trigger *some* burdens of proof, but the presumption of innocence, causing a macro-level burden, is attached to the original claim (the accusation) and does not shift during the discourse. While this connection of the principal burden to the original claim is relatively simple, it gets more complicated as different types of probative burdens clash over the course of the discussion (i.e., in positive claims of alibi, justification or norm conflict by the accused). These questions relevant for the details of some practical defenses in society, but fortunately they are not decisive for the establishment of the presumption of innocence at large, the topic of this essay.

Finally, some comments on the relationship of the set of arguments to each other. The arguments below are not organized in any linear manner (say, from strongest to weakest or most general to most specific), but they can all be located in a three-dimensional space. The respective dimensions are argumentative strengths, probative force, and situational trigger.

The first dimension is probably the easiest: Some arguments are more compelling than others, and some contain exceptions, conditions, or potential rebuttals while others do not. The arguments presented in the collection below occupy the full continuum from figurative analogy with little probative value to deduction from widely accepted philosophical principles. Because the arguments below serve a variety of theoretical and practical purposes, and because of the cumulative effect of some arguments with each other, I have opted for an inclusive approach to their selection.

The second dimension is the presumptive force each argument establishes or aims to establish. As indicated above, contrary to what Godden and Walton seem to suggest, I do not believe that the argumentative strengths (what they call the "foundation" of a presumption) and the presumptive force are strictly correlated. Yes, forceful presumptions require strong foundations, but the reverse is not true. Strong arguments can establish both, forceful and less forceful presumptions. In the first allegory above that means that weights of all sizes can be firmly attached to one particular scale. I believe the list below will give ample examples of both.

The third dimension is the most complex, and perhaps also the most contentious. I argue throughout this project and especially in this essay that a justified presumption of innocence is not limited to standard criminal trials. Instead, different flavors of the presumption are triggered by related but independent situational qualities, many of which happen to coalesce in the standard criminal case. Accordingly, the third aspect strictly speaking is more of a bumpy continuum than a precise mathematical dimension, but I think it serves the present purposes just as well. This continuum is inhabited by a rich set of relevant situations that range from 1: capital criminal trials (all principles of procedural justice are near universally accepted), 2: ordinary criminal trials (some jurisdictions reduce some principles) and 3: military or extraordinary trials (with frequently reduced or streamlined protections), via 4: minor trials and para-trials (formal setting, but commonly omitted procedural rights of the accused and weakened separation between accusers and decision maker), all the way to 5: informal three-party accusations (such as may happen in mediations or arbitrations, without preannounced accusation), 6: informal peer-to-peer accusations (two distinct parties remain, but the accuser and decision maker is the same agent) and even 7: informal self-accusations. Once again, this scale is far from perfect, but primarily an attempt to illustrate that as accusations get gradually more formal the presumption of innocence is more universally accepted and established. I believe there are good reasons why a friend who accuses a friend of lying to him has a duty to presume her innocent, but the respective elevated burden of proof is without doubt lesser than that for the prosecution in a murder trial. The analysis of the situational triggers that justify this distinction is one of the purposes of the following section.

1. Argument zero: A presumption on presumptions

The opening "argument" in this list isn't an argument in any critical sense of the term, yet it may well be the most frequently invoked justification for the presumption of innocence. Argument zero is a curious mix of appeals to the status quo and appeals to venerated authorities. One of Whately's most fundamental presumptions is the presumption in favor of the status quo, and I have already suggested above that this presumption is based on a sound principle of social economy: Any society has the right (and perhaps even duty) to avoid spending an avoidable amount of institutional, emotional, social, or financial capital on change without progress, in order to focus its finite capital on creating the best environment for its members instead. Yet, while this (weak) presumption of the status quo and its related deference to existing institutions may be a very well-founded principle, cases of the status quo being unjust are evidently legion. They cover nearly all interesting aspects of social progress: rights of women, minorities, personal liberties, kinds of governance and so on. Having said that, the examples above do not necessarily show that a weak presumption in favor of the status quo isn't principally justified, but rather that it is – and should be – easily overruled in the eyes of reasonable people by compelling arguments in favor of change.

With this caveat out of the way, it seems that an overwhelming agreement of the relevant authorities and institutions (in Aristotle's terms: a strong *endoxon*) does have some probative or at least some persuasive force in favor of the viability and desirability of a concept. Accordingly, introductions to the presumption of innocence frequently point to the authority of common legal principles, venerated authorities, or institutional manifestations. The Latin dicta *"In dubio pro reo"* (In doubt in favor of the accused), *"Ei incumbit probatio qui dicit, non qui negat"* (He who assert must prove, not he who denies) are famous slogans that cover aspects of the presumption of innocence. It is almost impossible to find an Anglo-Saxon discussion of this concept without running into either Blackstone's ratio *"It is better that ten guilty persons escape than that one innocent suffer."* (1769), or Benjamin Franklin's similar statement with a hundred to one ratio: *"That it is better 100 guilty Persons should escape than that one innocent Person should suffer, is a Maxim that has been long and generally approved."* (1795).

All these slogans and quotations are of course just paraphrases and affirmations of the presumption of innocence, rather than justifications for it. Yet, even while they do not add any argumentative value to the concept, they help perceive it as a well-established idea in society. Similarly, without independent argumentative force, but strong intuitive persuasiveness are references to the near universal endorsement of the presumption of innocence in contemporary national and international law. The Universal Declaration of Human Rights includes the presumption of innocence in Article 11 (*"Everyone charged with a penal offence has the right to be presumed innocent until proven guilty according to law in a public trial at which he has had all the guarantees necessary for his defense."*). Similar and related affirmations can be found in the European Convention on Human Rights (Art. 6.1), the Charter of fundamental rights of the European Union (Art 48.1), the American Convention on Human Rights (Art. 8.2), the Cairo Declaration on Human Rights in Islam (Art. 19e), the International Covenant on Civil and Political Rights (Art. 14.2), the Arab Charter on Human Rights (Art. 16), the ASEAN Human Rights Declaration (Art. 20.1) and many others (Clooney & Webb 2020, 200f.). Its manifestations and continuous confirmations in the laws and constitutions of civic law countries and precedent in common law countries are too numerous to count.

The near universal support of the presumption of innocence in contemporary positive law is a blessing and a curse at the same time: It is a blessing for the millions of accused people who benefit from it each year and for the countless convictions of innocents that it helps avoid. It is also a great sign of humanitarian progress and society's ability to agree on *"perhaps the most fundamental of fair trial guarantees, [...]"* (Clooney & Webb 2020, 199). At the same time, it is a curse because it largely prevents legal scholars from further arguing for a concept, that is already so wide accepted. This leaves defendants of the presumption unprepared in cases in which it does encounter resistance.[13]

Argument zero has two significant limitations. First, as already indicated, defenses of a concept with reference to the status quo are inherently systematically weak. What if Blackstone, Franklin, and the drafter of all those national and international laws, conventions and declarations simply erred in their acceptance of the presumption of innocence? Second, and more importantly for the purposes of this essay, argument zero establishes only a narrow presumption for criminal cases. The presumption of innocence is generally held to apply to the entire trials as well as to a limited set of pre-trial proceedings, and it is so fundamental that it does not permit derogations or reservations (Clooney & Webb 2020, 244ff.), but it only covers criminal cases tried in court. Argument zero has little or no supportive force for discourses of argumentation and defense with social, moral, or other non-criminal punishment at stake. In sum, argument zero creates a weak argument for a strong presumption in a narrow set of circumstances.

2. Argument one: Dialectical Fairness

Argument one is based on the discrepancy between an the discursive ideal of dialectical equality and the systematic deviations from this ideal in accusations and defenses. This discrepancy can be addressed by a weak presumption of innocence.

The argument from dialectical fairness requires two basic assumptions. First, differences of opinion should be solved in a reasonable manner by systematically equal parties. This argumentative egalitarianism finds an expression in a variety of dialectical systems (e.g. the Freedom rule of Pragma-Dialectics or Habermas' second presupposition, Bohman & Rehg 2017). As part of their systematic equality discussant should in principle have access to the same or similar argumentative moves and strategic choices.

Second, in discourses of accusation and defense, accusers impose a significant number of choices on their opponent. These choices cannot be altered, ignored, or opposed by the defendant without opting out of the discourse altogether and risking significant loss of face. The choices in question are the defining details of an accuser. A well-substantiated accusation is not limited to a structure of "I accuse you!", but rather contains a number of defining elements to create a "I accuse you of

---

[13] See e.g., former U.S. vice president Dick Cheney's attack on the presumption of innocence for Guantanamo detainees.
https://nymag.com/intelligencer/2014/12/cheney-alright-with-torture-of-innocent-people.html.

breaking norm a, at location b, at time c, against victim d" and so on. A typical example could thus be "I accuse Markus of stealing Karsten's encyclopedia from his home on August 23." Once leveled the accused has no option in altering the details ("Why don't we talk about how I *didn't* damage Dana's car in her driveway on May 1st?"). Changing any of the elements would be (rightfully) seen as evading the charge or trying to escape the discussion. Choosing and leveling this particular accusation out of a set of millions of other potential combinations gives the accusers a systematic advantage that has no equal on the defendant's side. One way to balance this dialectical difference is to confer a weak presumption of innocence onto the systematically disadvantaged side.

There are at least two possible objections to this argument: First, one may point out that *any* protagonist in *any* opening stage of a disagreement makes and imposes choices on the potential antagonist. A discussant adopting the standpoint of "The US should lower taxes for the households with significant student debt" also chooses to focus on one specific aspect of one law in one country. Yet the difference here lies in the strength of the imposition that this choice makes. The protagonist does not name his antagonist and force her to defend herself or else risk damage to her face. It is also quite possible for protagonist and antagonist to cooperatively alter the main standpoint during the discussion – a move that outside of trial avoiding plea deals would be systematically problematic in discourses of accusation and defense.

The second objection to the argument from dialectical fairness weighs heavier. Even assuming that the argument itself is sound, and the principle of argumentative equality requires a structural balancing in favor of the accused, there is still no indicator why this balancing should be done with the help of a weak presumption of innocence. Any other systematic advantage might be capable of addressing the imbalance: This could be taken from the typical list of procedural justice rights (requesting assistance from another discussant, requesting additional time to prepare, etc.), or even be as seemingly random as allowing the defendant to choose the venue of the discussion or require the accuser to provide food and drink for the duration of the disagreement. A presumption of innocence may be an intuitive plausible candidate to balance the dialectical disadvantage created by an accusation, but it is by no means the only possible choice. For this reason, the argument from dialectical fairness establishes only a weak argument in favor of a weak presumption of innocence, but it does so for a very wide set of circumstances (see also Konishi 2001; Goodwin 2001; Kauffeld 2002, 105–6). It is triggered by any accusation against a person, not just those that are tried in a formal criminal court.

3. Argument two: Walton's Loaded Gun

Argument two, Walton's Loaded Gun or the danger argument aims to establish a practical presumption based on the great potential harm a mishandling of a powerful instrument can create. This argument is little more than a figurative analogy – but a very persuasive one. Walton (1988, 238) compares the presumption of innocence to presuming any gun to be loaded: "*Suppose Larry picks up a gun, but has no information on whether the gun is loaded or not. It may make sense for him to presume that the gun is, or may be loaded, and to suit his actions of handling the gun to accord with that presumption.*" (Ibid.) Extending Walton's starting point, it may not only be a rule of prudence for the gun handler to presume any gun to be loaded, but also a justified imposition of bystanders and society at large to hold him to these standards even against his will. Continuing the same image illustrates that point further: Picture two people, A and B, in a park. A is juggling with a couple of plastic balls, and B is playing with his (potentially loaded) gun. It is intuitively immediately plausible that any third party has the right to impose a duty of care onto B, but not A. That seems to be the case because any mishandling of B can seriously harm, injure or even kill innocent bystanders (or himself), whereas A's mistakes can at worst embarrass herself. It is a small step from this plausible hypothetical to an argument from analogy in favor of the presumption of innocence.

Legal and moral norm systems are a magnitude more dangerous and harmful than any single loaded firearm. Criminal law alone is enforced by millions of people who are employed with the sole or principal purpose of enabling, supporting, or overseeing the intentional suffering or death of human beings. These people (parts of the police, guards, executioners, and other members of the criminal justice system) purposefully deprive humans of fundamental pleasures or impose suffering by locking them up, taking their property, limiting their rights or in some countries even beating, mutilating, or killing them. Most of this is done against convicted criminals under the assumption that the benefits of this suffering to society at large far outweigh the harm done to the individuals, but that by no means reduces the scope of the harm and suffering itself. If anything, it creates an even closer similarity between the criminal system and the gun: As most gun owners may firmly believe that their guns are mainly purchased for the purpose of justified self-defense or perhaps the minimum amount of necessary violence, so does the criminal system aim to use its awesome and terrifying power only for the defense of laws and public order. If the public was justified in imposing a special duty of care on the individual gun owner in the park, who at best could accidentally injure or kill a single person, then it is even more justified in imposing similar duties on those that handle an instrument that intentionally creates suffering and harm for millions. This special duty of care is expressed in the significantly elevated burden of proof for the accuser.

So far and for simplicity's sake, the danger argument has only compared the gun to the criminal system. The same argument can be made however for most non-criminal punishment. These punishments (loss of friends, positions, marriages, jobs, titles, memberships, public perception, income and so on) are on average less painful than the scale of criminal punishments, but at the same time they are more frequent and also capable of creating severe personal suffering and despair, causing some humans so much pain that they choose to end their lives.

The main weakness of the danger argument is the fact that it rests on the highly plausible, but not further justified distinction between the juggler and the gun owner. If that original distinction holds and society's right to impose a special duty of care based on voluntarily handling a very dangerous tool, then the same principle should also apply to accusers of any kind. The situational trigger for this presumption of innocence is the demand for punishment by the protagonist and the strengths of the presumption established is proportional to severity of the punishment at stake. In sum, the danger argument creates a fair support for a variable presumption of innocence in a wide set of circumstances.

4. Argument three: Perelman's 100

Argument three is one of the most complex, but I believe also one of the two most forceful arguments in favor of a strong presumption of innocence. It does however come with two clear rebuttal conditions that limit its applicability in certain extreme situations.

Argument three rests on two important, but very well supported and widely accepted assumptions. The first, it assumes with Chaim Perelman that abstract justice can be defined as "[...] *a principle of action in accordance with which beings of one and the same essential category must be treated in the same way.*" (Perelman 1963, 16).[14] The driving idea behind this definition of justice is of course not exclusive to Perelman or even creative in any meaningful way. It can be found in a variety of formulations of the golden rule throughout the history of ethics in Aristotle, Kant's categorical imperative and many others. Perelman's phrasing of this sentiment however lends itself particularly well for the present argument as well as its rebuttal conditions.

The second assumption concerns a Rawlsian starting point. In designing a fair society, the souls behind the veil of ignorance have the

---

14 Perelman and Olbrechts-Tyteca similarly define acting just as "[...] giving identical treatment to beings or situations of the same kind." (1969, § 52).

duty to obey the principles of justice such as the one above (which is ultimately the main reason why they are behind that veil in Rawls' thought experiment), but they also have the right to act self-interested within the restraints of that principle. The resulting principle of prudence will thus guide them to design the best society for all.

Uniting these two assumptions and applying them to the special case of procedural justice rules invites the just and prudent reasonable agent to engage in a hypothetical Blackstone calculus: In the position of the accused do I prefer to live in society A, where there are ten (or even one hundred) guilty criminals on the street that have committed my alleged crime (e.g., murder) or in society B where I am innocently convicted of murder and sentenced to life imprisonment while these are off the street. I argue that for most reasonable, just, and prudent people the decision in favor of society A – the society created by a strong presumption of innocence that is more likely to protect the innocent but also to let some guilty go free – is an easy one. The certain harm suffered by the innocently convicted that are punished for crimes they have not committed and ostracized by society without any guilt or personal agency would be perceived as far greater than the potential danger of encountering one of the free criminals at a later stage.

Once again, the key argument is illustrated on the example of criminal law, but its basic premises work similarly in all cases of accusation in which the severity of the punishment stands in a similar relationship to the severity of the social harm created by offense (i.e., at a minimum in all cases that contain elements of punishment as deterrent in favor of a norm): I would rather not be treated as a proven liar by my friends even at the cost that some lies in our group stay undetected.

As indicated above, argument three comes with two important rebuttal conditions that either undercut the force of the first assumption or the force of the second assumption and thus suspend the argument from Perelman's 100 under certain conditions.[15]

Undercutter one goes into effect in circumstances where those in the position to design and uphold the rules of procedural justice and those affected by these rules are perceived as beings of two essentially *different* categories. In a society that ascribes essentially different values to people based on their sex or gender, male witch hunters can suspend a presumption of innocence for the accused without violating either of the assumptions above, as long as they can make sure that only women will be innocently convicted. Similarly, in a society that treats nationality and/or ethnicity as essential categories, American military prosecutors can reasonably suspend the presumption of innocence against "enemy

---

[15] This absolutely does not mean that under those conditions the presumption of innocence is not justified, but rather that in those cases other arguments for its establishment are required.

combatants" as long as these are exclusively foreign, brown, Muslim, or otherwise qualifying as members of the essentially different group.

Undercutter two applies to cases in which the potential further harm done by the offenders who are acquitted because their guilt could not be established beyond a reasonable doubt is perceived as an existential threat to society at large. In cases in which a single witch is perceived as having the power to poison the wells, kills scores of infants or doom entire villages, or a single "enemy combatant" assumed to bring about incidents of international terrorism that kill thousands and terrify millions, [16] a reasonable, just, and prudent agent may believe that Blackstone's calculus swings in favor of society B, even at the cost of that agent's personal and unacknowledged martyrdom for the safety of that society. Instances of undercutters one and two sometimes correlate (as in the Spanish Inquisition and Guantanamo examples above), but they can also occur in isolation. The McCarthyan communist hunt is a good historical example of undercutter two appearing without undercutter one present.

Importantly, these two undercutters apply only to a very limited set of circumstances and they are in no way sufficient for establishing a strong case *against* a presumption of innocence even under those conditions. Yet their existence is a helpful illustration why a concept as essential as the presumption of innocence is best founded on a diverse set of independent reasons. Perelman's 100 in sum offers a strong argument in favor of a forceful presumption that applies to a wide variety of circumstances. Its triggering condition is the invocation of a deterrent punishment.

5. Argument four: Kant's Categorical Defense

Argument four is based on one two assumptions. The first is Immanuel Kant's most famous contributions to practical philosophy, his absolute prohibition to treat humans as mere means. In *Grundlegung zur Metaphysik der Sitten* he writes: "*Der praktische Imperativ wird also folgender sein: Handle so, daß du die Menschheit, sowohl in deiner Person als in der Person eines jeden anderen, jederzeit zugleich als Zweck, niemals bloß als Mittel brauchst.*" (The practical imperative will thus be the following: Act in such a way, that you use humanity in your own person, as well as in the person of every other, always at the same time as an end,

---

[16] This does not imply that terrorism is similarly illusionary as witchcraft, but rather that in both cases it is solely dependent on the perception of those in power to design the rules. It does not matter how much harm is done, but how much harm is feared in case too many guilty go free.

never merely as a means. 4:428-429). Like so much of Kant's work, many of the interpretative details of this moral imperative are intensely debated and this is not the place to reproduce the respective discussion. For the present purposes I will understand his position to demand that a morally relevant action which involves a human being must at least in part be motivated by a concern for that human.

The second assumption posits that while punishing an innocent person is inherently undesirable (i.e., it causes unnecessary suffering to the person and it is a wasted effort by the society), it may occasionally be necessary to uphold respect for the law and order in society. In other words, it acknowledges that punishing innocent people is an evil, but one that may occasionally be condoned on grounds of general deterrence.

It is easy to see how the two assumptions stand in immediate tension to each other. Punishing a person with the sole purpose of benefitting others (by deterring would-be offenders) is a paradigmatic case of using humans as mere means. This situation arises every time an innocent person is being convicted, and a moral society (in the Kantian sense) must thus do everything in its power to avoid these situations. A strong presumption in favor of the accused is one of the most effective tools to achieve this aim and may thus be morally required.

Argument four has two significant weaknesses. First, its complete dependance on one specific approach to moral duties. For anyone who does not support the sentiment expressed by Kant's formulation of the imperative, or is unconvinced of his reasoning for it and interprets it in a different way, this argument is moot. Second, its latter assumption introduces material that are dangerously close to circular reasoning. Properly speaking argument four is thus not even an independent argument in favor of establishing a presumption of innocence, but rather a counterargument against those that would oppose a strong presumption of innocence on grounds of general deterrence.

In sum, Kant's categorical defense provides some (if indirect) justification for establishing a forceful presumption that applies to a wide variety of circumstances. The triggering condition for the presumption thus justified is the presence of a demand for punishment.

6. Argument five: the Hyper-Villain

The remaining three arguments in favor of the presumption of innocence form a systematic group. They are all based on the danger of doing unnecessary or intolerable harm. The main difference between them are the potential perpetrators and respective victims of that harm. Argument five focuses on the potential harm done by society to the defendant. Argument six focuses on the relationship between accuser and society and argument seven on the relationship between accuser and defendant. Importantly, while this group of arguments thus covers the complete triangle between accuser, defendant, and society, and while they share one central assumption (i.e., intentionally creating preventable and unjustified suffering of others is immoral), they are completely independent from each other and create a multiple argument structure. The danger of creating unnecessary harm to either party is a sufficient condition for installing a preventative policy – independent of potential additional harm to others. This is highly relevant for the present purposes, because the remaining three arguments come in very different strengths. Argument five is – together with the argument from Perelman's 100 – perhaps the most forceful argument in favor of establishing a strong presumption of innocence for a wide set of circumstances, whereas arguments six and seven provide considerably weaker support for that concept.

Argument five is based on four assumptions that all seem very straightforward and individually easy to defend: 1) By default, the creating of unnecessary suffering is immoral, 2) Punishment is in essence the purposeful creation of suffering, 3) Judges about alleged offenses are (like all human actors) inherently fallible, and 4) of two alternative designs, the one that systemically leads to less harm is morally preferable. While each of these assumptions should – I hope – encounter little intuitive resistance, it will be necessary to unpack them in more detail to fully flash out the argument from the hyper-villain.

The first assumption probably needs the least explanation: Inflicting unnecessary and unjustified suffering on a conscious agent is perhaps the single best candidate for defining vice or evil that most moral systems share. It is the definition of "necessary", "justified" and "suffering" itself where they tend to disagree, rather than on this very basic assumption.

The second assumption is, when fully examined, equally close to a truism: Punishment is in its very essence the creation of suffering of those that society deems deserving of it. The main classes of known punishment are not only objectively designed with the suffering of their recipients in mind,[17] they also each resemble actions that, when committed by private actors without society's authorization, are usually classified as crimes or offenses. This is the case for many widely accepted forms of punishment: the deprivation of one's property (fines and seizures vs theft and robbery), the deprivation of one's freedom of movement (imprisonment and

---

[17] There might be a hypothetical case against this claim in which a society bases its punishment in theory solely on resocialization and/or incapacitation and in practice implements it in a way that is neutral or pleasant for the person being punished. To my knowledge such as society has never existed, and if it did, they would probably not use the concept of "punishment" for their corrective and preventative actions.

detention vs abduction or false imprisonment), threats against a person (probation sentences vs harassment or intimidation), harm to one's honor and image (guilty verdicts vs libel or slander), and even intentional killing (death sentence vs. murder or manslaughter). Other forms of punishment are more controversial, but also either condoned by some societies, condoned in the past or effectively tolerated by the criminal system. These include the infliction of physical pain (corporal punishment vs. assault), bodily mutilation (also classified as corporal punishment or assault respectively), and other forms of physical or sexual violence, which are not typically explicitly condoned, but tolerated by many prison systems to a degree that may well be considered complicity. All this is a long way of saying: Punishment by design resembles crime in an uncanny way. What differentiates the (criminal) actions of a villain and the (punitive) actions of a public representative is merely the justification of the latter: the claim that the recipient of the punishment deserves it based on her offense.

The third assumption acknowledges the fallibility of any human action – especially complicated processes like judging the competing merits of accusations and defenses under circumstances of limited time, knowledge, and cooperation. Given the inherent fallibility of these judgements, societies need to decide if errors in one direction, convicting innocents people or acquitting offenders, are significantly worse than in the other. Based on the second assumption this alternative can be further clarified: Any instance of failing to punish a guilty person for her offense lets a villain prevail. Any instance of punishing without justification, lets society behave like a villain itself – thereby creating a next-level villain.

This leads to the final assumption. If societies have a duty to choose that structure which is less likely to create unnecessary harm or suffering, which is more harmful, the influence of many unchecked ordinary villains or that of one unchecked hyper-villain. The harm of ordinary offenders to society is evident: they illegally or immorally disturb the social order with their actions. Depending on the offense, the harm caused can be relatively mild (lying or cheating) or quite severe (assault or murder). There are however two reasons why the harm caused by a hyper-villain is significantly worse: the hyper-villain prevents most forms of self-defense in a practical as well as a theoretical level, thus removing all agency and hope from the victim. Practically, victims of an ordinary offense often can defend themselves against theft, robbery, assault, or slander. The same kind of defense would be exponentially harder or entirely futile against the enforcement agents of an unjust guilty verdict. The criminal justice system has access to an overwhelming amount of power that makes individual resistance practically infeasible. The situation is even worse on a theoretical level. Where proportional defenses against ordinary would-be villains are permitted or even encouraged and victims of offenders have access to remedies if the defense fails, no such equivalent exist against the hyper-villain. Against a completed offense of the hyper-villain (i.e., a guilty verdict against an innocent person that has been confirmed and executed) any further resistance is prohibited. So even if the victim of the hyper-villain had the practical ability to resist the offense (e.g., by escaping from prison), that defense would not bring a resolution, but instead create even graver consequences. In sum, compared by similar offense, theft versus unjustified fine, abduction versus unjust prison sentence or manslaughter versus the execution of an innocent person, each action of a hyper-villain creates significantly more harm to the victim, because it deprives the person not only of the primary good, but also of her ability to defend herself, her hope to do so, any agency of preparing against or evading the offense as well as possible remedies or restitutions. From this follows that taken one by one convicting innocent people does significantly more harm than acquitting guilty ones, and societies have the moral duty to tilt the scale in favor of the accused by installing a strong presumption of innocence.

The trigger for this presumption is a significant power disparity between alleged offender and the party that executes a punishment. This disparity is evident in the criminal system, but also applies to most non-criminal forms of punishment that are enforced by organizations, institutions, or society at large. This places argument five as a strong reason for a forceful presumption in a wide set of circumstances.

7. Argument six: Leave me in Peace

The remaining two arguments in this group are considerably simpler, but also weaker than the argument from the hyper-villain. Argument six starts from an easy premise: In and off itself, moral and legal prosecutions and trials are a nuisance. They cost time, energy, and money, they pitch people against each other, and they disturb the social peace. In a nutshell they use up valuable and limited resources of a community. Using up these resources is a potential harm that accusers do to societies.

One way to discourage the overuse of these resources by any would-be accusers is elevating their burdens and making it harder for them to succeed. The presumption of innocence with its effect on the burden of proof is an excellent candidate to install this negative incentive.

The huge problem with this argument is immediately apparent: Imposing an elevated burden of proof on the accuser may indeed deter hasty accusations, disincentivize wasting resources on them and defend societal peace from being disturbed. Yet, given that successful offenders also drain limited resources and usually disturb the peace in a more serious manner, it is not clear that the net benefit for the community would be positive. While hasty accusations are indeed a problem for a society, so are structures that discourage the prosecution of actual offenses.

In sum, argument six probably provides the least support for the presumption of innocence of all reasons discussed. The presumption itself has relatively little force (it only needs to discourage the overzealous) and its application is narrow. The triggering condition is the involvement of institutional resources that are usually found in criminal trials and formalized non-criminal proceedings.

8. Argument seven: Just Slander

The final argument on this list focuses on the potential harm that the accuser does to the defendant. It is based on a single assumption: An unsubstantiated accusation by itself creates a threat to the accused's honor and face. In some instances, the very terms that describe a moral or legal offense are even used as ordinary language insults.

Under normal circumstances society does not tolerate grave threats to its members' honor and face. Accordingly, it installs a variety of criminal or moral rules against insults, slander, libel, and related offenses. If the default effect of an accusation is thus an offense to the accused, then society may and should impose a burden on the accuser to offer conclusive justifications for his actions. The elevated burden of proof caused by the presumption of innocence reflects this justified imposition.

I cannot see any evident weakness or rebuttal condition for this argument, but the presumption it establishes is rather weak. It needs to be just forceful enough to establish that the accusation is not used as a random attack, but that it withstands at least a prima facie testing. The triggering condition for this presumption is any potentially face-threatening accusation, which makes the area of applicability very wide.

Calling an academic project, a "work in progress" has become such a commonplace in scholarship, that I use this phrase only with great hesitation. Yet, that is what this endeavor truly is. I believe the list of seven arguments and one presumption above is in its sum sufficient for justifying a strong presumption of innocence in a wide set of circumstances. Some of the arguments are reserved only for criminal or highly codified accusations, but most of them are based on conditions that are also fulfilled by non-formalized extra-criminal accusations. I think that the argumentative foundation of the presumption of innocence is thus strong enough to ultimately carry the axioms of a modern stasis theory. Having said that, I hope and believe that there are additional reasons for this presumption out there, that I have not found yet and I am grateful for any suggestion of what this essay is missing.

References

Aikin, Scott. 2011. "A Defense of War and Sport Metaphors in Argument." *Philosophy & Rhetoric* 44 (3): 250–72. https://doi.org/10.5325/philrhet.44.3.0250.

Bodlović, Petar. 2020a. "Presumptions, Burdens of Proof, and Explanations." In *OSSA Conference Archive*. 17. Windsor: University of Windsor. https://scholar.uwindsor.ca/ossaarchive/OSSA12/Saturday/17.

———. 2020b. "On Presumptions, Burdens of Proof, and Explanations." *Informal Logic* 40 (2): 255–94. https://doi.org/10.22329/il.v40i2.6312.

———. 2021. "On the Differences Between Practical and Cognitive Presumptions." *Argumentation* 35 (2): 287–320. https://doi.org/10.1007/s10503-020-09536-w.

Bohman, James, and William Rehg. 2017. "Jürgen Habermas." In *The Stanford Encyclopedia of Philosophy*, edited by Edward N. Zalta. Metaphysics Research Lab, Stanford University. https://plato.stanford.edu/archives/fall2017/entries/habermas/.

Clooney, Amal, and Philippa Webb. 2020. *The Right to a Fair Trial in International Law*. First edition. New York: Oxford University Press.

Eemeren, Frans H. van, and Peter Houtlosser. 2002. "Strategic Maneuvering with the Burden of Proof." In *Advances in Pragma-Dialectics*, edited by Frans H. van Eemeren, 13–29. Amsterdam: Sic Sat.

Godden, David. 2017. "Presumption as a Modal Qualifier: Presumption, Inference, and Managing Epistemic Risk." *Argumentation* 31 (3): 485–511. https://doi.org/10.1007/s10503-017-9422-1.

———. 2019. "Analyzing Presumption as a Modal Qualifier." In *Presumptions and Burdens of Proof. An Anthology of Argumentation and the Law*, edited by Hans Vilhelm Hansen, Fred J. Kauffeld, James B. Freeman, and Lilian Bermejo-Luque, 206–20. Tuscaloosa: The University of Alabama Press.

Godden, David M., and Douglas N. Walton. 2007. "A Theory of Presumption for Everyday Argumentation." *Pragmatics & Cognition* 15 (2): 313–46. https://doi.org/10.1075/pc.15.2.06god.

Goodwin, Jean. 2001. "Reply to Takuzo Konishi: A Generalized Stasis Theory and Arguers' Dialectical Obligations." In *Proceedings of the 2001 Meeting of the Ontario Society for the Study of Argumentation*, edited by Hans Vilhelm Hansen, Christopher W. Tindale, J. Anthony Blair, Ralph Johnstone, and Robert Pinto. Windsor: University of Windsor.

Hansen, Hans V., ed. 2019. *Presumptions and Burdens of Proof: An Anthology of Argumentation and the Law*. Rhetoric, Law, and the Humanities. Tuscaloosa: University of Alabama Press.

Hoppmann, Michael J. 2014. "Preciseness Is a Virtue: What Are Critical Questions." In *Proceedings of the 10th International Conference of the Ontario Society for the Study of Argumentation (OSSA)*, edited by Dima Mohammed and Marcin Lewiński. Windsor: University of Windsor.

Kant, Immanuel. 2019. *Werke [in zwölf Bänden]. 7: Kritik der praktischen Vernunft. Grundlegung zur Metaphysik der Sitten*. Translated by Wilhelm Weischedel. 23. Auflage. Frankfurt am Main: Suhrkamp.

Kauffeld, Fred J. 1998. "Presumptions and the Distribution of Argumentative Burdens in Acts of Proposing and Accusing." *Argumentation* 12 (2): 245–66. https://doi.org/10.1023/A:1007704116379.

———. 2002. "Pivotal Issues and Norms in Rhetorical Theories of Argumentation." In *Dialectic and Rhetoric*, edited by Frans H. van Eemeren and Peter Houtlosser, 97–118. Dortrecht: Kluwer Academic.

———. 2003. "The Ordinary Practice of Presuming and Presumption with Special Attention to Veracity and the Burden of Proof." In *Anyone Who Has a View*, edited by Frans H. van Eemeren, J. Anthony Blair, Charles A. Willard, and A. Francisca Snoeck Henkemans, 133–46. Dordrecht: Springer.

Konishi, Takuzo. 2001. "A Generalized Stasis Theory and Arguers' Dialectical Obligations." In *Proceedings of the 2001 Meeting of the Ontario Society for the Study of Argumentation*, edited by Hans Vilhelm Hansen, Christopher W. Tindale, J. Anthony Blair, Ralph Johnstone, and Robert Pinto. Windsor: University of Windsor.

Krabbe, Erik C. W., and Jan Albert van Laar. 2011. "The Ways of Criticism." *Argumentation* 25 (2): 199–227. https://doi.org/10.1007/s10503-011-9209-8.

Perelman, Chaïm. 1963. *The Idea of Justice and the Problem of Argument*. London: Routledge & Paul.

Perelman, Chaïm, and Lucie Olbrechts-Tyteca. 1969. *The New Rhetoric*. Translated by John Wilkinson and Purcell Weaver. Notre Dame, IN: University of Notre Dame Press.

Rawls, John. 1999. *A Theory of Justice*. Rev. ed. Cambridge, Mass: Belknap Press of Harvard University Press.

Rescher, Nicholas. 1988. *Rationality: A Philosophical Inquiry into the Nature and the Rationale of Reason*. Clarendon Library of Logic and Philosophy. Oxford [England] : New York: Clarendon Press ; Oxford University Press.

———. 2006. *Presumption and the Practices of Tentative Cognition*. Cambridge: Cambridge University Press.

Ullman-Margalit, Edna. 1983. "On Presumption." *The Journal of Philosophy* 80 (3): 143. https://doi.org/10.2307/2026132.

Walton, Douglas. 1988. "Burden of Proof." *Argumentation* 2 (2): 233–54. https://doi.org/10.1007/BF00178024.

Walton, Douglas 2014. *Burden of Proof, Presumption and Argumentation*. New York, NY: Cambridge University Press.

Whately, Richard. 2010. *Elements of Rhetoric Comprising an Analysis of the Laws of Moral Evidence and of Persuasion, with Rules for Argumentative Composition and Elocution*. Paperback ed. Landmarks in Rhetoric and Public Address. Carbondale, Ill: Southern Illinois University Press.

Yoffe, Emily. 2017a. "The Uncomfortable Truth About Campus Rape Policy." The Atlantic. September 6, 2017. https://www.theatlantic.com/education/archive/2017/09/the-uncomfortable-truth-about-campus-rape-policy/538974/.

———. 2017b. "The Bad Science Behind Campus Response to Sexual Assault." The Atlantic. September 8, 2017. https://www.theatlantic.com/education/archive/2017/09/the-bad-science-behind-campus-response-to-sexual-assault/539211/.

———. 2017c. "The Question of Race in Campus Sexual-Assault Cases." The Atlantic. September 11, 2017. https://www.theatlantic.com/education/archive/2017/09/the-question-of-race-in-campus-sexual-assault-cases/539361/.

What Justifies the Presumption of Innocence? Commentary on Hoppmann's "Mapping Arguments in Favor of the Presumption of Innocence"

PETAR BODLOVIĆ
*Institute of Philosophy,
Ulica grada Vukovara 54, Zagreb, Croatia, EU*
pbodlovic@fcsh.unl.pt

Abstract

What justifies the presumption of innocence? Once the challenger doubts it, how can the proponent defend it? And what are the potential weaknesses of her arguments? In his (2024) paper "Mapping arguments in favor of the presumption of innocence," Michael Hoppmann tentatively answers previous questions. In this commentary, I offer several remarks, ranging from clarificatory suggestions to identifying weaknesses and missed opportunities.

1. Introduction

What justifies the presumption of innocence (henceforth, POI)? What, exactly, makes it acceptable? Once the challenger doubts it, how can the proponent defend it? And what are the potential weaknesses of her arguments? In his (2024) paper "Mapping arguments in favor of the presumption of innocence," Michael Hoppmann answers previous questions. He identifies, interprets, and evaluates eight independent (but mutually connected) arguments supporting POI. The paper belongs to the broader project that aims to develop "a modern stasis model", so Hoppmann's motivation is primarily practical: the stasis model should "equip critical citizens with a tool to judge the fairness of judicial and para-judicial proceedings and to actively speak up for justice".

In this commentary, I summarize Hoppmann's views and offer several remarks, ranging from clarificatory suggestions to identifying gaps and missed opportunities. I believe that the greatest potential for improving the author's research lies in studying more closely the dialectical statuses and logical relations between eight (allegedly) different arguments that support POI. However, Hoppmann has already written an inspiring, well-motivated, and much-needed article.

2. Preliminaries

2.1. Presumption of innocence in discussion

The presumption of innocence instructs agents to proceed as if the accused is innocent—i.e., not to punish the accused—if available evidence does not prove guilt, and there is pressure to decide whether to punish the accused or not. Since it guides agents' actions (instead of beliefs), it is a paradigmatic case of practical presumption (Rescher, 2006; Bodlović, 2021). Importantly, Hoppmann (2024) emphasizes that POI operates in a discussion — typically a "two-party disagreement" — and that its applicability and strength may vary depending on different audiences. To illustrate the discussion's competitive nature, as well as POI's role in the competition, he uses the Apothecary Scale Analogy.

> In the present allegory, these presumptions work like weights that have been screwed to the right arm of the scale, thus tilting it against the protagonist before the discussion even begins. (Hoppmann, 2024)

This analogy nicely explains the central deontic feature of POI: according to traditional accounts, presumptions asymmetrically allocate the burden of proof. Since the weights are initially screwed to the antagonist's (e.g., the defendant's) right arm of the scale, the protagonist (e.g., the accuser) carries the heavier burden of proof: to win the discussion (to tilt the scale towards the left), she must provide stronger arguments (add more weights) than the antagonist. Put technically, a presumption's challenger has the so-called "burden of persuasion" (Prakken and Sartor, 2009) or "macro-level burden of proof" (Hoppmann, 2024).

Comment 1: Competition and evidence accessibility

It is natural to analyze POI in competitive dialogical contexts. But since Hoppmann aims to categorize arguments supporting POI and also evaluate POI's strength, it must be acknowledged that competition includes various aspects and comes in different degrees.
    The criminal trial is extremely competitive. The stakes are high and the cooperation between the defense and prosecution is low. The prosecutor will exploit any weakness in the defender's argument (even by

violating the rules of charitable interpretation), and the defender will not share her evidence if it potentially incriminates her client. As Goodwin (2000) puts it, instead of Grice's Cooperative Principle, discussants follow the Principle of Zeal.

But consider the next example. Jim was recently employed in a delivery center, and some order went wrong: the package was not delivered to the correct address. The manager is looking for an explanation, but she is not planning to punish anyone. She says: "This was a minor mistake without serious consequences. I just want to avoid the same mistake happening again in the future." The order was taken by telephone call. Since this happened during Jim's shift and he is new on the job, the manager accuses him of writing down the incomplete address.

This challenges his reputation as a reliable employee, so the manager must justify her accusation. Jim believes he is innocent, but there is no bad blood between the parties and he trusts the manager's intentions. Crucially, since the manager is not after punishment, Jim is more likely to cooperate with the accuser than the defender in the criminal trial. For instance, he might elaborate on the telephone calls received that day, show where the address was originally written down, or even offer his handwriting sample. In principle, this evidence might help the manager to show she was right, i.e., defeat the presumed claim ("Jim is innocent").

To summarize, Hoppmann's further research might benefit from acknowledging that, in addition to (1) the standard of proof (determined by costs and risks), the presumption's strength also depends on (2) the accessibility of evidence (determined by various factors) (Bodlović, 2022, p. 140-142), and (3) the cooperativeness in dialogue (Goodwin, 2000; Macagno, 2012; Bodlović, 2022, p. 157). In the Delivery case, POI is weaker than in the criminal trial because (1) the manager must meet a lower standard of proof (the stakes are lower), and (2) the potential evidence is more accessible to her due to the (3) dialogue's cooperative nature.

2.2. Motivation for evaluating the presumption of innocence

Hoppmann's research is primarily motivated by the following discrepancy: the presumption of innocence is an old, popular, and highly influential concept, but its justifications—especially those concerning POI's non-standard applications—are difficult to come by. Hoppmann's paper aims to fill this gap. In addition, he distinguishes between three different kinds of reasons motivating his research: (1) theoretical, (2) political, and (3) practical.

For the most part, theoretical reasons concern the relationship between justification and strength. Put simply, to understand why POI has a unique dialogical effect—e.g., why does it place the heavier burden of proof on the challenger's side—we must understand the arguments supporting it. Furthermore, in para-courts, POI is usually just another tool in the politician's hands, and we might want to know if they are using it properly. For instance, the rules of prosecuting transgressions in higher education (associated with the U.S. American Title IX proceedings) change from one administration to another: Obama's administration weakened POI (lowered the standard of proof needed to support accusations), Trump's strengthened it, and the Biden's administration is weakening it, once again. Finally, from the practical viewpoint, studying the foundations of POI might enable citizens to arrive at better judgments about the fairness of various legal and para-legal procedures (Hoppmann, 2024).

Comment 2: Clarifying the dialectics

I believe there is a slight discrepancy between Hoppmann's overall motivation and some of the arguments presented in the second part of his paper. Namely, his research is primarily motivated by the lack of arguments for POI's non-standard applications. In Hoppmann's words:

> [The presumption of innocence's] justification is frequently questioned, and its applicability challenged. These challenges are rarely aimed at paradigmatic cases – say the respectable citizen standing in an ordinary court being accused of murder – but instead focus on chipping away around the corners: Is this presumption really required in minor cases? Or outside of formal courts? For foreigners? And terrorist? ... The answer to all these questions is a resounding "yes!". But unfortunately, the reasons for that answer are not that easy to come by. ... This essay is an attempt to fill part of that gap. (Hoppmann, 2024)

However, many arguments presented by Hoppmann represent POI's initial, general, default justifications, and seem naturally related to standard applications.

For instance, according to the Dialectical Fairness Argument, the accuser must carry a heavier burden to compensate for gaining the dialectical advantage: by making an accusation, the accuser frames the discussion, sets the agenda, and selects the questions that the accused must answer. Thus, it seems fair to restore dialectical balance by burdening the accuser. But rather than a response to some exceptional case, the Dialectical Fairness Argument seems like a general reason that initially justifies POI. In other words, "What if the accused is a potential terrorist?" is more plausibly understood as a response to the Dialectical Fairness Argument than vice versa. I believe that a similar concern applies to the Danger Argument (Walton's Loaded Gun Argument), Veil of Ignorance Argument (Perelman's 100 Argument), Leave me at Peace

Argument, and Just Slander Argument. So, it is questionable whether arguments presented by Hoppmann fill the gap they were meant to fill.[1]

The potential discrepancy highlights the importance of more precise dialectical modeling. It is essential to identify the time when the alleged lack of arguments appears. Is it initially, at t_3, just after the pure challenge:

t_1 Proponent: The accused is presumed innocent until proven guilty.
t_2 Challenger: Why should we presume innocence?
t_3 Proponent: *no initial arguments.*

Or is it later on, at t_5, when some initial argument is already provided, and the Challenger refers to an exception:

t_1 Proponent: The accused is presumed innocent until proven guilty.
t_2 Challenger: Why should we presume innocence?
t_3 Proponent: Because: INITIAL ARGUMENT.
t_4 Challenger: But the accused is a potential terrorist. So, why: INITIAL ARGUMENT?
t_5 Proponent: *no additional arguments.*

Moreover, an analyst should clarify the stage at which arguments (identified by Hoppmann) naturally appear. This is crucial for another goal of the paper: evaluating the strength of arguments that justify the presumption. Take the argument that letting a hundred guilty criminals free is less harmful than convicting one innocent person. Clearly, this argument is more convincing at t_3 than at t_5. It works well in standard cases, but the risk assessment dramatically changes after t_4: if someone is accused of terrorism, not convicting the guilty person might become more harmful than convicting an innocent one.

Accordingly, identifying a dialectical stage where argumentative gaps occur, and specifying the exact roles of available arguments in filling these gaps seems essential to Hoppmann's analysis.

---

[1] In this context, it is somewhat indicative that Hoppmann's later analysis starts with presenting arguments supporting POI, and then continues with discussing defeaters or exceptions. It does not start with exceptions to standard arguments, and continue with discussing further arguments dealing with exceptions.

3. Justifications of presumption of innocence

3.1. Argument analysis: three dimensions

Before exploring the main arguments justifying the presumption of innocence, Hoppmann identifies three dimensions of argument analysis, or "three-dimensional space" (2024) where these arguments can be located.

1. *Argumentative strength*: How compelling is the argument given the kinds of principles it rests upon, as well as associated exceptions and defeaters?
2. *Probative (presumptive) force*: How does the argument influence the dialogical obligations? Should the Challenger carry a lighter or heavier burden of proof to successfully eliminate the presumptive status of the argument's conclusion?
3. *Situational trigger*: What is the corresponding institutional context? For instance, does some argument justify the application of POI in a capital criminal trial, ordinary criminal trial, para-trial, informal three-party accusation, informal peer-to-peer accusation, self-accusation, etc.

Comment 3: Additional aspects of argument analysis

Hoppmann's three-dimensional approach is intuitive, adequate, and well-established in the theory of presumptions (see Godden and Walton, 2007). In future research, however, the proposed analytic matrix might become more sophisticated and detailed.

First, in his recent analysis of justificatory factors, Bodlović distinguishes several considerations affecting the presumption's strength: a) a kind of uncertainty (e.g., expert vs. public disagreement) triggering "Presumably, p," b) the structure of presumptive rule (rebuttable vs. irrebuttable presumptions), c) rule's formal status (is POI an institutional requirement in a criminal trial, or as a regular inferential principle in informal deliberation), d) rule's further justification (does POI protect someone's life, reputation, or mere convenience?), and e) available defeaters (see Bodlović, 2022, pp. 160-167). Furthermore, he explains that, from a dialectical viewpoint, an argument's strength also depends on f) adequacy (does the challenger accept the premises, even if they are objectively weak?), and g) legitimacy (does the relevant majority accept the premises, even if the challenger rejects them?) (2022, p. 136). These considerations should be incorporated in a more refined analysis of the argumentative strength.

Second, various deontic factors affect the presumption's probative force. Clearly, the Challenger's dialectical obligation becomes more demanding if she is a) required to meet the higher standard of proof. But presumption's probative force also increases if the Challenger becomes b) more strictly committed to presenting an argument, [2] c) faces time limitations (it is heavier to defeat "Presumably, p" within minutes rather than weeks), or d) the evidence for defeating "Presumably, p" is not accessible (e.g., since acquiring it would violate moral norms) (Bodlović, 2022, pp. 167-169). Such considerations influence POI's dialogical status and corresponding dialectical obligations.

Finally, Hoppmann's treatment of relevant situational triggers—ranging from formal to informal; from monologues to polylogues (Lewinski and Aakhus, 2014)—is illuminating. But other contextual features must also be included in future research. For instance, Bodlović distinguishes between a) the dialogue's collective goal (POI is strong in the decision-making dialogues, but weak or even inapplicable in epistemic ones), b) the dialogue's cooperativeness (see Comment 1), c) background information (POI's strength varies depending on stakes and background information, related to the different stages of the dialogue, see Comment 2), and d) the intended audience (POI is stronger if challenger shares fundamental values underlying it, despite rejecting the conclusion) (2022, pp. 155-160). Admittedly, Hoppmann's analysis implicitly recognizes many of these factors but might benefit from a more systematic approach.

Comment 4: Is argumentative strength correlated to probative force?

Hoppmann's present paper does not investigate categorical relations between argumentative strength, probative force, and situational triggers. Still, on two occasions, it challenges the usual Correlation Thesis according to which a strong argument for "Presumably, p" implies the high probative force of "Presumably, p."

> [C]ontrary to what Godden and Walton seem to suggest, I do not believe that the argumentative strengths (what they call the "foundation" of a presumption) and the presumptive force are strictly correlated. Yes, forceful presumptions require strong foundations, but the reverse is not true. Strong arguments can establish both, forceful and less forceful presumptions. (Hoppmann, 2024)

---

[2] Being more strictly committed to present an argument is conceptually different from being (equally) committed to present a stronger argument (for details, see Bodlović 2022, pp. 142-145).

> I think a relatively weak presumption (such as the presumption in favor of the status quo) could be established by equality compelling arguments as a very strong presumption (such as the presumption of innocence). (Hoppmann, 2024)

This is an intriguing challenge since the correlation between strength and probative force is taken for granted in the case of practical presumptions (see Godden and Walton, 2007; Ullmann-Margalit, 1983; Rescher, 2006; Godden, 2017; Bodlović, 2022).[3] Unfortunately, Hoppmann does not further develop his intuition in the present paper. In what follows, I set the stage for the criticism of the Correlation Thesis by identifying several conditions that a good counterexample should satisfy.

First, the Correlation Thesis (CT) should be unpacked. In its standard formulation, it comes down to:

> CT: Iff p is uncertain and, potentially, acting on p reduces (in)significantly greater harm than acting on non-p [justification / foundations / argument strength], then the Challenger's standard of proof for defending non-p is (in)significantly higher than the Proponent's standard of proof for defending "Presumably, p." [probative force].

Put simply, challengers carry heavier burdens than the proponents of practical presumptions because, potentially, their mistakes cost more: convicting the innocent is more harmful than not-convicting the guilty. But how strong Challenger's arguments must be depends on the harm assessment in a particular case. For instance, stronger arguments (meeting a higher proof standard) are required in the criminal trial—where the wrong conviction takes away a person's freedom—than in the Delivery Case (where the wrong conviction only endangers a person' reputation). Since Hoppmann challenges the relation from left to right, he is expected to offer the following type of counterexample:

> Potentially, acting on p reduces significantly greater harm than acting on non-p [strong justification], but, nevertheless, the Challenger should meet the weak standard of proof to defend non-p [weak probative force].

It is difficult for me to construct such an example. But my essential question is: If such cases of practical presumptions exist, what explains, then, the low standard of proof? If the presumption's justification in terms

---

[3] Recently, Bodlović (2022, pp. 167-171) argued that this correlation does not work in the case of cognitive presumptions, but did not challenge its plausibility in practical cases.

of uncertainty, cost considerations, and value hierarchies is not the only explanans, then what other variables explain the standard of proof allocation? Sure, there are various situational triggers, and features of a particular dialogue, that might make the Challenger's job easier: more time at her hands, accessible evidence, cooperative interlocutor, etc. But do these features really reduce the *standard of proof* or just an overall difficulty of satisfying it? To use an analogy, many external features might make somebody's run easier—from better weather conditions to weak competitors—but none of them make the final distance any shorter.

In sum, although Hoppmann's rejection of the Correlation Thesis is inspiring, it is currently unsupported. Further research requires a detailed analysis of a concrete counterexample [4] and an explanation of the alternative mechanisms for allocating the standard of proof.

3.2. Arguments justifying the presumption of innocence

In what follows, I briefly present Hoppmann's eight arguments that support POI. I will not systematically present their limitations at this point.

Status Quo Argument

The status quo presumption states that established institutions, beliefs, policies, and practices should not be abandoned without a good reason. It is based on a principle of social economy: societies should avoid

---

[4] Hoppmann briefly compares the presumption of innocence and the status quo presumption. In his opinion, although they are justified by equally compelling arguments, POI is very strong (defeating it requires proof "beyond a reasonable doubt") and the status quo presumption is relatively weak (e.g., defending it requires satisfying the "preponderance of evidence" standard). But this comparison faces many difficulties. First, the two presumptions might have different types of justifications. While cost-management associated with POI includes comparing two distinct values (personal freedom vs. social safety), the cost-management associated with some status quo presumption (e.g., keeping the current tax regulation) can rest on only one value (e.g., avoiding the so-called "transition costs"). It is difficult to compare two practical arguments if they address different issues and are based on distinct considerations. Second, to compare strengths of POI and status quo presumption, an analyst should first instantiate them. What POI, exactly, do we have in mind: the strong POI from the murder trial, or the weak POI from the Delivery case? And is POI compared to the weak status quo presumption concerning current environmental policies ("We should continue treating our environment in the same way") or the strong one concerning the education system ("We should keep investing in our universities"). Finally, it is difficult to compare the two presumptions as independent units since the status quo presumption, to some extent, justifies POI (as Hoppmann acknowledges).

unnecessary "transition costs" (Nebel, 2015), i.e., "spending an avoidable amount of institutional, emotional, social, or financial capital on change without progress" (Hoppmann, 2024). Since POI is a well-established legal rule, the burden is on those trying to initiate an institutional change.

Dialectical Fairness Argument

The Dialectical Fairness Argument rests on the idea that reasonable discussions should promote dialectical equality: in principle, discussants must "have access to the same or similar argumentative moves and strategic choices" (Hoppmann, 2024). But by making an accusation the accuser limits the defendant's options and gets the initial advantage: she sets the tone and frames the discussion by selecting the questions that the accused must answer. POI is justified as a means to restore the dialectical balance: if the accuser gets the initial advantage by setting the agenda, then the accused must get the advantage of, ultimately, carrying the weaker burden of proof.

Comment 5: Fighting for dialectical balance with dialectical means

One objection to the Dialectical Fairness Argument stresses the existence of alternative means. According to Hoppmann, we may restore the balance by allowing the accused to get "assistance from another discussant," have "additional time to prepare," "choose the venue of the discussion," or even by something completely random like making the accuser obliged "to provide food and drink" (Hoppmann, 2024). Then, the question is: Why should we choose POI to promote dialectical equality instead of other means?

This is a fundamental and complex question, but there are, I believe, three general reasons. First, POI provides a reliable and consistent advantage. By contrast, some other means might provide unreliable (e.g., what if the accused chooses the incompetent assistant who gives her bad advice?), or inconsistent advantage (e.g., having more time helps the accused, but, in some circumstances, it also helps the accuser). Second, POI provides a systematic advantage: it applies to all dialogue stages, including the so-called "concluding stage" where parties determine the outcome of the deliberation (after evaluating arguments previously introduced to the discussion) (van Eemeren and Grootendorst, 2004, pp. 61-62). Here, POI resolves the potential uncertainty in the defendant's favor: after argument evaluation, if the guilt is still relatively uncertain, we proceed as if the accused is innocent. Other means—like getting assistance or additional time—might restore dialectical equality during the *exchange* of arguments, but they do not help the accused at the final stage when the decision is being made.

Finally, while various factors may influence the discussion, POI appeals to reason and is argument-centered. On the one hand, it is supported by arguments and, on the other, determines how strong the accuser's arguments must be to justify the accusation. Unlike the rule "The accuser must provide food and drinks," POI restores dialectical equality by dialectical means: requesting stronger arguments from the accuser, partially, improves the discourse. This seems suitable since the accuser's initial advantage is also dialectical: initially, the accuser determines the claims and questions, but, ultimately, the accused can offer weaker arguments in response to these questions and still win the discussion. The accuser gets the advantage of initial selection, the accused gets the advantage of ultimate evaluation criterium.

The Danger Argument

According to the Danger Argument, when we are not sure how to proceed (due to epistemic uncertainty of some claim), we should choose an action that avoids greater potential harm. As Walton illustrates in his well-known example, if we are uncertain whether the gun is loaded or not, we should practically presume (act as if) the gun is loaded. In normal circumstances, proceeding as if the gun is loaded (when it is empty) is less dangerous than proceeding as if the gun is empty (when it is loaded): it avoids the risk of fatal injuries. The presumption of innocence has a similar justification: if we are uncertain whether the accused is innocent or guilty, we should proceed as if she is innocent. Convicting the innocent is worse than not-convicting the guilty, so POI avoids greater harm.

Comment 6: Is the Danger Argument just a figurative analogy?

Hoppmann admits that the Danger Argument is persuasive, but states that it "is little more than a figurative analogy" (2023, p. 18) with "little probative value" (p. 12). I believe this interpretation is misleading.

As I see it, the Loaded Gun example and many others[5] are not meant to provide a foundation of an analogical argument. Rather, there are meant to illustrate a piece of practical reasoning. Accordingly, POI is not (un)justified *because* the Loaded Gun inference is (un)justified: both POI and Loaded Gun are (un)justified because they instantiate a special form of practical reasoning, a particular argument scheme—probably some variation of the "negative practical reasoning" or "practical ad ignorantiam" (Walton, Reed, and Macagno, 2008, pp. 99-100). In effect, the probative value of POI will depend on different elements of this scheme—e.g., evidential uncertainty, time pressure, the precautionary principle,

---

5 For instance, the Ladder example (Dare and Kingsbury, 2008), the Coil of Rope example (Walton, 2014), the Party case (Bodlović, 2020), the Umbrella case (Bodlović, 2021), etc.

policy of avoiding costly errors, normative inequality, etc. (Bodlović, 2021, pp. 294-295)—once they are applied to the special case of POI.

In short, the Loaded gun example might help us to understand POI's justification in simple and everyday terms but is not directly relevant to assessing POI's force. Rather than a weak figurative analogy, it provides a good illustration of a plausible argument scheme.

The Veil of Ignorance Argument

The Veil of Ignorance Argument—or, as Hoppmann calls it "Perelman's 100 Argument"—shows that POI satisfies Rawls' famous fairness criterium. Hypothetically, when accused and under the veil of ignorance, a reasonable decision-maker will choose to live in society A (where innocent defendants are protected, but many guilty escape punishment) rather than society B (where many innocent defendants are punished, but guilty rarely escape punishment).

Kant's Categorical Argument

Kant's categorical imperative is a famous (deontological) principle of moral action. It demands that our actions always treat other people as ends, and never merely as means. It coheres well with POI because the latter protects innocent people and ignores considerations that, from some consequentialist perspective, might justify punishing innocent defendants.

Hyper-Villain Argument

Hoppmann's Hyper-Villain Argument explores the relationship between society and the accused. Societies must follow procedures that systematically avoid unnecessary suffering, e.g., reduce the number of undeserved punishments. Choosing such procedures is challenging since, sometimes, we face both moral and epistemic uncertainty: social institutions must assess different moral costs (convicting the innocent vs. not-convicting the guilty) without knowing which cost is (significantly) more likely (is the accused more likely innocent, or guilty). Hoppmann believes that society avoids greater harm by choosing POI. Choosing the alternative rule—that, in uncertainty, systematically punishes innocent defendants—would turn the society into Hyper-Villain: the society would cruelly punish innocent citizens without good reasons, and endanger some reasonable and moral practices (e.g., self-defense).

Leave me at Peace Argument

The Leave me at Peace Argument focuses on the relationship between the accuser and society. It stresses that accusers disturb social peace and spend social resources: e.g., processing accusations in courts costs money and time. Since POI makes it difficult to prove guilt, it is a good means to deter hasty accusations and reduce social costs.

Just Slander Argument

Just Slander Argument focuses on the relationship between the accuser and the accused. By accusing, the accuser damages the defendant's reputation and automatically generates harm. Hence, the defendant must be protected by the presumption: unless the accuser provides very strong reasons to the contrary, she must be considered innocent.

4. Comment 7: Bringing it all together

Due to his practical motivation, Hoppmann treats previous arguments as independent units and does not attempt to closely study their mutual relationships. However, it is surely practically beneficial to investigate broader categories these arguments belong to, as well as their logical connections. For instance, critical citizens might want to know that POI has two main sources of justification:

(1) fairness (justice, moral) considerations,
(2) pragmatic (cost-oriented) considerations.

Also, it is important to know that these considerations are not independent. As the red lines in Figure 1 illustrate (see below), some moral (fairness) considerations strengthen pragmatic arguments, and vice versa. On the one hand, Kant's Categorical Argument might provide a reason why it is morally costlier to punish the innocent (than let the guilty person free): it may justify the basic premise of the Danger Argument. On the other hand, the moral cost of punishing the innocent—the crucial premise of the Danger Argument—may explain why reasonable deliberators choose POI under the veil of ignorance, thereby becoming relevant to the argument concerning social fairness.

Also, Figure 1 clearly shows that the Danger argument lies at the heart of POI's justification, providing the foundation for the Hyper-Villain Argument and the Veil of Ignorance Argument. These arguments have distinct flavors and emphasize different aspects, but they are logically intertwined and should be analyzed accordingly. Studying the logical connections between various arguments does not threaten Hoppmann's pragmatic agenda. In my opinion, it can only help critical citizens who wish to evaluate POI or similar justice-related practices.

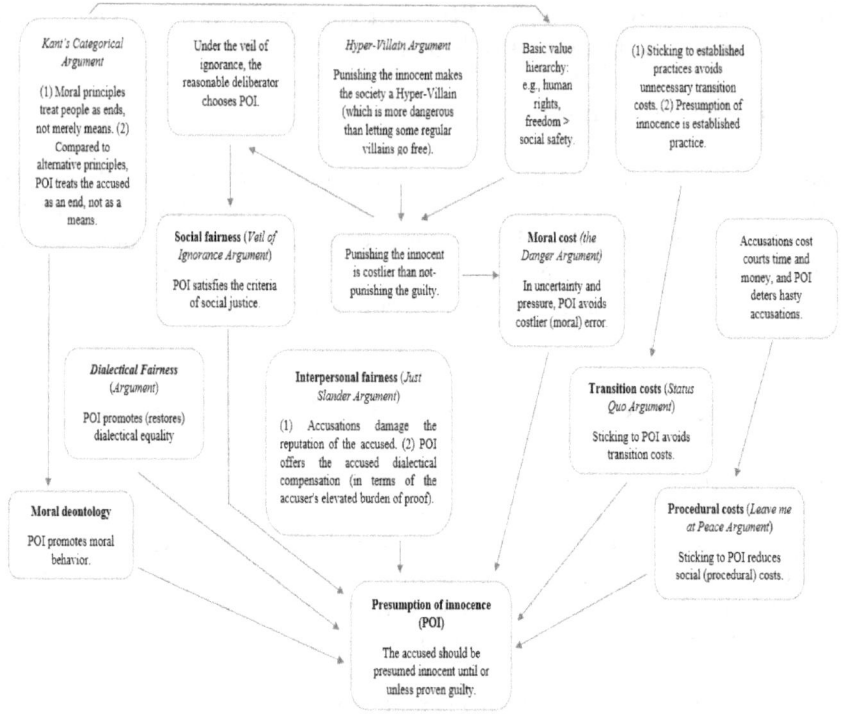

Figure 1: Hoppmann's arguments supporting POI (overview and connections)

5. Conclusion

I tried to identify some gaps and weaknesses, in order to offer suggestions on how to improve the analysis. I believe that the greatest potential lies in (1) clarifying the dialectics (at what dialogical stage does some argument occur?), (2) studying more closely the logical connections between arguments, (3) re-assessing the nature of particular examples (e.g., Walton's Loaded gun example), and (4) further justifying theoretical claims (e.g., the lack of correlation between strength and probative force). Nevertheless, even if all these suggestions are on point, Hoppmann has already written the important paper that anyone interested in the study of presumptions should read.

References

Bodlović, P. (2020). On Presumptions, Burdens of Proof, and Explanations. Informal Logic, 40(2), 255-294.
Bodlović, P. (2021). On the differences between practical and cognitive presumptions. Argumentation, 35, 287-320.
Bodlović, P. (2022). Presumptions in Argumentation. A Systematic Analysis. University of Groningen (doctoral thesis).
Dare, T., & Kingsbury, J. (2008). Putting the Burden of Proof in Its Place: When Are Differential Allocations Legitimate? The Southern Journal of Philosophy, XVI, 503-519.
Godden, D. (2017). Presumption as a Modal Qualifier: Presumption, Inference, and Managing Epistemic Risk. Argumentation, 31(3), 485-511.
Godden, D., & Walton, D. (2007). A theory of presumption for everyday argumentation. Pragmatics & Cognition, 15(2), 313-346.
Goodwin, J. (2000). The noncooperative pragmatics of arguing. In E. Nemeth (Ed.), Pragmatics in 2000: Selected papers from the 7th International Pragmatics Conference (Vol. 2) (pp. 263-277). Antwerp: International Pragmatics Association.
Hoppmann, M. (2024). Mapping arguments in favor of the presumption of innocence. Conference paper, European Conference of Argumentation (ECA), 28-30 September 2022, Rome, Italy (pp. 1-30).
Lewinski, M., & Aakhus, M. (2014). Argumentative Polylogues in a Dialectical Framework: A Methodological Inquiry. Argumentation, 28, 161–185.
Macagno, F. (2012). Presumptive Reasoning in Interpretation. Implicatures and Conflicts of Presumptions. Argumentation, 26, 233-265.
Nebel, J. (2015). Status Quo Bias, Rationality, and Conservatism about Value. Ethics, 125, 449–476.
Prakken, H., & Sartor, G. (2009). A Logical Analysis of Burdens of Proof. In H. Kaptein, H. Prakken, & Verheij, B. (Eds.), Legal Evidence and Proof: Statistics, Stories, Logic (pp. 223-253). Farnham: Ashgate Publishing (Applied Legal Philosophy Series).
Rescher, N. (2006). Presumption and the practices of tentative cognition. Cambridge: Cambridge University Press.
Ullmann-Margalit, E. (1983). On presumption. Journal of Philosophy, 80(3), 143-163.
van Eemeren, F.H., & Grootendorst, R. (2004). A Systematic Theory of Argumentation. The Pragma-Dialectical Approach. Cambridge: Cambridge University Press.
Walton, D. (2014). Burden of proof, presumption and argumentation. New York: Cambridge University Press.
Walton, D., Reed, C, & Macagno, F. (2008). Argumentation schemes. New York: Cambridge University Press

DEVELOPING AN INVITATIONAL APPROACH TO PRAGMA-DIALECTICS

BROOKE HUBSCH
The Pennsylvania State University
bstanley5@wisc.edu

Abstract
In this article, I argue that the potential value of invitational rhetoric to argumentation studies has been overshadowed by the introduction of invitational rhetoric as a discrete alternative to persuasion. Invitational rhetoric offers a practical guide for discursive participants to fully articulate and understand each other's beliefs, perspectives, and standpoints. Rather than considering this understanding as the end in and of itself, invitational rhetoric might be better positioned as a supplement to the confrontation and opening stages of Pragma-Dialectics. Extended confrontation and opening stages reconceptualized through the guidelines of invitational rhetoric would allow for greater externalization of standpoints by participants, aiding both in fully understanding the stance of their opposition prior to a move to the argumentation stage. I consider potential tensions between the norms of invitational rhetoric and the rules of Pragma-Dialectics and how they might be resolved through modifications to the original theory of invitational rhetoric. While this move runs contrary to the motivations of many invitational rhetoric's advocates and requires the abandonment of some of its tenets that violate Pragma-Dialectics' commitment to reasonableness, it offers an avenue for invitational rhetoric to offer broader practical, discursive value.

1. Introduction

In this article, I argue that invitational rhetoric should be reconsidered as an addition to the confrontation and opening stages of Pragma-Dialectics not as an alternative to persuasive rhetoric. Invitational rhetoric was introduced by Foss and Griffin (1995) as a new strategy for ethically engaging in discourse that was grounded in feminist theory. However, invitational rhetoric has not been widely adopted in rhetorical studies due to resistance from both traditional rhetoricians and critical theorists. By introducing invitational rhetoric as an alternative to persuasion and positing persuasion as inherently (or largely) violent, Foss and Griffin (1995) faced significant backlash due to their characterization of the history of persuasive rhetoric. On the other hand, some theorists critiqued invitational rhetoric for not pushing far enough, advocating norms of civility and openness that were unlikely to result in material change due to structural inequities in power among discursive participants. While there continues to be a small, dedicated group of scholars examining the value of invitational rhetoric, particularly with regards to studies of pedagogy, support groups, and research focus groups, invitational rhetoric has had limited uptake in modern communication theory.

However, I argue that the potential value of invitational rhetoric to argumentation theory has been overshadowed because Foss and Griffin (1995) introduced invitational rhetoric as a discrete alternative to persuasion. Invitational rhetoric offers a practical guide for discursive participants to fully articulate and understand each other's beliefs, perspectives, and standpoints. Rather than considering this understanding as the end in and of itself, invitational rhetoric might be better positioned as a supplement to the confrontation and opening stages of Pragma-Dialectics. Pragma-Dialectics was introduced by van Eemeren and Grootendorst (1992) as a model for practical argumentation aimed at resolving differences of opinions through a dedication to reasonableness. The confrontation and opening stages of Pragma-Dialectics are dedicated to determining whether there is a difference of opinion between discursive participants and establishing shared premises and rules. Only after the conclusion of the opening stage are participants asked to defend and attack standpoints. Extended confrontation and opening stages reconceptualized through the guidelines of invitational rhetoric would allow for greater externalization of standpoints by participants, aiding both in fully understanding the stance of their opposition and determining whether a move to the argumentation stage is necessary. I consider potential tensions between the norms of invitational rhetoric and the rules of Pragma-Dialectics and how they might be resolved through modifications to the original theory of invitational rhetoric. While this move runs contrary to the motivations of many invitational rhetoric's advocates and requires the abandonment of some of its tenets that violate Pragma-Dialectics' commitment to reasonableness, it offers an avenue for invitational rhetoric to offer broader practical, discursive value.

2. A Brief Overview of Invitational Rhetoric

Sonja Foss and Cindy Griffin (1995) first introduced invitational rhetoric as a response to perceived patriarchal bias in rhetoric, specifically the definition of rhetoric (wholly or primarily) as persuasion. Invitational rhetoric was meant to be a feminist alternative to persuasion in which participants offered perspectives with the goal of inviting others to understand them rather than attempting to "dominate" and change an audience. This alternative rhetoric was based in feminist values of equality (the elimination of dominance and elitism), immanent value (the inherent worth of every being), and self-determination (the freedom of all to decide for themselves how to believe and act). It offered potential to communication theory as a new avenue for understanding ideas and people through the creation of an atmosphere in which all participants felt welcome and free from judgment, encouraging greater self-disclosure. However, while invitational rhetoric as a practice and a theory has been picked up over the last several decades in some spheres, including research on focus groups (Johnson et al., 2021), feminist blogs (Hayden, 2021), and parent-adolescent communication (Pariera and Turner, 2020) (for a comprehensive overview of modern invitational theory, see Foss and Griffin, 2020), it has also faced significant criticism and failed to gain a larger audience within rhetorical theory. Some barriers against a wider adoption of the theory have been Foss and Griffin's (1995) account of the history of rhetoric and invitational rhetoric's narrow applicability and force.

While Foss and Griffin (1995) do admit that persuasion is often necessary, this statement is often overshadowed by their introduction to "Beyond Persuasion: A Proposal For An Invitational Rhetoric," in which they frequently rely on Gearhart (1979) to argue that "embedded in the efforts to change others is a desire for control and domination" (p. 3), that changing others "devalues the lives and perspectives of those others" (p.3), and that the traditional conception of rhetoric writ large is characterized by patriarchy and violence. Through this introduction, Foss and Griffin (1995) set up an opposition between feminist invitational rhetoric and violent, patriarchal persuasion which became the subject of significant criticism for defenders of persuasion. Pollock et al. (1996) argued that Foss and Griffin (1995) and Gearhart (1979)'s definition of persuasion is essentialized, stating that superior and oppressive attitudes are neither necessary nor inherent to persuasion and offering counterexamples of persuasion motivated by care and love. Condit (1997) took a more forceful approach, stating that the 1995 essay "condemns the entire history of the discipline of rhetoric in a single page" that the "reductive account does not adequately represent the diversity of rhetorical theory" (p. 100-101). Condit (1997) supported this argument with examples from eras of rhetorical theory in which persuasion was seen as the "feminine" alternative to "masculine" violence. Jørgensen (2007) critiqued both Gearhart (1979) and Foss and Griffin (1995), stating that the linkage of persuasion and violence only held if we depict the rhetor as all an powerful force speaking to an audience without free will, a "'hypodermic theory' of communication that we all are supposed to have left behind us" (p. 166). These critiques reflect a broader resistance to invitational rhetoric among rhetoricians who view persuasion as a valuable resource that lacks an inherent violent quality.

However, even among those more sympathetic to a violent characterization of persuasion, further critique came due to invitational rhetoric's narrow applicability and passivity. Pollock et al. (1996) argued that giving up on advocacy (as is required by invitational rhetoric) "requires a luxury of time, status, and circumstance not always available to those who are underresourced, underrepresented, and oppressed" (p. 150). Lozano-Reich and Cloud (2009) charged Foss and Griffin (1995) with prioritizing understanding and dialogue over material change for the oppressed, arguing that incivility and confrontational (or even violent) tactics (including persuasion) are the greatest resource for combatting discrimination and achieving justice. This argument was furthered by Chick (2021), who claimed that invitational rhetoric privileges the already privileged, placing an undue burden on the already marginalized to suppress feelings and maintain civility. He, like Lozano-Reich and Cloud, critiqued Foss and Griffin's (1995) moralizing of invitational rhetoric, a move that he believed excluded violent and confrontative rhetoric from being an acceptable response to injustice. Meyer (2007), while seeing the benefits of a feminist rhetoric, argued that replacing forceful terms such as "persuasive" or "convincing" with "invitation" and "offering" introduced a passivity in women's rhetoric that "robs women of their power of intent in rhetorical construction" (p. 10). Dow (1995) cautioned against a utopian rejection of tools of power, privilege, and patriarchy such as persuasion, warning that this allows men to monopolize more forceful strategies and prevent feminists from creating institutional change. This line of criticism is best summarized by Bruner (1996), who says, "if feminist argumentation theory assumes that one cannot constrain and enable at the same time, or nurture and at the same time seek to change the perspective of another, then feminist argumentation is limited to a very narrow range of argumentative situations" (p. 187).

Across critiques both from traditional rhetoricians defending persuasion and critical rhetoricians advocating for occasionally violent tactics to achieve social justice is a rejection of invitational rhetoric's goal of understanding for understanding's sake and its position as a replacement for persuasion. From the feminist side, one attempt to bolster invitational rhetoric's value came from Ryan and Natalle (2001). They argued that a fundamental weakness in invitational rhetoric was the contradiction between the feminist values of self-determination and imminent value and invitational rhetoric's goal of dialogue. According to Ryan and Natalle (2001), the subjective epistemological stance of self-determination and imminent value is that "knowing" is an internal process, coming solely from self-reflection and entails a wholesale rejection of external authority. This position is decidedly at odds with Foss and Griffin's (1995) use of invitational rhetoric as a way to know and understand new perspectives through engagement with others. To resolve this tension, Ryan and Natalle (2001) fuse invitational rhetoric with feminist standpoint theory (the belief in structural power differentials between people that create situated knowledge, to be understood by reflecting on one's positionality) and philosophical hermeneutics (understanding through the questioning the external world and its implicit traditions) to create a new, less contradictory, feminist rhetoric.

Rather than continue this project, I argue that invitational rhetoric can also be recovered to support the non-violent characterization of persuasion supported by traditional rhetoricians by binding it to Pragma-Dialectics. The challenge with a practical theory of invitational rhetoric (as was noted above) is that a rhetoric whose end goal is understanding for its own sake lacks applicability beyond a narrow set of situations. However, while Foss and Griffin (1995) treat invitational rhetoric as a discrete alternative to persuasion, invitational rhetoric is in no way bound to that distinction. In Bone, Griffin, and Scholz's (2008) response to critics of invitational rhetoric, they stated that, "like all forms of argument, [...] invitational rhetoric often is woven into a public argument that contains moments of informing, persuading, and inviting" (449). I seek to weave invitational rhetoric into public argument a bit more tightly than was perhaps intended by Bone, Griffin, and Scholz (2008) by theorizing its contributory value for the confrontation and opening stages of Pragma-Dialectics.

3. Similarities Between Invitational Rhetoric and Pragma-Dialectics

One of the obvious hurdles to this project is the seemingly mutual opposition between invitational rhetoric and Pragma-Dialectics. Invitational rhetoric is based in feminist theory, rejects any and all deliberate attempts at changing the mind of others, prioritizes relationality over reason, and seeks only the understanding and recognition of inherently valuable others. In contrast, Pragma-Dialectics finds its bearings in critical-rationalism, prioritizes depersonalized, reason-based argumentation, and is oriented wholly towards the testing ideas based on their merits to determine whether an opponent's mind needs to be changed. Pragma-Dialectics is unlike the traditional rhetoric of persuasion targeted by Foss and Griffin (1995) because the intention of Pragma-Dialectics is not to change an opponent's mind through rhetoric, but rather to test the viability of a standpoint through argumentation. However, if the merit is determined to align with the protagonist, this does entail that the antagonist must then change their mind.

Despite these differing orientations, there are threads of connective tissue between both communicative theories. Perhaps the most promising potential for connection between invitational rhetoric and Pragma-Dialectics is the fact that in their original article, Foss and Griffin (1995) specifically call out the similar aims between their theory and the new model of argumentation introduced only three years prior by van Eemeren and Grootendorst (1992): Pragma-Dialectics. In their discussion of implications, Foss and Griffin (1995) laud van Eemeren and Grootendorst's (1992) model of argumentation that "is designed to create an open and free exchange and responsible participation in cooperative, dialogic communication" (p. 15). Implicit in this suggestion is the view of Foss and Griffin (1995) that invitational rhetoric and Pragma- Dialectics are not so mutually opposed as to prevent the possibility for collaboration between the two theories. However, Foss and Griffin (1995) suggest that the value invitational rhetoric has to offer Pragma-Dialectics is not in aiding the oppositional model van Eemeren and Grootendorst (1992) created, but rather as a way to abandon models of communication characterized by opposition and competition. This potential for collaboration between the two models was not picked up in the decades following Foss and Griffin's (1995) original article.

Both invitational rhetoric and Pragma-Dialectics are concerned with theorizing and guiding practical conversations between parties who hold different perspectives. Bone, Griffin, and Scholz's (2008) inform that "when rhetors use invitational rhetoric their goal is to enter into a dialogue in order to share perspectives and positions, to see the complexity of an issue about which neither party agrees" (p. 436), a statement that is similar to van Eemeren and Grootendorst's (2004) description of argumentation as an invariable "part of an exchange of views between two parties that do not hold the same opinion" (p. 59). In order to facilitate this exchange of views, Foss and Griffin (1995) state that "no subject matter is off limits, and all presuppositions can be challenged" (p. 12) and that speakers using invitational rhetoric "communicate a willingness to call into question the beliefs they consider most inviolate and to relax their grip on those beliefs" (p.7). Despite invitational rhetoric's opposition to intentions to change, this standard closely mirrors Pragma-Dialectics' first rule for critical discussion, which grants an unconditional right to discussants, stating: "standpoints can refer to anything, and that in principle, every standpoint can be called into question" (van Eemeren and Grootendorst, 2004, p. 165). Both invitational rhetoric and Pragma-Dialectics impose rules to ensure equal treatment of discussants, with Foss and Griffin (1995) rejecting traditional rhetoric for its supposed superiority of a rhetor over the audience, instituting in its place "a commitment to the creation of relationships of equality and to the elimination of [...] dominance and elitism." van Eemeren and Grootendorst (2004) hold in rule 1 that "no special preparatory conditions apply to the position or status of the speaker or writer and listener or reader" (p. 136) and outlaw any directives such as orders or prohibitives that would allow one participant to restrict the other's ability to engage freely. While Pragma-Dialectics does ultimately prioritize argumentation and change (in cases where a standpoint is successfully defended or refuted), two of the four stages of Pragma- Dialectics are not meant to contain any argumentation or attempts to critique the ideas of one's opponent, but rather are meant to establish whether two (or more) participants have different views, what common ground exists between participants, and which conventions they will use and positions they will adopt to engage in a dialogue to test their differing views (van Eemeren and Grootendorst, 2004).

Beyond comparing the guidelines of each theory, there is one particularly useful example of what invitational rhetoric looks like in the real world that is offered by Bone, Griffin, and Scholz (2008). As one of their four case studies meant to counter claims that invitational rhetoric is "utopic" or "does not exist in any meaningful way in the 'real world'" (p. 448), Bone, Griffin, and Scholz (2008) describe a series of dialogues between the pro-life and pro-choice movements conducted by the Public Conversations Project following a Boston-area incident where John Salvi opened fire on two Planned Parenthood clinics. Participants established conventions and ground rules covering what they would like to be called and how to label their positions, "challenged each other to dig deeply, defining exactly what we believe, why we believe it, and what we still do not understand" (p. 453), and agreed "to listen openly and speak candidly and to allow their ideas to be 'challenged, but not attacked'" (p. 453). While Bone, Griffin, and Scholz (2008) use this case study to detail a successful case of invitational rhetoric, it could just as easily be used as a series of examples of the opening and confrontation stages of Pragma-Dialectics in which participants establish their differences of opinion, set conventions for how to resolve the discussion, define their terms, and discuss their views to determine what common ground exists between the participants.

4. What Does Invitational Rhetoric Offer Pragma-Dialectics?

When establishing the guiding principles for Pragma-Dialectics, van Eemeren and Grootendorst (2004) argue that the rules for critical discussion must be conducive [...] above all to the optimal externalization of differences of opinion" (p. 156). The importance of externalization (one of the four meta-theoretical principles of Pragma-Dialectics defined by the creation of a public commitment through speech acts) is repeatedly addressed by van Eemeren and Grootendorst (2004). They state that "a difference of opinion that is only partly externalized, or not externalized at all, does not make having a discussion superfluous, but it does make it difficult" (van Eemeren and Grootendorst, 2004, p. 135). In Pragma-Dialectics, the rules each participant is bound by depend on the speech acts each participant performed. Participants are only bound to commitments to defend their views after making an explicit statement regarding their standpoint or challenge. Without clear and explicit speech acts, it becomes difficult to determine which rules and obligations bind each participant. Further, van Eemeren and Grootendorst (2004) state that "a difference of opinion can never be resolved if it is not clear to the parties involved that a difference of opinion exists and what that difference entails" (p. 190). Unless discussants clearly express their intentions and try to interpret those of their opposition accurately, faulty interpretation can lead discussants to either establish a pseudo-difference or a pseudo-solution (van Eemeren et al., 2014). Discussants are always granted the right in every stage of discourse to use or request that their opposition use usage declaratives (such as definitions, specifications, amplifications, and explanations) in order to lessen the likelihood that discussants will misinterpret each other's speech acts or speak at cross-purposes by revealing spurious differences or clarifying rules or vague premises (van Eemeren and Grootendorst, 2004).

However, despite the importance of optimal externalization in ensuring that differences of opinion are substantive and understood by both parties, there is potential for these differences to remain implicit during the first two stages of Pragma-Dialectics. In the confrontation stage, the primary goal is to establish whether there is a difference of position between two (or more) parties. However, as van Eemeren and Grootendorst (2004) note, while discussants can explicitly state their difference of opinion, in practice, discussants may proceed only on the assumption that a difference of opinion exists (or that there is a possibility of difference), never outright stating their positions. Further, in the opening stage, discussants are meant to identify what common ground they share relevant to the debate (including rules for the discussion as well as facts, truths, norms, values, and value hierarchies that can act as shared premises). Shared rules and premises are vital to a successful debate, without which van Eemeren and Grootendorst (2004) argue there is "no point in venturing to resolve a difference of opinion through an argumentative exchange" (p. 60). However, at the same time, they state that in a "great many cases, the opening stage of an argumentative exchange of views will remain largely implicit, because it is generally tacitly assumed that the required common ground exists" (van Eemeren and Grootendorst, 2004, p. 60-61). Taken together, this means that in at least some argumentative exchanges following Pragma-Dialectics' model, discussants proceed with argument based only on presumed or potential differences of opinions and with assumptions of common ground regarding what they know, value, and believe that may not, in reality, be shared grounds. Implicit differences or grounds may not be large hurdles for some discussants, but when considering (for example) trying to tackle a reasonable approach at resolving differences between the pro-life and pro- choice movements during the dialogues following the Planned Parenthood shootings, the need for a clear, painstaking assembly of facts, truths, norms, values, and value hierarchies subscribed to by participants and the need for the identification of how those premises align and differ cannot be overstated.

While van Eemeren and Grootendorst (2004) attempt to lessen discussants' reluctance to commit to certain premises or standpoints by instituting the rule that "participants in a discussion may not prevent the other party in any way (verbal or non-verbal)" (p. 165) from advancing or questioning standpoints, the reality of practical disagreement is that the fear of stigma, ridicule, or isolation (from threats real or imagined) will prevent many potential discussants from expressing standpoints or making explicit the facts and norms they believe in. Be it in a classroom, political debate, or conversation with friends and family, differences of opinion that might otherwise be resolved through reasonable argumentation are buried or only partially externalized due to (often well-founded) concerns about how an honest account of one's position will influence their image in the eyes of others, leading to unrealized differences, pseudo-solutions that don't actually address the difference, or unproductive argumentation hindered by a failure to understand what the other person believes, says, and is committing to. It is in these cases that invitational rhetoric may have something to offer to Pragma-Dialectics.

Invitational rhetoric is specifically tailored to guiding discussants to disclose their views "as carefully, completely, and passionately as possible to give them full expression and to invite their careful consideration by the participants in the interaction" (Foss and Griffin, 1995, p. 7). Because invitational rhetoric is framed as pursuing understanding for its own sake, it encourages self-disclosure without advocating support or adapting one's disclosure to the resistance anticipated from the audience. For differences of opinion in which participants are likely to anticipate great resistance or backlash (even that backlash unexpressed by their opponent), invitational rhetoric can offer tools to Pragma-Dialectics for the confrontation and opening stages that encourage more complete disclosure. Invitational rhetoric can assist in creating a more significant divide between the first and second halves of a dialectic, where the confrontation and opening stages are devoted to understanding and externalization (in a more invitational fashion) and the argumentation and concluding stages are devoted to changing the mind of one's opponent. This differentiation may facilitate a more optimal externalization and a more productive resolution of differences.

Invitational rhetoric offers several strategies to Pragma-Dialectics that may foster more complete confrontation and opening stages of discourse. These strategies come in two categories: those intended to guide how individuals disclose their perspective and those intended to foster an environment that encourages individuals to disclose (Foss and Griffin, 1995). The first category is described as offering perspectives. In offering perspectives, speakers are asked to share their point of view in a longer form during with they "tell what they currently know or understand; they present their vision of the world and show how it looks and works for them"

(Foss and Griffin, 1995, p. 7). Rather than engaging in a persuasive case for their point of view, rhetors offering their perspective are asked to articulate their viewpoint in a way that is maximally comprehensible by those that differ from them. A speaker is only considered to be engaging in offering if they are "sharing what they know, extending one another's ideas, thinking critically about all the ideas offered, and coming to an understanding of the subject and of one another" (Foss and Griffin, 1995, p. 8). Audience members are welcome to ask questions or make comments with the goal of "learning more about the presenter's ideas, understanding them more thoroughly, nurturing them, and offering additional ways of thinking about the subject" but they are not allowed to ask questions or make comments designed to show the stupidity or error of the perspective (Foss and Griffin 1995 p. 8). Finally, when a participant does offer their perspective, "the perspective presented through offering represents an initial, tentative commitment to that perspective – one subject to revision as a result of the interaction" (Foss and Griffin, 1995, p.8).

Offering perspectives is a different rhetorical framing to the kind of dialogue one might hope to see during the confrontation and opening stages of Pragma-Dialectics. Participants clearly and completely articulate their views, invite others to ask questions or engage from different perspectives to reveal the complexity and range of standpoints one might have on an issue, and by articulate their perspectives, and through this process, they are committing to the standpoints expressed (though in a less rigorous fashion than is required by Pragma-Dialectics). However, the foregrounding of understanding within offering perspectives is meant to prioritize disclosure over argumentation or anticipated resistance, which could promote greater externalization during the confrontation and opening stages.

However, the greater factor influencing whether this offering of perspectives happens is the atmosphere developed by participants. Foss and Griffin (1995) argue that for invitational rhetoric to facilitate mutual understanding, it must create three external conditions: safety, value, and freedom. For speakers and audience members to feel safe, they must believe that their ideas and feelings will be received with respect and care without being subject to harm, degradation, belittlement, retribution, or rebuttal. These beliefs can be fostered through mutual commitment by participants to these behaviors (and repeated rounds of successful adherence). However, a clear issue with this guideline is that, within Pragma-Dialectics, no standpoint is protected from rebuttal. A resolution between these two standpoints is to give discussants freedom from rebuttal at this time. Discussants should be granted the room to express themselves fully and without critical response when first establishing the difference of opinion and common ground to encourage maximal disclosure. Other participants are allowed to express different opinions during this time, but a full rebuttal should be suspended until after the opening stage as ended, something already intended by the pragma-dialectic model.

The second external condition is value, derived from Benhabib's (1992) principle of universal moral respect. Participants must be treated as having intrinsic worth and recognized as unique and irreducible by their audience (Foss and Griffin, 1995). One strategy for fostering this condition is through absolute listening, in which "listeners do not interrupt, comfort, or insert anything of their own as others tell of their experiences" (Foss and Griffin, 1995, p. 11). By granting speakers both the respect and space to share their view without interruption from potential opposition, those speakers may feel less risk of backlash from other discussants and be willing to disclose more fully due to their felt value. Further, the audience is encouraged to engage with Benhabib's (1992) "reversibility of perspectives" (p. 145), where "they reason from the standpoint of others" (Foss and Griffin, 1995, p. 12). By teaching an absolutely listening audience to reason from the other's perspective, this external condition facilitates potential opposition to fully hear the disclosures that the discussant does make and understand them more completely, minimizing the potential for misunderstanding.

The final external condition is freedom, which carries with it a few aspects. First, as with Pragma-Dialectics, discussants must be free to engage with any and all matters, any standpoint can be challenged, and there is no special status granted to the discussant over the audience. A second aspect of invitational freedom is encouraging "audience members to develop the options that seem appropriate to them, allowing for the richness and complexity of their unique subjective experiences. Perspectives are articulated as a means to widen options – to generate more ideas than either rhetors or audiences had initially" (Foss and Griffin, 1995, p. 12). Beyond merely making participants comfortable with expressing the standpoints they already believe, the value of a prolonged confrontation stage facilitated by invitational rhetoric is its potential for unearthing richer differences of opinion to engage with. Rather than proceeding from the first difference discovered, an extended opening stage might allow both discussants to reveal deeper and more engaging differences that neither had independently identified. It is the third aspect of the freedom condition that is a bit more difficult to incorporate: it should make no difference to the discussant whether the audience accepts or rejects their standpoint. For Pragma-Dialectics, it is precisely the externalized acceptance or rejection of a standpoint that structures the rest of the dialogue, at the end of which the protagonist must retract their standpoint or the antagonist must accept the standpoint (van Eemeren and Grootendorst, 2004). However, when describing the standard of "makes no difference" (p.12), Foss and Griffin (1995) seem to prioritize the relationships between a speaker and the audience. The speaker is meant to give unconditional positive regard to the audience, continuing to respect and engage with them regardless of their adherence to the speaker's standpoint. For the first two stages of Pragma-Dialectics, meeting this threshold causes no tension, as neither stage entails a forced change in perspective when the antagonist does not adhere to the protagonist. Both participants can treat each other with unconditional positive regard while expressing differing standpoints. It is only after leaving the opening stage that the protagonist or antagonist must endeavor to test the viability of a standpoint to determine who in the exchange must change their mind, at which point (under this model) both parties switch from an invitational frame of understanding to an argumentative frame of viability.

5. Challenges to Incorporating Invitational Rhetoric with Pragma-Dialectics

Despite the potential for contribution invitation rhetoric has to Pragma-Dialectics, there are some aspects of invitational rhetoric that cannot be incorporated easily with Pragma-Dialectics' guiding standard of reasonableness, namely the feminist principles of equality, immanent value, and self-determination that serve as the theoretical core of invitational rhetoric. Equality rejects any efforts to dominate or gain power over others (which Foss and Griffin (1995) believe persuasion and argumentation are intended to do). Immanent value requires that every individual is treated as a "unique and necessary part of [...] the universe" which entails "the eschewal of forms of communication that seek to change that individual's unique perspective" (Foss and Griffin, 1995, p. 4). Self-determination requires that speakers allow audiences to "constitute their worlds as they choose" without interference from the speaker (Foss and Griffin 1995, p. 4). Adherence to these principles of feminism appear to be at odds with the critical-rationalist view adopted by Pragma-Dialectics, which "proceeds on the basis of the fundamental fallibility of all human thought. To critical rationalists, the idea of a systematic critical scrutiny of all fields of human thought and activity is the principle that serves as the starting point for the resolution of problems" (van Eemeren and Grootendorst, 2004, p. 131). This approach all but requires the engagement with others to uncover error, to challenge their beliefs, and (when a standpoint is successfully defended or refuted) to require the opponent to change their stance. If the use of invitational rhetoric I am advocating is intended to remain faithful the feminist theory that inspired Foss and Griffin (1995), this presents a significant problem for an invitational addition to Pragma-Dialectics' model. For this, I have two responses.

The first response is that while Pragma-Dialectics does at times require discussants to change in response to a successful defense or refutation of a standpoint, it does still align with many of the standards embedded in these feminist principles. Regarding the tenet of equality, Pragma-Dialectics rejects efforts for control or domination through (among other examples) a commitment to rule 1 which denies special status to any speaker or position that would advantage one discussant over another, a rejection of directives that grant or use authority to prohibit or order, and mutual agreement and adherence by the discussants to the same set of rules (van Eemeren and Grootendorst 2004). Each of these rules are specifically intended to grant equal status to interlocutors in order to facilitate the testing of an idea solely on its merits.

Regarding the tenet of immanent value, Pragma-Dialectics does not specifically contribute to the treatment of individuals as unique or irreducible, but it also does not contradict or interfere with this standard. In Pragma-Dialectics, the identity of individual speaking holds has no bearing on how an idea is meant to be evaluated. Further, Pragma-Dialectics does not seek to change an individual's perspective as persuasive rhetoric does, but rather to test an idea to determine whether a perspective ought be changed. Regarding the final tenet of self-determination, the right of an individual to constitute their own beliefs is specifically practiced through their engagement with opposition during a dialectic. van Eemeren and Grootendorst (2004) state that while discussants have an unconditional right to advance and challenge standpoints (which would then entail an obligation to defend), this right does not impose any obligation to advance a standpoint or challenge a standpoint. Any individual that does not wish to have their perspective challenged is entitled to the right to refrain from engaging in Pragma-Dialectics. Further, for those that choose to engage in a dialectic, rule 11 grants the right to the protagonist (the discussant whose standpoint is being challenged) to retract an argument they made and replace it with a new argument, which removes the protagonist's obligation to defend the prior argument (van Eemeren and Grootendorst, 2004). This allows the protagonist to reshape their arguments to align with their new beliefs as their perspective changes as a result of their engagement in a dialectic. This is only prohibited for the shared premises established during the opening stage, which only become free to be challenged or retracted after the final stage has concluded. Finally, to engage with the language used by Foss and Griffin (1995), the right of individuals to "constitute their worlds as they choose" might very well be exercised by an individual choosing to constitute their world view through an engagement with others in reasonable argumentation.

However, the fact remains that Pragma-Dialectics does at times require a protagonist to retract their initial standpoint or an antagonist to accept that standpoint (which entails the antagonist changing their mind). In addition, while there is significant room to argue that Pragma-Dialectics does uphold principles of equality and self-determination without interfering with immanent value, the spirit in which Foss and Griffin (1995) invoke these feminist principles differs significantly from the practice with which Pragma-Dialectics supports them. This is apparent in Foss, Griffin, and Foss's (1997) defense to critics of invitational rhetoric, in which they say,

> We do not believe that we, as rhetors, have the right to suggest to others which of their beliefs are integral or not, egregious or not, or correct or not. We believe that individuals should be allowed self-determination so they may make their own decisions about what they want to believe and how they want to live. Individuals make choices in these areas for reasons that make sense to them, and our initial inclination is to try to understand and respect others' beliefs rather than to impose our judgment on them and to attempt to change them (Foss et al., 1997, p. 123).

The fundamental tension between the feminist principles supported by Foss and Griffin (1995) and those supported by van Eemeren and Grootendorst (2004) is whether a discussant has the right to suggest to another that their standpoint is incorrect. This is where I offer my second response. There are already uses of invitational rhetoric faithful to Foss and Griffin's (1995) original model which retain their feminist principles as well as adaptations of invitational rhetoric that seek to strengthen its feminist ties (Ryan and Natalle, 2001). This model is intended to serve as one avenue for utilizing invitational rhetoric that answers the critiques of traditional rhetors defending the value of persuasion. Merely because invitational rhetoric derived its strategies for fostering understanding through its invocation of feminist principles does not require every iteration of its usage to retain a feminist approach. As such, I suggest that Pragma-Dialectics adopt the tools that invitational rhetoric offers for encouraging maximal disclosure without adopting the feminist principles that inspired the creation of those tools. Fostering understanding can be a powerful feminist end in its own right, but it can also be used as just one part of a longer dialogue that uses understanding to achieve a different end, and a non-feminist use of invitational rhetoric in no way invalidates the value feminist invitational rhetoric provides in other spheres.

In practical terms, there is still a question regarding how invitational rhetoric is meant to be incorporated into Pragma-Dialectics' model. I am not suggesting that Pragma-Dialectics adds (for example) an eleventh commandment that requires participants to offer perspectives and foster atmospheres based in safety, value, and freedom. The use of invitational rhetoric is not necessary to engage in a reasonable exploration and resolution of differences of opinion, nor am I recommending it be used in all cases. In many dialogues that follow Pragma-Dialectics' model, differences of opinion or common ground can remain only partially externalized without interfering with the achievement of a successful resolution. Instead, I propose that the practices of invitational rhetoric could function more similarly to the conventions described by van Eemeren and Grootendorst (2004) in rule 5. Per this rule, the protagonist and antagonist create and agree to a set of rules regarding how standpoints are to be defended and attacked and what constitutes a successful defense of attack. These rules only take on status as a convention that must be followed after they have been agreed to by both parties. This process takes place after a difference of opinion has been identified and roles and standpoints have been adopted, but it can function as a model for how invitational rhetoric might be incorporated prior to the discovery of differences. Discussants that anticipate polarizing, stigma-laden, or otherwise challenging differences that would discourage participants from fully externalizing might agree to the guidelines of invitational rhetoric and set them as conventions for the opening and confrontation stages of their dialectic prior to engagement. In essence, invitational rhetoric would exist as a set of guidelines that participants in a dialogue can choose to adhere to only in those cases where an implicit difference of opinion or set of shared premises would be likely to result in a failure to externalize and resolve those differences due to the anticipation of backlash or of scarce common ground.

6. Conclusion

Foss and Griffin (1995) built useful tools for engaging in practical dialogue that help participants achieve the worthy goal of understanding one another. Incorporating invitational rhetoric into Pragma-Dialectics offers new strategies for encouraging optimal externalization and ultimately facilitates a more productive testing of ideas and resolution of differences. This article is meant to both serve as an introduction as to how these two models of discourse might work together rather than functioning as discrete alternatives and as an attempt to resolve some of the potential tensions that exist between invitational rhetoric and Pragma-Dialectics. However, there are still more questions regarding what challenges this incorporation might face, particularly if one considers modifications and additions made to invitational rhetoric after Foss and Griffin's (1995) original article. Beyond the potential value that invitational rhetoric offers Pragma-Dialectics, I also intend for this article to function as an example of how we might use other rhetorical models (even those grounded in theory that seem to run counter to the goals of reasonable argumentation) to develop new strategies for fostering productive and reasonable resolution of differences of opinion. Finally, I offer this article as a strategy for engaging in and with feminist theories of rhetoric that retains many of these theories' ethical goals without isolating them from the power and value of argumentation.

References

Benhabib, S. (1992). Situating the self: Gender, community, and postmodernism in contemporary ethics. Routledge.

Bone, J. E., Griffin, C. L., & Scholz, T. M. L. (2008). Beyond Traditional Conceptualizations of Rhetoric: Invitational Rhetoric and a Move Toward Civility. Western Journal of Communication, 72(4), 434–462. https://doi.org/10.1080/10570310802446098

Bruner, M. L. (1996). Producing Identities: Gender Problematization and Feminist Argumentation. Argumentation and Advocacy, 32(4), 185–198. https://doi.org/10.1080/00028533.1996.11977994

Chick, D. M. (2021). Memorializing Senator McCain's "Uncivil Tongue": Invitational Rhetoric and the Problem of Confrontation and Violence. Southern Communication Journal, 86(4), 320–334. https://doi.org/10.1080/1041794X.2021.1933154

Condit, C. M. (1997). In Praise of Eloquent Diversity: Gender and Rhetoric as Public Persuasion. Women's Studies in Communication, 20(2), 91–116. https://doi.org/10.1080/07491409.1997.10162405

Dow, B. J. (1995). Feminism, difference(s), and rhetorical studies. Communication Studies, 46(1–2), 106–117. https://doi.org/10.1080/10510979509368442

Foss, S. K., & Griffin, C. (Eds.). (2020). Inviting understanding: A portrait of invitational rhetoric. Rowman & Littlefield.

Foss, S. K., & Griffin, C. L. (1995). Beyond persuasion: A proposal for an invitational rhetoric. Communication Monographs, 62(1), 2–18. https://doi.org/10.1080/03637759509376345

Foss, S. K., Griffin, C. L., & Foss, K. A. (1997). Transforming Rhetoric Through Feminist Reconstruction: A Response to the Gender Diversity Perspective. Women's Studies in Communication, 20(2), 117–136. https://doi.org/10.1080/07491409.1997.10162406

Gearhart, S. M. (1979). The womanization of rhetoric. Women's Studies International Quarterly, 2(2), 195–201. https://doi.org/10.1016/S0148-0685(79)91809-8

Hayden, W. (2021). From Lucifer to Jezebel: Invitational Rhetoric, Rhetorical Closure, and Safe Spaces in Feminist Sexual Discourse Communities. Rhetoric Society Quarterly, 51(2), 79–93. https://doi.org/10.1080/02773945.2021.1877797

Johnson, J., Friz, A., Randall, C., & Vitolo-Haddad, C. (2021). "Keep Talking, I Keep Changing My Mind": The Value of Invitational Rhetoric for Focus Group Research. Western Journal of Communication, 85(3), 381–399. https://doi.org/10.1080/10570314.2020.1829024

Jørgensen, C. (2007). The Relevance of Intention in Argument Evaluation. Argumentation, 21(2), 165–174. https://doi.org/10.1007/s10503-007-9044-0

Lozano-Reich, N. M., & Cloud, D. L. (2009). The Uncivil Tongue: Invitational Rhetoric and the Problem of Inequality. Western Journal of Communication, 73(2), 220–226. https://doi.org/10.1080/10570310902856105

Mathison, M. A. (1997). The Complicity of Essentializing Difference.: Complicity as Epistemology: Reinscribing the Historical Categories of "Woman" Through Standpoint Feminism. Communication Theory, 7(2), 149–161. https://doi.org/10.1111/j.1468-2885.1997.tb00146.x

Meyer, M. D. E. (2007). Women Speak(ing): Forty Years of Feminist Contributions to Rhetoric and an Agenda for Feminist Rhetorical Studies. Communication Quarterly, 55(1), 1–17. https://doi.org/10.1080/01463370600998293

Pariera, K. L., & Turner, J. W. (2020). Invitational Rhetoric between Parents and Adolescents: Strategies for Successful Communication. Journal of Family Communication, 20(2), 175–188. https://doi.org/10.1080/15267431.2020.1729157

Pollock, M. A., Artz, L., Frey, L. R., Pearce, W. B., & Murphy, B. A. O. (1996). Navigating between Scylla and Charybdis: Continuing the dialogue on communication and social justice. Communication Studies, 47(1–2), 142–151. https://doi.org/10.1080/10510979609368470

Ryan, K. J., & Natalle, E. J. (2001). Fusing horizons: Standpoint hermeneutics and invitational rhetoric. Rhetoric Society Quarterly, 31(2), 69–90. https://doi.org/10.1080/02773940109391200

van Eemeren, F. H., Garssen, B., Krabbe, E. C. W., Henkemans, A. F. S., Verheij, B., & Wagemans, J. H. M. (2014). The Pragma-Dialectical Theory of Argumentation. Handbook of Argumentation Theory (pp. 517–613). Springer Netherlands. https://doi.org/10.1007/978-90-481-9473-5_10

van Eemeren, F. H., & Grootendorst, R. (2004). A systematic theory of argumentation: The pragma-dialectical approach. Cambridge University Press

REFRAMING IN DISPUTE MEDIATION: AN UMBRELLA TERM ENCOMPASSING FOUR ARGUMENTATIVE PHENOMENA

CHIARA JERMINI-MARTINEZ SORIA
Università della Svizzera italiana
chiara.jermini@usi.ch

Abstract

This paper aims at providing an analysis of the phenomenon of reframing in dispute mediation from an argumentative point of view. In order to find out how reframing actually works in terms of inference, an argumentative analysis of mediation sessions transcriptions' in different languages was carried out – employing the extended pragma dialectical model of a critical discussion (Van Eemeren, 2010) and the Argumentum Model of Topics (Rigotti and Greco, 2019). Reframing has long been considered a useful tool that fosters conflict resolution (see for e.g. Donouhue et al., 1988) but its functioning had not been fully elicited yet.

1. Introduction

Dispute mediation is a dialogical and communicative process in which a neutral third party, a mediator, helps conflicting parties modifying the way in which they interact with each other so that they become able to co-construct a solution to their conflict that is mutually satisfactory. Mediators cannot impose solutions to the parties, yet they play a fundamental role in conflict resolutions as they enable people to be in the right mindset to explore options for win-win solutions by themselves.

Mediators are able to perform this task employing different techniques; it has been observed several times, both in scientific research and in professional practice, that reframing is an important communicative competence of mediators (see e.g. Donohue, Allen and Burrell, 1988). However, research has not elicited yet how reframing actually works, i.e. what a mediator should – or should not – say in order to do a reframing. As Atran and Axelrod note, "surprisingly little has been written on how the process actually works" (Atran and Axelrod, 2008, p. 222). The RefraMe project's team members started studying reframing relying on the assumption that reframing has an inferential dynamic (see Hoffmann

2011; Greco 2016): when a mediator asks conflicting parties to leave behind their interpretation of the conflict(or of parts of it), s/he is asking them to make an inference; and therefore s/he must provide them with elements (arguments) to reason in a specific way. Different scholars describe the kinds of reasoning enacted by reframing in terms that are very similar to what argumentation scholars would call argument schemes (e.g. Rigotti and Greco, 2019). For example, reasoning from analogy is mentioned (Putnam and Holmer, 1992, p. 140), as well as cause-effect relations (Putnam and Powers, 2015, p. 386).

Starting from this assumption, complementary theoretical frameworks were employed to carry out an analysis of reframing instances in mediation cases. Firstly, a definition of reframing provided by communication scholar Putnam (2004) was employed to code instances of reframing in two corpora of mediation sessions' transcriptions. Secondly, an analysis of these cases was carried out employing the extended pragma-dialectical model of a critical discussion (van Eemeren et al. 2004) – in particular the notion of strategic maneuvering (van Eemeren, 2010), which focuses on the analysis of argumentation, and the Argumentum Model of Topics (AMT, Rigotti and Greco, 2019), which looks at how inferences are drawn. But before discussing in more detail how the analysis was carried, it is worth discussing the concept itself of reframing, as it has been defined (like the related concepts of "frames" and "framing")differently in several disciplines and by different scholars.

2. What is reframing? Fillmore and Putnam's approaches

Before I discuss what reframing is and, more specifically, how it is defined in this work, a premise on the related concepts of frames and framing is needed. According to Dewulf et. al (2009), it is possible to see two broad research streams around frames: one views them as cognitive structures that help people making sense of their experience (Dewulf et al., 2009) whereas the other focuses on the communicative aspect of frames, i.e. "on how parties negotiate meaning in interactions" (Dewulf et al., 2009, p. 156). Although Dewulf et al. (2009) stresses the differences between these two approaches, they both bring interesting insights for the study of frames. For example, in a conflictual situation, to find out what is a party's understanding of the conflict, a mediator needs to rely on what is expressed by that party through communication (e.g. how does s/he depicts the other party, what words s/he chooses to describe the facts), and this understanding changes in the interaction (for e.g. a party can shift from characterizing an incident as a "personal attack" to describe it as a "misunderstanding"). So, the mediator focuses on the communicative aspects of frames, however the way parties will verbally describe their conflict and the way they will behave is linked to the way in which they make sense of that conflict – i.e. to frames as cognitive structures.

An approach to frames that takes into account both their communicative and their cognitive dimension isthe one developed by linguist Charles Fillmore. In fact, he explains that"[...] the meaning of words may, more than we are used to thinking, depend on contexted experiences; that is, the contexts within which we have experienced the objects, properties or feelings that provide the perceptual or experiential base of our knowledge of the meaning of a word (or phrase or grammatical category) may be inseparable parts of those experiences." (Fillmore, 1976, p. 24). If we keep this way of thinking about frames in mind and we consider now what reframing does, or, as Boulle et al. (2008) put it, what the goal of reframing is – namely "to change [parties'] frame[s] of reference in order to get the parties think differently about things, or at least to get them to see things in a different light" (Boulle et al. 2008, p. 128) –, we can see that it can be achieved thanks to "the fact that the language we use affects how we perceive the world [therefore] by changing language we can change perceptions, and [...] changed perceptions can lead to changed behaviour" (Boulle et al. 2008, p. 128). According to Fillmore,"[p]articular words or speech formulas, or particular grammatical choices, are associated in memory with particular frames in such a way that exposure to the linguistic form in an appropriate context activates in the perceiver's mind the particular frame – activation of the frame, by turn, enhancing access to the other linguistic material that is associated with the same frame"(Fillmore, 1976, p. 25). Fillmore makes the example of the frame of a "commercial event" (Fillmore, 1976, p. 25), that is activated by any of the words belonging to that semantic domain, such as "buy"; "sell"; "pay"; "cost" and so on(Fillmore, 1976, p. 25). He then points out that "each of these (words) highlights or foregrounds only one small section of the frame" (Fillmore, 1976, p. 25). This research shows that it is indeed in this way – by activating new frames in the parties' minds through uttering specific words – that it is possible for mediators to reframe.

Fillmore's approach has not been applied to conflict resolution studies so far, but it can serve as a background to understand how frames are activated in discourse through the use of specific words and how an analyst can reconstruct them. An interpretation of reframing compatible with Fillmore's approach is the one proposed by communication scholar Putnam and colleagues (see Putnam, 2004; Putnam and Holmer, 1992): they view framing and reframing as communicative processes and she defines reframing as "shifts in levels of abstraction" that foster conflict transformations (Putnam, 2004). This means that reframing always includes a shift from one category to a different one. More precisely, as based on her work in organizational conflict, Putnam provides different categories of shifts that in the present work will be used to identify instances of reframing. In this work, starting from Putnam's definition of

reframing as shifts in levels of abstraction, the author has systematically categorized instances of reframing present in mediation sessions' transcripts, adding also some new categories that emerged
empirically during the analysis (see Greco and Jermini-Martinez Soria, 2021; Jermini-Martinez Soria 2021).

3. Methodology: data collection

To study the functioning of reframing in dispute mediation, the present work takes an empirical focus, like several other studies of argumentation in mediation activity (see for example Jacobs, 2002; Jacobs and Aakhus, 2002; van Eemeren et al., 1993; Aakhus, 2003; Janier et al., 2016; Cisterna Rojas, 2007; Greco Morasso, 2011; Janier and Reed, 2015; Vasilyeva, 2012; 2015; 2017). Two corpora of mediation sessions' transcriptions were analyzed. They consist together of 26 transcriptions of mostly role-played mediations sessions relative to different kinds of interpersonal conflicts (total number of words: approximately 217000). The choice to include cases of mediation from different domains is justified by the hypothesis that reframing does not change in its core functioning depending on the context or type of conflict. Of course the reframing's "content" will vary in each case, but its inferential dynamic will always be the same. For this reason, it is appropriate to compare cases of mediation practice applied to different "interaction fields" (Rigotti and Rocci, 2006) in order to verify reframing's functioning in as many settings as possible.

Corpus I is composed of 17 transcriptions [1] of mostly role-played mediation cases in English, mediated by 13 different mediators. All cases are either teaching materials used for the training of mediators, or anyway demonstrations of what is considered good mediation practice, organized by mediation institutions or similar organizations (such as Universities offering mediation-related courses, like the Program on Negotiation – Harvard University).

Corpus II is composed of nine transcriptions of role-played mediation sessions (two in Italian and seven in French) organized by the author in different areas of Switzerland between 2017 and 2018 within the context of the RefraMe project. The role-play scenarios were selected by the researcher, combining scenarios proposed online by an institution that promotes mediation with scenarios created within the RefraMe project to guarantee for enough variety of conflict types – analogously to Corpus I. In each role-play the mediator is a certified professional mediator working

---

[1] Within the RefraMe project, part of corpus I (six cases) was already transcribed and analyzed for other purposes in Greco Morasso (2011).

in Switzerland at the time of data collection, and the conflicting parties are played by volunteers.

In conclusion, considering both corpus I and corpus II for the present study The author took into consideration 26 mediation cases of interpersonal conflicts of various kinds in 3 different languages (English, French and Italian)[2]. As for the use of role-plays, the choice was made because of the impossibility of acquiring real data due to the confidentiality of the process, it is a standard way of gathering data on mediation (see for example Jameson et al., 2014; Janier and Reed 2017; van Bijnen 2020).

4. Data analysis: Theoretical tools

4.1. Pragma-dialectical approach: argumentation as a critical discussion

The pragma-dialectical approach, which is based on a dialogical view of argumentation as a critical discussion (van Eemeren and Grootendorst, 2004), has already been employed for studying argumentation in dispute mediation (see for example Greco Morasso, 2011; Vesper, 2015). In the present work, the combination of the AMT and the pragma-dialectical model allows considering the specific inference moves regarding mediators' reframing in the broader context of the argumentative discussion in which the mediator and the parties are involved. In particular, reframing is considered as being part of a mediator's strategic maneuvering, the core concept of the extended pragma-dialectical model (van Eemeren 2010). Strategic maneuvering has three interrelated aspects (topical potential, adaptation to audience demand and presentational devices) and it refers to "the continual efforts made in all moves that are carried out in argumentative discourse to keep the balance between reasonableness and effectiveness"(van Eemeren, 2010, p. 40).

4.2. The Argumentum Model of Topics (AMT)

In order to bring forward the argumentative interpretation of reframing, the tool that I will use to find out which is the reasoning involved in each reframing, i.e. to make its argumentative dimension explicit, is the Argumentum Model of Topics (AMT), developed by Rigotti & Greco (2010,

---

[2] The fact that mediation cases the author analyzes were recorded in different countries is not an issue for the development of this research. In fact, professional mediation is a standardized practice at the international level (see for e.g. Menkel-Meadow 2005).

2019), a tool that "allows to reconstruct the inferential configuration of arguments schemes (Rigotti & Greco Morasso 2010, p.502)"(Greco, 2016, p. 358; for a detailed description of the model see section 4.2 below). In fact, in order to understand how the shift in levels of abstraction happens and how it is justified, it is important to use an argumentation model that focuses on inference.

5. Findings: reframing corresponds to four different phenomena in which argumentation plays a different role

In the 26 mediation sessions that constitute corpus I and corpus II, I found 59 instances of reframing, the majority of which (37 cases) finds itself in the empirical counterpart of the argumentation stage. Of these, 19 are to be considered counterarguments that attack one or both parties' standpoints, whereas the other 18 cases are changes in the discussion issue proposed by the mediator and supported by arguments. Of the remaining 22 cases, 6 are cases of reframing take place at the empirical counterpart of the confrontation stage, as they consist of proposals of new discussion issues that are not supported by arguments, and 16 cases are reframing as reformulations. The following sections illustrate examples of each of these cases.

5.1. Reframing as a change of discussion's issue

In any argumentative exchange, parties' support standpoints (or doubt other parties' standpoints) regarding an issue, i.e. "a proposition which can be doubted by the interlocutors (Stump, 1978, p. 33)"(Schär, 2021, p. 19; see also the discussion on the concept of issue in Goodwin, 2002). If two or more people have entered mediation, it is likely that there will be a number of issues around which they are not able to get to an agreement without help. Mediators employ many different techniques in order to help parties solve their conflict and obviously simply always changing a discussion's issue would not be fruitful, as some issues need eventually to be addressed as they are core ones in the conflict. However, sometimes the conversation between the parties gets stuck as they quarrel about issues that do not necessarily need to be addresses (or perhaps not at that stage of the discussion). So, mediators may intervene by introducing a different issue in the discussion (one that is, to a degree that may vary, related to the previous one). Perhaps parties will not hold opposing standpoints around this new issue, or, if they do, they might find easier to find a compromise on this issue than on the previous one.

Notably, however, that not all changes of discussion issues proposed by mediators are indeed reframing instances. In order for a reframing to be present, there should also be a shift in levels of abstraction that fosters positive conflict transformation. This kind of reframing can either be supported by one or more arguments or not, as we will see in sections 5.1.1 and 5.1.2 below.

5.1.1 Reframing as a change of discussion's issue non supported by arguments

I will now discuss, through the analysis of an illustrative example, reframing cases that simply consist of a change of issue proposed by the mediator and not supported by arguments. Let us consider the excerpt below (example 1), taken from a mediation session between two business partners that own a Bagels' shop. In the excerpt below, they are discussing about the possibility of investing in machines for mechanizing part of the bagel's production and one of the parties' worries that it could be too expensive and that they would have to spend their private money to face the investment. At turn 187, the mediator asks them a question involving a reframing:

181 M Ah (.) okay (.) so you're probably looking [then at about a
182 D [About a thousand bucks a [month I expect=
183 M [about fifteen hundred a month (Robert nods his head)
184 D Sure↓
185 M Payment correct↑
186 D Well (.) give or take (.) yeah (.) so that's seven hundred and fifty bucks each (.) of which I think we
should be paying a little more than [I should
187 M [let's let's let's hold on about how much it is each each for a moment and think about it in terms of the BUSINESS perspective (.) what (.) could you if you had (.) if you did make that investment (.) could the business pay a mortgage of fifteen hundred a month↑

Here the mediator introduces a new discussion issue:

Could the business pay a mortgage of fifteen hundred a month?
And he does so in order to shift the discussion away from a very conflictual topic, namely how much of their personal money David and Robert should spend in order to mechanize the production. If they manage to think only in terms of what would be possible or not to do using only the business's money they might be able to find a mutually satisfactory solution. In interventions of this kind the mediator does not advance standpoints nor supports the change of issue he proposes by arguments. Thus, these types of reframing are not per se argumentative, although one

might argue that implicit argumentation always lies behind them, in a structure that could be:

Should you discuss about Y instead of X?
(1 Yes, you should)
Because it would help you more to solve your conflict)

However, this general argumentation is implicit in all mediator interventions and is not specific of this case of reframing, which does not present an explicit justification. Therefore, in cases like the one we are considering, reframing does not concern the argumentation stage but finds itself in the empirical counterpart of the confrontation stage of the critical discussion, as the mediator expresses that s/he has a difference of opinion with the parties concerning what needs to be the topic of their discussion. In fact, "in the confrontation stage the parties establish that they have a difference of opinion"(van Eemeren and Snoeck Henkemans, 2017, p. 20). In this case, an implicit standpoint of the parties' (i.e. "I want to discuss about X") "meets with doubt or criticism" (van Eemeren and Snoeck Henkemans, 2017, p. 21) on the part of the mediator.

Whether a mediator decides to justify reframing as changes of discussion issues with one (or more)argument(s) or not depends on each situation, on whether, arguably, he or she senses it would be beneficial in terms of adaptation to audience demand and/or presentational devices. In the next section, I will discuss about cases in which mediators decide to support reframing with arguments.

5.1.2 Reframing as a change of discussion's issue supported by arguments

There are cases in which a reframing instance enacts a change of the discussion issue that is supported by one or more arguments. This has usually the function to move the discussion in a way that is more productive for conflict resolution. Let's consider now example 23. This example is taken from a mediation case between a woman and her brother-in-law about a complex family conflict involving a money loan and other issues. In the next excerpt, where Eloise (E) is talking about the problems she was having with her husband before their divorce, one of the mediators (M1) enacts a reframing involving a meta-pragmatic shift (turn 98):

---

3 For space reasons, only the English translation of the excerpt is reporter here, however the original language of this mediation session's video-recording is French.

97 E in that moment one is not well one needs support things like that and and when I asked him ( ) (.) well
nobody was there anymore (.) when I told him now you have to choose between his ideals his activism things like that and really focus on his FAMILY for once ↓ well there was nobody there anymore his ideals are more important than anything else (.) well listen it's not really me the one who decided eh (she is almost crying while she pronounced these lasts words) I still loved him

98 M1 so it's a little complicated because about what exactly would you like to discuss ↑ because actually it bounces a bit everywhere there is this money that you lended that you said was part was the the family's support for your own family project (.) your family project didn't work out (.) after that you also have an opinion about what happens in in their family (.) it's a bit tangled this (.) what could you do that that would allow you to put the pieces in order ↑ (4) actually about what shall we discuss ↑

The mediator notices that the parties' conflict seems to involve several issues that they are trying to address in an untidy fashion therefore he asks them "what do you want to talk about?" introducing in the discourse the following issue:

Should we discuss about what topic to address first?

In this sense, also this case of reframing involves a change of issue, which according to Aakhus (2003) would go under the name of redirection; however, as I will show in what follows, this change of issue is justified. We can see that the mediator's preferred answer to this question is "yes"; the mediator provides an argument for his standpoint, as can be seen in the following reconstruction:

1 Yes, we should discuss about what topic to address first
1.1 Because it's a bit tangled
(1.2 And this is not helpful for conflict resolution)

Note that the most important part of the argument advanced by the mediator remains implicit: he points out that at the current moment the discussion is unfolding in an untidy way, which he imagines the parties will agree is not a good thing for the conflict resolution, based on what we could call an endoxon, i.e. the fact that in order to find a solution to a complex problem one should proceed in a tidy fashion. Notably, this endoxon remains implicit.

In terms of strategic maneuvering, in this type of reframing the selection of the topical potential is clearly a prominent aspect, because the mediator asks directly to the parties to choose which topic they feel as the most important and urgent to address. Moreover, in terms of presentation devices, the mediator stresses the fact that their discourse is not unfolding around a specific issue by stating that it is "emmêlé" (i.e."tangled") and by using the metaphor that "ça fait un peu de ricrochets partout" ("it bounces a little bit everywhere"): this metaphor perhaps can help the disputants understand better how the mediator is perceiving their interaction, and therefore hopefully encourages them to change it by focusing on one specific topic.

Both unjustified and justified reframing perform the function of changing the discussion issue in a way that possibly steers conflict resolution or at the very least does not let it degenerate even further. As the pragma-dialectical reconstructions shows, they consist of mediators' standpoints disagreeing with parties' usually implicit standpoints regarding the need and/or importance of addressing a discussion topic instead of another one. These kinds of reframing fulfil an important function in the crafting of the dialogue space, as they select topics for discussion that should help conflict resolution. Reframing as changes of discussion issues tend to usually be accepted[4] by the parties because they expect mediators to guide their discussion, thus redirection (Aakhus, 2003) is usually accepted.

5.2. Reframing as counterargument

Let us consider now an example of reframing as counterargument. In example 35, the two parties are working together in a development project for an NGO. One of them is the NGO's leader and the other one is a major benefactor who asked to be involved in decision-making processes. In the following excerpt, they are addressing the fact that they disagree over the reasonableness of having weekly reports sent to Alec by Jember. According to Jember, Alec's request of her sending him weekly reports is "unreasonable" because it requires a considerable amount of time that she would rather spend doing other tasks useful for the project. Alec explains why in his opinion weekly reports are reasonable and useful.

> 223 M ok (.) you don't get the sense that weekly reports are unreasonable you said they were normal
> 224 A oh well ehm definitely I all of my businesses are ehm (.) based on weekly reporting ehm and in fact I have in many cases when the business has been in trouble or in a in a start-up phase which I think is analogous to to jember's organization I've had daily reports
> 225 M mmh

---

[4] On the acceptance of reframing by the parties, see discussion in Jermini-Martinez Soria 2021 (unpublished PhD Dissertation).
[5] The author would like to thank the Program on Negotiation – Harvard University for allowing her to transcribe and use this teaching material for research purposes.

REFRAMING IN DISPUTE MEDIATION

226 A from my key ( ) in fact in the case of the ( ) a real estate business that I was asked to take over I would meet with the key managers at 7 a.m and we'll met again at the end of the day eh for exceptional reports not a regular thing but if anyone had anything exceptional to report we'd get together in the evening (.)and so
227 M these are computerized reports↑
228 A well ehm in in in most cases yes those would be (.) a stock control report sales reports marketing reports those sorts of things financial statements and so on
229 M I mean one of of the things I could imagine (.) ehm is just there is a difference in time (.) if you're dealing in the developed world at at the speed of new york and london stock exchange a lot happens in the day (.) you've got the telephone you've got computers you've got all kinds of large scale business going on being accounted for in in electronic ( ) has to be at this point (.) a week a lot happens in a week
230 A yes
231 M if you're dealing if I'm understanding it if you're dealing where you probably have no motorized
transport (.) for most things /
232 J that's correct

At turn 224, Alec provides the following standpoint and argument in favor of having weekly reports:

1 Asking for weekly reports to Jember is reasonable
1.1 Because it works for analogous businesses and projects in a start-up phase

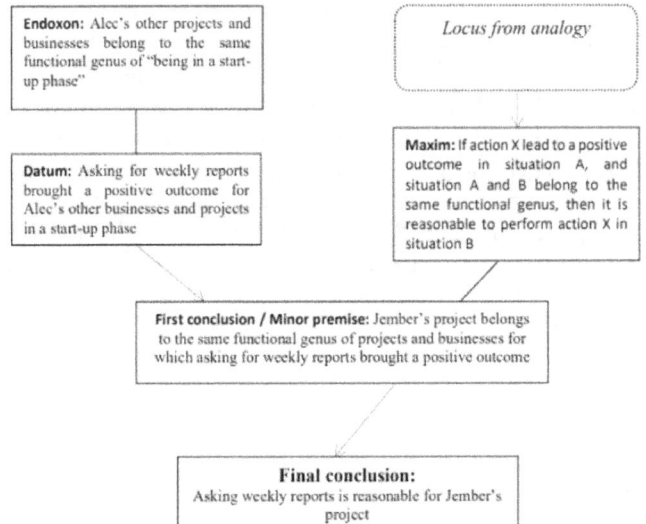

Figure 1. AMT reconstruction of Alec's reasoning.

Alec's reasoning can be reconstructed with the AMT model (figure 1). For Alec, the most salient trait of this project is its "being in a start-up phase", the way he frames the situation is by comparing it to other projects "in a start-up phase", taking for granted that the potential problems will be similar, and therefore applying a measure such as having weekly reports that has proven effective in the past.

Alec's reasoning involves a locus from analogy, as he compares this project to other ones that share a trait with it, namely the fact of being in a start-up phase. Rigotti and Greco (2019) observe that "arguably, argumentation from analogy is not effective if the functional genus is ill-designed" (Rigotti and Greco, 2019, p. 262) – which is indeed what the mediator will try to point out. In fact, the mediator attempts a reframing (at turn 229) involving a shift from quality to external condition by drawing Alec's attention to the fact that there is indeed also a big difference between his other businesses and project and this particular one, namely the country in which they are based. The mediator expresses an argument (1.1) in favor of the following implicit standpoint:

(1 You cannot carry out this project in the same way as your other businesses in a start-up phase = you
should not ask for weekly reports)
1.1 Because there is a difference in time between Ethiopia and the US

By 1.1. he means that the time required in Ethiopia to carry out tasks, communicate with other people and so on is much more, due to lack of infrastructure, internet. It is something that Jember has explained before in the mediation session. What it means can be inferred from his following words at turn 229: "(.) if you're dealing in the developed world at at the speed of New York and London stock exchange a lot happens in the day (.) you've got the telephone you've got computers you've got all kinds of large scale business going on being accounted for in in electronic ( ) has to be at this point) [...]" and at turn 231: "If you're dealing if I'm understanding it if you're dealing where you probably have no motorized transport (.) for most things / [...]". The mediator's reframing can be reconstructed with the AMT in figure 2:

REFRAMING IN DISPUTE MEDIATION

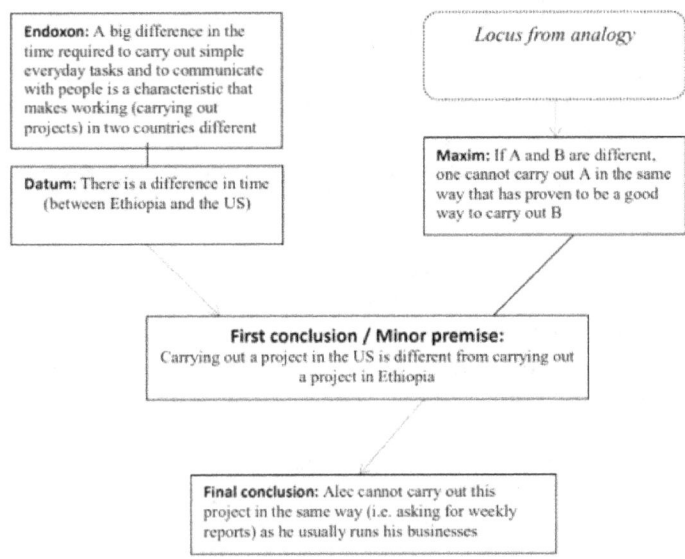

Figure 2. AMT reconstruction of the mediator's reasoning.

The mediator's final conclusion is oppositional to Alec's standpoint and it coincides with Jember's point of view. In this case, the mediator advances an argument in support of a standpoint regarding the issue the parties are already arguing about – there is no change in discussion issue.

Note, however, that the fact that the mediator advances a counterargument probably does not undermine the mediator's impartial role. In fact, the mediator only explicitly states a datum at turn 229 ("there is a difference in time") that performs the function of argument and Alec accepts it (turn 230). For the rest of the mediation session, they no longer discuss whether they will have or not weekly reports in the future, but it is most probably that they will not, as Alec accepts to go and spend some time in Ethiopia to see for himself how things are getting done on the field. In terms of adaptation to audience demand the mediator does not openly express disagreement with Alec, he simply draws his attention to an aspect of the situation that he might not have thought of but that is important to take into consideration and that contests the functional genus Alec has created. However, it is clear that the mediator expects Alec to come to his same conclusion on his own. The mediator would not introduce in the discourse his observation that "there is a difference in time" if he had no intention of supporting a standpoint and using this as an argument supporting it [6]. Considering now presentational devices in this

---

[6] In order to infer that the mediator is actually supporting a standpoint – more precisely, one that honours his/her neutral

intervention, notably, the mediator emphasizes the factual, objective nature of his claim by stating "there is a difference in time" (my emphasis) but at the same time he acknowledges that the relevance of his claim might be challenged by Alec by saying "I mean one of the things I could imagine ehm is just there is a difference in time" (my emphasis). This case shows how a reframing targets and criticizes a standpoint expressed by one of the parties.

5.3. Reframing as reformulation

Finally, I would like to discuss what can be considered yet another type of reframing, one that does neither involves a change of issue nor attacks parties' standpoints, but one in which the mediator rephrases something a party has said giving it a label that in the mediator's view will be more useful for clarity and/or conflict resolution. This type of reframing, thus, has a different role in the discussion than (justified) redirection or counter-argumentation; it is a reformulation of the parties' positions. Although the role of (re)formulations in mediation has been connected to argumentation (van Eemeren et al., 1993), it seems to us that the argumentative function of this type of reframing is less direct than the other cases discussed; of course, however, a positive formulation of the parties' statement is, indirectly, subservient to the goal of conflict resolution. For instance, Van Eemeren et al. (1993) observe: "[...] mediators will often formulate the standpoints and arguments of the disputants and the possible options for moving along with the discussion. Officially, such formulations have straightforward informative and procedural functions [...]. In performing these functions, however, formulations can also give substantive shape and direction to an argument, and in that respect formulations can serve as a technique for mediators to manage the substantive character and argumentative force of a discussion without entering into the discussion as an advocate" (van Eemeren et al., 1993, p. 120). Let us now consider the following excerpt (example 4) from a mediation case between two business partners that follows a part where one of the parties has just explained why he behaved in a certain way in the workplace:

---

role – I follow this consideration by Van Eemeren and Snoeck Henkemans: "If speakers or writers do not explicitly express
their standpoint, as a rule they expect the listener or reader to be able to infer this standpoint from the arguments put forward. Why would they otherwise bother to present argumentation? When an argument lacks an explicit standpoint, it is not only immediately obvious that the standpoint is missing, but it is also easier than in the case of other indirect speech acts to figure out what is really meant. This is because there is an extra tool available, namely logic" (Van Eemeren and Snoeck Henkemans, 2017, p. 49).

377 P Ehm (.) I just wish he'd said it to me it would have been no big deal
378 M If if Joe had said this to you (.) ehm /
379 P Well I would have just read the things myself and (.) you know signed it or whatever put it off for a   week till I came back or ( ) he wasn't telling me and then (.) I thought it was something else I thought he was too( ) over you know said the wrong thing or something I I mean I hadn't a clue
380 M So you've been making many assumptions about Joe (.) in the last while↑ Eh that may not actually be(.) valid↑
381 P Probably yeah

At turn 380, the mediator summarizes what Pat (=P) has said in the previous turn by stating "you've been making assumptions about Joe". This involves a shift in levels of abstraction from concrete – what concretely happened in their situation (I thought it was something else...) to abstract (the idea of making assumptions about someone's behaviour). Here the mediator reformulates what Pat has stated in a way that makes clear that all the parties involved agree on the standpoint "Pat has made assumptions about Joe", perhaps with the intention of establishing a clear common ground. This type of shift allows to "creat[ing] a level of meaning together" (Putnam, 2004, p. 208) that in turn makes it possible for parties to "redefine issues and alter the naming of a conflict" (Putnam, 2004, p. 280). This technique is not always successful; therefore, each mediator should decide carefully when to employ it. It clearly fulfils an important function within the discussion, namely to have clear labels to define events the parties are referring to, avoiding in this way to give room to possible misunderstandings and ambiguities. Moreover, as Jacobs notes, "statements of disputant positions (i.e. that summarize or rephrase what a party has said) can also give direction to a line of reasoning and add weight to an argument, and, in that respect, they can manipulate the substantive character of a discussion and push disputants towards settlements they might not ordinarily accept" (Jacobs, 2002, p. 1414).

5.4. Conclusions

In conclusion, this analysis revealed that reframing is a term that includes different phenomena in which argumentation plays different roles. Understanding this enables us to better grasp different implications of the use of this technique: for example, understanding that reframing can correspond to a counterargument imposes a reflection on mediators' neutrality and its limits; or, reflecting upon in which cases it is advisable for mediators to support the changes of issues they propose with arguments can deepen the understanding of mediation practice. This work opens interesting avenues for future research, among which are: exploring the way in which reframing is carried out differently in terms of presentational devices in different languages; continue research on potential correlations between characteristics of reframing and other aspects of each mediation session such as the type of conflict, its use in caucuses versus joint sessions, the evolution in its use in the first mediation session versus successive meetings with the same parties and so on; continue to study reframing in correlation with the concept of "argumentative style" and, in particular, of "reconciliatory style" (van Eemeren 2019), as the RefraMe Project's members have recently started to work on this (Greco & Jermini-Martinez Soria, 2021).

References

Aakhus, M. (2003). Neither naïve nor critical reconstruction: Dispute mediators, impasse, and the design of argumentation. Argumentation, 17(3), 265-290.

Atran, S., and Axelrod, R. (2008). Reframing sacred values. Negotiation Journal, 24(3), 221-246.

Cisterna Rojas, V. (2007). Pragma-dialectical analysis of two family mediation cases in Chile. Proceedings of the Sixth Conference of the International Society for the Study of Argumentation. Amsterdam: Sic Sat.

Donohue W. A., Allen M., and Burrell N. (1988). Mediator communicative competence. Communication monographs, 55(1), 104-119.

Eemeren, van, F.H., Grootendorst, R., Jackson, S., and Jacobs, S. (1993). Reconstructing argumentative discourse. Tuscaloosa: The University of Alabama Press.

Eemeren, van, F.H., and Grootendorst, R. (2004). A systematic theory of argumentation. Cambridge: Cambridge Universtiy Press.

Eemeren, van, F.H., (2010). Strategic maneuvering in argumentative discourse: Extending the pragmadialectical theory of argumentation. Amsterdam: John Benjamins.

Eemeren, van, F.H, and Snoeck Henkemans (2017). Argumentation: analysis and evaluation. New York: Routledge Taylor & Francis Group.

Eemeren, van, F.H. (2019). Argumentative style: A complex notion. Argumentation, 33, 153-171.

Goodwin, J. (2002). Designing issues. In van Eemeren, F. H. & Houtlosser, P. (Ed.), Dialectic and rhetoric: The warp and woof of argumentation analysis (pp. 81–96). Dordrecht: Kluwer Academic Publishers.

Greco Morasso, S. (2011). Argumentation in dispute mediation: A reasonable way to handle conflict. Amsterdam: John Benjamins.

Greco, S. (2016). Framing and reframing in dispute mediation. In M. Danesi & S. Greco (Eds.), Case studies in discourse analysis (pp. 353-379). Munich: Lincom.

Greco, S. and Jermini-Martinez Soria, C. (2021). Mediators' reframing as a constitutive element of a reconciliatory argumentative style. Journal of argumentation in context 10 (1), 73-96.

Hoffmann, M. H. G. (2011). Analysing framing processes in conflicts and communication by means of logical argument mapping. In W. A. Donhoue et al. (Eds.), Framing matters: Perspectives on negotiation research and practice in communication (pp. 136-164). New York: Peter Lang.

Kovach, K.K. (2005.) Mediation. In The handbook of dispute resolution, ed. M.L. Moffitt & R.C. Bordone, 304–317. San Francisco: Jossey-Bass.

Jacobs, S. (2002). Maintaining neutrality in dispute mediation: Managing disagreement while managing not to disagree. Journal of Pragmatics, 34(10-11), 1402-1426.

Jacobs, S., and Aakhus, M. (2002a). What mediators do with words: Implementing three models of rational discussion in dispute mediation. Conflict Resolution Quarterly, 20(2), 177-203.

Janier, M., and Reed, C. (2015). Towards a theory of close analysis for dispute mediation discourse. Argumentation, 31, 45-82.

Janier, M., Aakhus, M., Budzynska, K., & Reed, C. (2016). Modelling argumentative activity in mediation with Inference Anchoring Theory: The case of impasse. In D. Mohamed & M. Lewiński (Ed.), Proceedings of the first European Conference on Argumentation, 9-12 June 2015, Lisbon.

Jermini-Martinez Soria, C. (2021). Reframing as an argumentative competence in dispute mediation. Unpublished PhD dissertation, USI – Università della Svizzera italiana, Lugano, Switzerland.

Menkel-Meadow, C. (2005). Roots and inspirations. A brief history of the foundations of dispute resolution. In M. L. Moffitt and R. C. Bordone (Eds.), The handbook of dispute resolution (pp. 13- 31). San Francisco: Jossey-Bass.

Putnam, L. (2004). Transformation as a critical moment in negotiations. Negotiation Journal, 20(2), 275-295.

Putnam, L. L., and Holmer, M. (1992). Framing, reframing and issue development. In L. L. Putnam and M. E. Roloff, Communication and negotiation (pp.128-155). Newbury Park: Sage.

Putnam, L., and Powers, S. R. (2015). Developing negotiation competencies. In A. F. Hannawa & B. H. Spitzberg (Eds.), The handbook of communication science: Communication competence (22) (pp. 367-395). Berlin: De Gruyter.

Rigotti, E., and Greco, S. (2019). Inference in argumentation: A topics-based approach to argument schemes. Cham: Springer.

Rigotti, E., and Rocci, A. (2006). Towards a definition of communication context. Studies in Communication Sciences, 6(2), 155-180.

Rigotti, E., and Greco Morasso, S. (2010). Comparing the Argumentum Model of Topics to other contemporary approaches to argument schemes: The procedural and material components. Argumentation, 24(4), 489-512.

Saposnek, D. T. (1983). Strategies in child custody mediation. A Family system approach. Mediation Quarterly, 2, 29-54.

Schär, R. (2021). An argumentative analysis of the emergence of issues in adult-children discussions. Amsterdam : John Benjmins Publishing.

Stump, E. (Ed.). (1978). Boethius's "De topicis differentiis." Ithaca, NY: Cornell University Press.

Van Bijnen, E. (2020). Common ground in conflict mediation: An argumentative Perspective. Unpublished PhD dissertation.

Van Slyck, M. R., Newland, L. M. and Stern, M. (1992). Parent-child mediation: Integrating theory, research, and practice. Mediation Quarterly, 10 (2), 193-208.

Vasilyeva, A.L. (2012). Argumentation in the context of mediation activity. Journal of Argumentation in Context, 1, 209–233.
Vasilyeva, A.L. (2015). Identity as a resource to shape mediation in dialogic interaction. Language and Dialogue, 5(3), 355-380.
Vasilyeva, A. L. (2017). Practices of topic and dialogue activity management in dispute mediation. Discourse Studies, 19(3), 341-358.
Vesper, A. (2015). A mediator as a pragma-dialectical critical designer of acceptance. In B. J. Garssen et al. (Eds.), Proceedings of the Eighth Conference of the International Society for the Study of Argumentation, 1451-1458. Amsterdam: Sic Sat.
Yarn, D. (1999). Dictionary of conflict resolution. San Francisco: Jossey-Bass.

COMMENT ON PAPER: REFRAMING IN DISPUTE MEDIATION: AN UMBRELLA TERM ENCOMPASSING FOUR ARGUMENTATIVE PHENOMENA

ELENA MUSI
University of Liverpool
elena.musi@liverpool.ac.uk

1. Introduction

This paper focuses on the argumentative underpinnings of reframing in the context of discourse mediation. Its merits are both theoretical and empirical: through a corpus-based study of 26 mediation transcripts, the author unravels a set of argumentative functions displayed by reframing. As to analytic frameworks, strategic manoeuvring and the Argumentum Model of Topics are successfully combined, revealing a high degree of complementarity. Overall, this study constitutes a valuable contribution to the investigation of reframing from an argumentative point of view. In what follows, I propose a few suggestions/challenges to further improve it:

2. What is reframing?

Besides the cognitive and the communication perspectives on frames as intended by Fillmore and Putnam, I would suggest discussing the concept as discussed by Goffman (frames are "principles of organization which govern events – at least social ones – and our subjective involvement in them" (1974: 10-11) since it played a major role in media research. Furthermore, the connection identified by Greco (2012) between edoxa and frames shall be accounted for: "Frames provide the implicit material premises (endoxa) which are at the basis of argumentations through which newspapers interpret and comment on the news (Greco 2012: 197)"

Finally, since the author adopts Putnam's definition of reframing, the expression "level if abstraction" shall be explained. My guess is in fact that the author has identified the presence of reframing in correspondence of shifts in the levels of abstraction. However, there is a clear reference to such a shift only in section 5.3.

3. Methodology

The choice of not constraining mediation topic is promising, but the fact that the majority of the sample is constituted by role-plays might constitute an issue for generalization of the results. I would suggest to keep detached the analysis of the results in the two corpora. It would be interesting to see whether the typed of reframing identified are equally distributed among the role-played mediation corpus and the 'natural' one.

4. Data analysis

It would be worth explaining the difference between endoxon and datum according to the Argumentum Model of Topics since readers might not be familiar with it.

5. Findings

It is important to explain how the instances of reframing have been empirically identified: this entails ---I imagine --- explaining what is meant by levels of abstraction.

From the sentence "In order for a reframing to be present, there should also be a shift in levels of abstraction that fosters positive conflict transformation" it seems that the outcome of the discourse moves affects its categorization as an instance of reframing (?)

I would also suggest to visualize the typology through a table to make it clear 1) what is the argumentative stage the reframing belongs to ii) what is the argumentative component the proposition expressing the reframing is part of iii) what is the mediation role played by reframing iv) what is the general topic of the mediation iv) what is the source (corpus I or 2).

5.1.1 This is a very interesting case. I would stress that the reframing brings to zoom out from individual instances adopting a cooperative perspective (from the "my problem" to the "our problem" approach).

5.1.2 I am not convinced that this in an instance of change of discussion issue since what is advocated for is first of all the identification of an issue. In other words, the mediation has the scope over the argumentation structure rather than the content.

5.2 What are other types of argument schemes you encountered in the analysis of cases where reframing works as a counterargument? It would be highly interesting to see some trends ...

5.3 This type stems from different criteria since a reformulation can in principle propose a change in discussion issue or advance a counterargument. I would keep distinct in your typology argumentative functions from formal aspects (reformulations, rephrases etc.).

THE CONCEPT OF ARGUMENT: REVISITED AND REENGINEERED

MARCIN LEWIŃSKI
Nova Institute of Philosophy
Nova University Lisbon, Portugal
m.lewinski@fcsh.unl.pt

Abstract
Part of the business of argumentation theory is a conceptual dispute over what argumentation and argument are in the first place. The goal of this paper is to contribute to this dispute in four steps. First, I go back to the classic work of Frege as a fruitful background to better navigate the conceptual map of language, thought, and argument. This "semantic housekeeping" allows me, second, to critically engage some of the recent philosophical literature on the concept of argument. Third, I present a positive proposal of a minimal, contrastivist concept of argument as *a set of reasons advanced to support a conclusion C_1 rather than any other conclusion C_n*. Finally, I draw three consequences of the concept: a) argument does have a proper function, that of support; b) there is one concept of argument, so there is no need to conceptually distinguish, following O'Keefe et al., arguments$_1$ and arguments$_2$; c) there are not so called "conductive" arguments or, instead, all arguments are conductive

1. Introduction

What is an argument? That is the question explored in this paper.

The term "argument" is used to refer to many things. A recent *Oxford English Dictionary* records five senses of the term:

1. an exchange of diverging or opposite views, typically a heated or angry one.
2. a reason or set of reasons given in support of an idea, action or theory.
3. *Mathematics & Logic* an independent variable associated with a function or proposition and determining its value.

4. *Linguistics* any of the noun phrases in a clause that are related directly to the verb, typically the subject, direct object, and indirect object.
5. *archaic* a summary of the subject matter of a book.
(*OED*, digital edition, 2021)

We thus face a classic case of an ambiguous, homophonic *term* of a natural language which denotes, in this case, no less than five different *concepts*. Leaving aside the specialized and well-defined concepts of mathematics and linguistics – as well as the archaic use[1] – we are left with "two concepts of argument", namely (1) and (2) above. There's been much ado about these two in argumentation theory, with O'Keefe's (1977) paper of that very title – "Two concepts of argument" – being the celebrated classic. My goal here is not to retrace the fine details of the nearly 50 years (and counting) of the debate on the topic – although the references listed below would allow a more diligent scholar to do that. Instead, I engage in a conceptual investigation. It progresses in four basic steps. First, I go back to the classic work of Frege as a fruitful background to better navigate the conceptual map of language, thought, and argument. This "semantic housekeeping" allows me, second, to critically engage some of the recent philosophical literature on the concept of argument. Third, I present a positive proposal of a minimal, contrastivist concept of argument as *a set of reasons advanced to support a conclusion C_1 rather than any other conclusion C_n*. Finally, I draw three of the not-so-minimal consequences of the concept: a) argument does have a proper function, that of support; b) there is one concept of argument, so from the point of view defended here there is no need to conceptually distinguish, following O'Keefe et al., arguments[1] (see sense (2) above) and arguments[2] (see sense (1) above); c) there are not so called "conductive" arguments or, if one so wishes, all arguments are conductive.

2. Semantic housekeeping: Frege as an argumentation scholar

As a useful background to the analysis of the paper, I briefly revisit the conceptual ontology of the uses of language laid out in Frege's classic *Über Sinn und Bedeutung* (1892/1948) and *Der Gedanke: Eine Logische Untersuchung* (1918-1919/1956). I rely on these two texts, but in the

---

[1] Other European languages such as Portuguese and Spanish, however, use "argumento" still today to denote a plot, story, scenario, or screenplay of a film.

context of the broader discussion of Frege's philosophy of language, especially in Dummett's (1993) interpretation.

Frege famously distinguished between three "realms", namely

External objective world:	the realm of *things* that we "see" (thanks to the inherent "intentionality" of perception, that is, its "directedness" or "aboutness" towards things)
Internal subjective world:	the realm of *ideas* / representations / sensations that we "have"
"The third realm":	the realm of *thoughts* about the other two realms that we "apprehend" (grasp, entertain)

Frege's "thoughts" belong to a world not unlike the world of Platonic forms.

They are eternal and unchangeable, free of the causal relations of the physical and psychological worlds. While they are immaterial (like private ideas), they exist objectively for everyone to take notice (like things). As such, they are publicly shared and thus communicable. This communication happens via sentences of a language, capable of expressing thoughts. *Sentences* are linguistic constructions which express a *sense* (*Sinn*) containing the thought. They are then assigned a truth-value, that is, the *reference* (*Bedeutung*) via an *utterance* with an assertive force (as opposed to interrogative or directive force). Internally, what corresponds to the assertion is a *judgement*, a mental act of deeming the thought true or false. We thus arrive at the following schematization:

External world: things, events

Mental world: thinking: ideas, representations of grasped thoughts, judgments

Linguistic world: (thanks to the facts of communication, it belongs to "the common outside world", Frege, 1956, p. 310) utterances with, notably, assertoric forces attached to sentences that express a specific sense, and ascribing to them a reference (a truth value)

Platonic world: abstract objects: thoughts, senses, propositions

For anyone interested primarily in the questions of truth and meaning – as Frege arguably was (Dummett, 1993) – at the very center of philosophical investigations lies the world of meaningful communication. Communication is not merely a display mechanism for the ideas we have and thoughts we grasp, but an indispensable tool to understand, express, and share thoughts. As a centerpiece, communication directly relates to the world of our mental activities and the world of thoughts:

> Grasping, entertaining, judging
>
>
> Communicating by uttering sentences with *force* and *content* (*sense* and, in assertions, also *reference*)
>
> **F(c)**
>
> which express:
>
>
> Thoughts

Now, this schematic representation of how human psychology and linguistic communication relate to the abstract world of thoughts, applies quite well to the world of argumentation:

> Grasping, entertaining, judging
> *Reasoning, inferencing (Argument$_0$)*
>
>
> Communicating by uttering sentences with *force* and *content* (*sense* and, in assertions, also *reference*)
>
> **F(c)**
>
> *Speech act of argumentation (Argument$_1$ + Argument$_2$)*
>
> which express:
>
>
> Thoughts
> *Arguments as abstract inferences (Argument$_L$)*

Against this background, I can begin ploughing through my conceptual arguments (more below):

(A1) Logicians' concept of argument – argument$_L$ [my invention] – is, in the end, sustainable only for *types* of inference such as *modus ponens* but unsustainable for *tokens* of arguments containing concrete premises and conclusions. Concepts such as

INFERENCE RULE, SET, or PROPOSITION can quite well capture all this.

(A2) Argument$_0$, a term coined by Hample and defined as "the cognitive dimension of argument – the mental processes by which arguments occur within people" (1985, p. 2), is a concept validated by psychological facts. But it's simpler to use the terms "thinking", "reasoning", or "inference" for the concepts of THINKING, REASONING, or INFERENCE.

(A3) Argument$_1$ and argument$_2$ collapse into one concept – term: "argument", concept: ARGUMENT – despite the overwhelming practice of distinguishing them. Here's how they collapse: Argument$_1$, argument *made*, inherently, even if sometimes implicitly, involves contrast and thus disagreement with other "diverging or opposite views" (*OED*, see above). It thereby encroaches on the territory of what traditionally belonged to Argument$_2$, argument *had*. Whatever remains in that territory of argument$_2$ – people having a verbal fight, a "heated or angry exchange" (*OED*, see above) – but is not support- or justification-relevant simply belongs to another concept, for example that of a VERBAL FIGHT or ANGRY EXCHANGE.

As a result, we remain with one and only one concept of argument. How so? In my attempt to answer this open question, I will not argue for (A2), it being mostly a terminological point. The (heavy) burden of proof lies with (A3), which requires a positive account of the concept of ARGUMENT. But before moving there, let me first go through the critical (A1). What's wrong with treating arguments as abstract objects, similarly to the way Frege (and, by extension, the entire mainstream philosophical tradition after him) treated thoughts? My argument rests on two simple moves: the question of a category mistake and of individuation.

3. Concrete abstract arguments

3.1. Between the world and the mind

In the logical tradition, arguments are understood as sets of propositions organized in such a way that some of these propositions are a conclusion and others are premises giving inferential support to the conclusion. As such, arguments are typically conceived of as abstract, atemporal, mind-independent, and "otherworldly" objects akin to Plato's forms whose use in communication is a negligible side-issue. So delimited, logical study of argument is a consistent and highly successful endeavor necessary in the broader theory of argumentation. Yet, it is in and of itself not sufficient. This, somewhat paradoxically, is revealed in philosophers' discussions over the exact status of arguments as abstract objects (for a recent line of discussion see Hitchcock, 2007; Goddu, 2009, 2011a; Simard-Smith & Moldovan, 2011; Goddu, 2011b; Goodman, 2018; Svoboda, 2020). Protagonists in these discussions all agree that individual propositions, their sets, and some basic rules of inference such as *modus ponens*, exist independently of human intervention. Yet key questions regarding human-argument relations remain.

Goddu (2009, 2011a) defends the concept of arguments as mind-independent platonic objects that human activity does not produce, but merely discovers, either in the dialogical process of argumentation or otherwise, for instance through solitary reflection or storytelling. But this position, beyond what's discussed below, needs to overcome two important challenges. First, on such a view only good arguments are arguments, so fallacious and otherwise flawed arguments are no arguments at all (as observed by Goodman). Second, and even more importantly, it requires a very complex metaphysics: "the third realm" inhabited not just by basic mathematical formulas, logical axioms and rules of inference, and physical laws, but also by all propositions about natural and social world, and, most suspiciously, by all possible relations and combinations of these propositions, grasped – in the case of argumentation – in the particular inferential units of an individual argument. Such metaphysics is, at best, "mythological" (Dummett, 1993). If Searle's arguments against strong AI would in fact always exist mind-independently, atemporally, and non-spatially, then it was Plato's fault to never just see them.

Others, alive to such challenges of metaphysical absolutism about arguments, defend the idea that it is human mind that creates arguments, similarly to the way languages, games, or musical compositions are created (Simard-Smith & Moldovan, 2011; Goodman, 2018). It is human intention (Simard-Smith & Moldovan) or belief (Goodman) to treat one member of the set of propositions (conclusion) as being inferred from other members of the set (premises) that turns these unorganized sets into arguments. While fine details vary – Do humans wholly invent or rather take active effort to discover pre-existing arguments? Once "entertained" by a human mind, do arguments stay with humanity forever, or disappear when falling into oblivion? Do humans properly infer a conclusion in hypothetical or devil's advocacy arguments? – some sophisticated relation between arguments as abstract objects and human intentionality is sought for in such logical analyses.

And yet, the dominant posture in these debates is stipulative hand-waving, well instantiated in Goodman's approach. Referring to Walton's (1990, p. 400) work, Goodman (2018, p. 591) disambiguates between "two fairly-well-delineated senses of 'argument'", namely, as "sets of propositions" [arguments$_1$ - ML] and "as a rule-governed kind of discussion entered into by two or more parties in order to resolve a conflict of opinions" [arguments$_2$ - ML]. For Goodman, those who focus on the latter sense – Walton, pragma-dialecticians – are "simply interested in a different topic" than logicians, such as himself, exclusively studying the former, argument-as-object sense.[2] But is Goodman really interested in a different topic? Having very carefully argued – in the sense he ostensibly is not interested in – against competing approaches to the nature of logical arguments, he ends up with the following definition:

> Take any graspable set of propositions. Any such set will have the following property: *being entertainable by an agent (or group of agents) who further believes there is a relation of support among all the members of the set, save one, and that other member*. Sets of propositions may never be entertained by anyone who believes there to be such a support relation among their members, but if and only if they *have been so-entertained at some time or other*, they are arguments. (Goodman, 2018, p. 596).[3]

Further, "[t]o entertain a set of propositions inferentially, on my view, is to believe there exists objective, a mind-independent support relation among them, whether such a relation exists or not" (Goodman, 2018, p. 598, fn. 14). Arguments$_{Goodman}$, then, are grounded in our beliefs about arguments$_{Goddu}$ – and so they move from Frege's third realm of thoughts to the realm of mental processes. Like sense-impressions they cannot quite exist without the objective, mind-independent world (other than in hallucinations and other delusions) but in the end they are psychological phenomena. Thus Goodman overlooks Frege's key lesson: one cannot do away with language and communication. "Entertaining", "apprehending", or "grasping" a thought – here, a set of thoughts – is an open act that includes questioning or hypothesizing the thought, pondering over it, and

---

2 Most argumentation scholars, such as Walton discussed here by Goodman, might agree that various concepts of argument constitute "a different topic", yet a topic they are nonetheless interested in, due to the important interrelations with "their" topic. Philosophers are far graver offenders and Goodman is just one among many. See, for one extra example, Parsons' mention of Hamblin's work on fallacies: "Hamblin's book has inspired a substantial literature that investigates fallacies defined within two-person dialogues. I ignore that literature because of my focus on argument (which I do not see as dialogue)" (Parsons, 1996, p. 174, fn. 13).

3 As a result of this: "a set becomes an argument when entertained with support relations in mind, so in a sense there exists a set that at first is not an argument, but then it becomes one due to something one of us does. [...] What we are responsible for is choosing to entertain some such sets with support relations in mind. When that happens, a set that was not an argument becomes one given something we've done. So, in that sense, an argument is a product of our activity." (Goodman, 2018, pp. 600-601, fn. 17). Simard-Smith and Moldovan similarly argue that "a set of propositions S becomes an argument just when some agent intends to infer some member of S from the other members of S in accordance with a rule of inference" (2011, pp. 254-255).

also believing it to be true or false. But in the latter case we rather speak of "judging" rather than merely "entertaining". It seems perfectly possible, both psychologically and conceptually, to entertain (e.g., to question) an argument we believe to be invalid. But on Goodman's definition, unless someone else first believed this argument to be valid, we are not entertaining an argument just yet: we merely think we are. And we would never know: maybe someone on the other side of the world thought this argument to be valid just 5 nanoseconds before I, independently of her, thought it to be utter rubbish. Enter the world of communication: without them being publicly asserted (a counterpart of judging) or at least questioned or hypothesized (counterparts of entertaining without judging), we would never know how, nor even that, humans come to "so-entertain" sets of propositions as arguments.

For instance, how did Anselm grasp or entertain his *Ontological Argument* for the existence of God, or how did John Searle entertain his *Chinese Room Argument* against the possibility of strong Artificial Intelligence? Historians of philosophy would tell us they did so exactly the way Goodman came to "entertain" the arguments for his concept of argument, namely, by being part of a larger intellectual conversation of the time. In this conversation, they encountered important problems the extant solutions to which they found objectionable. Through various intellectual activities – no doubt including individual reflection and thought experiments, but also direct spoken and written exchanges with colleagues – they eventually "grasped" their respective arguments and wrote them down in the form known to us today. Shortly, they produced their arguments via argumentation, something Goodman (2018), Goddu (2011a), and Simard-Smith & Moldovan (2011) all staunchly deny as a viable theoretical component of the concept of argument.

Now, one might resort to the classic distinction and claim that the context of discovery should not be confused with the context of justification. The way arguments might come into existence is a curious historical and empirical question, but that doesn't undermine the study of arguments *for their own sake*, as mind-independent abstract objects. However, arguments are never quite *for their own sake,* the way the Pythagorean theorem can be. This is clear in the way Goodman's arguments require an arguer's beliefs to *be* arguments and Simard-Smith & Moldovan's arguments require an arguer's intention to *be* arguments – while the Pythagorean theorem is simply out there to be discovered.

The way I see such confusions is that they owe to, often implicit, category mistakes and to the inadequate attention to the problem of individuation of arguments.

3.2. Category mistakes

Frege's distinctions capture very well what happens in which realm, and how they are separated (that also explains the enduring appeal of these distinctions, especially in speech act theory of Austin, Searle, Bach & Harnish, and others).[4] Notably, utterances are social phenomena of human communication, with all the properties of such phenomena: they are indexed to a specific time, space, and speaker; they are products of intentional human action; they enter into causal relations with other phenomena (uttering a question normally issues in an answer, calling someone names normally results in the addressee's distress, etc.); etc. Similarly for the psychological phenomena of reasoning, with the important difference that the latter are inherently private and, as such, non-communicable (they can nevertheless be communicated via thoughts about them). By contrast, "the thought, for example, which we expressed in the Pythagorean theorem is timelessly true, true independently of whether anyone takes it to be true. It needs no bearer. It is not true for the first time when it is discovered, but is like a planet which, already before anyone has seen it, has been in interaction with other planets" (Frege, 1956/1918, p. 302). The third realm of abstract objects – thoughts – is thus entirely distinct from human endeavors made possible by mental acts and by speech acts. Yet confusions abound. A striking example is Hitchcock's concept of argument:

> A premiss is an assertive, conceived as not necessarily asserted by anyone, and a conclusion is a speech act of any type, conceived as not necessarily performed by anyone or urged upon any addressee. In other words, arguments are abstract structures. When expressed, whether in language or in images or in physical behaviour, an argument invites its addressees to accept each conclusion on the basis of the acceptance of the assertives in its immediately supporting reasons. (Hitchcock, 2007, p. 121).

On this approach, "arguments are abstract structures" (elsewhere also characterized by Hitchcock as "abstract objects") that by his definition are composed of "speech acts", namely those of "premising" and of "concluding".[5] Yet, on the most straightforward interpretation of Frege, this simply can't be. Most importantly, as just seen in Frege's own words, assertion – taking something to be true and expressing it linguistically – cannot have any effect on the timelessly true abstract object, a thought.

---

4 See Textor (2021) for a recent defense of Frege's force / content distinction.
5 See Bermejo-Luque (2011) for a similar account of argument as a complex speech act composed of "the speech-act of adducing reasons" and "the speech-act of concluding."

On the other hand, contrary to what Hitchcock claims, any speech act IS "necessarily performed" by someone and it has a goal, generically, that of urging the addressee(s) to accept it. Here lies the category mistake that undermines Hitchcock's otherwise most insightful discussion of the concept of ARGUMENT. While he is exactly right to think of arguments as speech acts, he attributes to speech acts some qualities of abstract objects, and to abstract objects some of the qualities of speech acts.

3.3. Individuation problems

There do exist, in the Fregean sense of abstract thoughts, types of arguments. Modus ponens and tollens, hypothetical and disjunctive syllogism, (constructive/destructive, simple/complex) dilemma, reductio ad absurdum, etc., and their various combinations, are all good examples, indeed specimens, of abstract, immaterial, human-independent, and atemporal entities. This is the traditional realm of logical inquiry, not much different from inquiry into mathematical formulae and theorems, just the way Frege presented it. Such formal, content-free types of inference indeed constitute the object of logical inquiry by the very definition of modern formal logic. (Of course, troubling questions can immediately be asked: Are these exclusively deductive inferences? Can we treat inductions and abductions in the same way? How about arguments from analogy or causal arguments? And, by a slippery slope, all the 278 argumentation schemes of Walton? Assuming logical pluralism, objects of which logic do exist really, and which are merely a logician's inventions? Etc. While I have no time to discuss these questions here, they only reinforce my worry: it's hard to keep this concept neat, contrary to philosophers' best efforts.)

But how about concrete *tokens* of arguments, such as Judith Jarvis Thomson's *Violinist Argument* and John Searle's *Chinese Room Argument*? Recognizably, both are thought experiments that instantiate the reductio ad absurdum and the argument from analogy. But as concrete arguments, do they belong to the "timeless, eternal, unchangeable" (Frege, 1956/1918, p. 309) third realm, alongside the Pythagorean theorem? Could these arguments exist *before* humans invented a violin and Artificial Intelligence? Better: can these arguments exist *outside of a human history* in which the violin and Artificial Intelligence were invented? Can scientific and ethical arguments about the man-made climate change exist independently of humans? Could Plato or Frege somehow "grasp" or "entertain" arguments against Putin's aggression of Ukraine? (This applies to the theoretical and practical arguments alike.) As these instances of arguments clearly have human provenance, they cannot be "thoughts" understood as the abstract object of logical inquiry. Overall, the problem over what counts as an individual argument in the philosophically interesting sense looms large. Aren't all arguments reconstructible as modus ponens just the same argument, differing only in the contingent content of their premises and conclusions? Further, we can have various versions of Searle's *Chinese Room Argument*, but then they are just *versions* of the argument, which itself is an instance of the reductio and analogy. Where to draw the line between the abstract and mind-independent and the concrete and man-made?

4. Contrastivism about reason and about argument

In this section, I sketch a positive proposal on how to conceptualize ARGUMENT.[6] I hope to show how this concept avoids some of the puzzles and inconsistencies discussed above. In the following section, I will draw its further consequences.

Most importantly, argument is a contrastive concept. Philosophical *contrastivism* was first developed in the epistemological literature in the 1970s (Dretske, 1970, 1972) and then exported to other areas of philosophy such as practical reasoning,[7] philosophy of science, and philosophy of language (see Blaauw, ed., 2012; Schaffer, 2004; and Snedegar, 2015, for an overview). In the simplest of formulations, contrastivism holds that "reasons are always reasons for one thing as opposed to another" (Sinnott-Armstrong, 2008, p. 257). In Dretske's original account: "the proposition on which we operate must be understood as embedded within a matrix of relevant alternatives. We explain why P, but we do so within a framework of competing alternatives A, B, and C" (Dretske, 1970, p. 1021). More precisely:

> Epistemological contrastivism holds that reasons and grounds for belief, and hence knowledge and justified belief, are relative to contrast classes because a reason to believe a proposition is best understood as a reason to believe that proposition out of a set of conflicting propositions. In particular:
>
>> Someone, S, is justified out of a contrast class, C, in believing a proposition, P, when and only when S is able to rule out all other members of C but is not able to rule out P.
>
> (Sinnott-Armstrong, 2008, p. 259)

---

6 For further details, see Ch. 6 of Lewiński & Aakhus (2023).
7 Here, often under the term "comparativism" (see Chang, 2016; Temkin, 2012; for discussion, Lewiński, 2017a).

Two important factors should be stressed here: First, the contrastive reason is based on a relative (rather than absolute) contrast against "a" contrast class. Second, justification is in fact a falsificationist rather than verificationist activity: its key achievement is critically "ruling out" all other members, while not being able to rule out P. While this might sound like a simple version of dialectical rationality, whereby an argument is built in the process of a critical testing against an opponent's objections, it is not so.[8] Dretske's classic example is the one of a father taking his son on a trip to a zoo and showing him a zebra. Does he *know* it is a zebra? What reasons can he give to support his position before his inquisitive son? Well, that depends on a certain normative *standard* of knowledge required in this context, a standard encapsulated by the so-called *contrast class* against which the father's position is being defended. If the possible alternatives are an ostrich, elephant, or snake, the father can confidently claim this *is* a zebra, period. However, if employees at this zoo are known for clever tricks such as painting mules in white stripes (Dretske, 1970) or producing near-perfect robot replicas of zebras (Sinnott-Armstrong, 2008), then the father's burden of proof for warranting his assertion is much higher.[9]

To further appreciate the force of the contrastive view of argumentation, one can benefit from Schaffer's (2004) argument that *to know* is a contrastive predicate, not unlike *to prefer*. Saying simply that "John knows that Mary is home" is incomplete, much in the same way that saying "John prefers that Mary is home" is. What is missing is the "rather than" complement: properly speaking, "John knows that Mary is home rather than at work" and "John prefers that Mary is home rather than at work." Truth conditions of such statements depend, thus, on the contrast class: John might know that Mary is home rather than at work – for he just got her message "finished work, almost home now" – but not that she is home rather than stuck in the elevator of their apartment block, which never broke in the last 10 years. Similarly, he might prefer her being home

---

[8] Siegel & Biro (2008) criticize the dialectical model of argumentation on epistemic grounds. (Pragma-)dialectical "critical discussion" is only as good as the quality of arguers, their starting points, arguments, and critical questions. Hence, two stupid or biased arguers will most likely reach stupid or biased conclusions, even if their moves comply with all the moves and rules of an ideally reasonable dialectical procedure. Instead, as Siegel & Biro posit, a fixed logical or epistemic standard of rationality is needed for arguments to be sound. These criticisms do not easily apply to a contrastivist concept of argument, originally developed precisely to guard against skeptical challenges to belief and knowledge ascriptions via the recognition of legitimate contrastive restrictions on such ascriptions (see esp. Dretske, 1970; Sinnott-Armstrong, 2008).

[9] Sinnott-Armstrong (2008) distinguishes between modest, rigorous, and extreme contrast classes and thus standards of justification.

rather than at work, but not her being home rather than with him in the cinema.[10]

To argue admits of a very similar analysis – and, indeed, in a more straightforward and stronger sense. *To argue* or *to defend* that something is the case is even more obviously contrastive than *to know* (yes, that's a contrastive statement). In some restricted cases, one can argue that A rather than not-A (contradiction) or rather than? A (doubt), but normally one would argue that A, rather than B, C, or D or defend A against B, C, or D – and do so for specific reasons R_1 and R_2 rather than R_3 and R_4. For instance, in making a case that Bernie Sanders should be the next U.S. President, the Sanders supporter depends in intricate ways on the relevant contrast class in play for the discussion at hand. Is the Sanders supporter making a case relative to all the presidential candidates, including Republicans and independents, to all alternatives in the Democratic primary, to the progressive alternatives in the Democratic primary, to the viable alternatives for winning the general election, and so on? The contrast class is also a resource for formulating criticisms such as whether a relevant alternative has not been considered, whether a different contrast class must be considered, and so on.[11]

While further details of the contrastivist view need to be filled elsewhere (see Aikin, 2021; Aikin & Casey, 2022; Lewiński & Aakhus, 2023), this brief discussion lets me formulate a minimal, contrastivist concept of argument:

ARGUMENT = a set of reasons $\{R_1, R_2, R_3, ..., R_n\}$ advanced to support conclusion C_1 rather than any other conclusion C_n.

Strictly speaking, the set of reasons $\{R_1, R_2, R_3, ..., R_n\}$ is also contrastive, as another set of reasons $\{R_4, R_5, R_6, ..., R_n\}$ can instead be presented to support the very same conclusion C_1. Agreeing on the same conclusion but for divergent reasons is as much an argumentative situation as disagreeing over different conclusions; in both cases, the goal

---

10 Schaffer's (2004) original example is a simple statement: "Mikey knows he is drinking Coke." Does he? It depends on the contrast class, thus revealing the incompleteness of the knowledge statement. Compare: (1) "Mikey knows he is drinking Coke rather than Sprite" which sounds perfectly reasonable with (2) "Mikey knows he is drinking Coke rather than Pepsi" which sounds highly suspicious.

11 "... notice something important about participating in the dialogue. First, what reasons we give depends on the breadth of the options under consideration. So, for example, my reasons for supporting a particular candidate for office will differ depending on whom I am addressing, because I will be offering reasons for my choice as opposed to, for example, the far-right gun-nut my uncle prefers, the centrist caretaker my father likes, the crypto-Maoist my daughter likes, or just opting out of the process altogether, which my anarchist wife prefers" (Aikin, 2021, p. 840).

is to defend the contrastive bestness of one's argument, understood as a unique *case*: a unique arrangement of reasons and a conclusion (see Lewiński, 2019; Lewiński & Aakhus, 2023).

In the definition presented, I outsource the discussion over the meaning of "reasons" and "conclusion" to the mainstream philosophical literature (reasons are, of course, "justifying" reasons, not "motivating" reasons; conclusion is a generic term covering various conceptions of "claims", "standpoints", "positions", etc.). But it's important to notice that the contrastive "rather than" works very well with the concept of SUPPORT or JUSTIFICATION. In the sense relevant to this discussion, the verb "to support" means "suggest the truth of; corroborate" as in *"the studies support our findings"* (*OED*, 2021). The stress on "corroboration" rather than "proof" seems about right: inconclusive but presumably "supportive" arguments are still arguments. At the same time, "support" semantically presupposes that the thing supported is in need of help, that it cannot stand on its own. And so we make active effort to make it stand rather than collapse, or rather than let others stand in its place. Finally, this "active effort", in this case, is one of "advancing" reasons to support: support is aimed at but not guaranteed. Bad arguments are arguments too.

Now, the key point here is that support in this definition is a man-made, mind-dependent support that is "advanced". In the natural world, a giant stone can support a slab of rock, thus forming a waterfall without any human intervention: it is simply there for us to discover. But in the social world – as in *"the dome was supported by a hundred white columns"* (*OED*, 2021) – the support is in the first place constituted by a human intervention. And it is so in all the other senses of "to support", including the one of "suggesting the truth of; corroborating". Arguments – I take it from Simard-Smith & Moldovan and Goodman – belong to the social world. Each and every original argument is a unique arrangement of reasons (columns) to support a conclusion (dome). This uniqueness is a result of the creativity of human mind which, from among a class of contrastive alternatives entertained, picks one as the best of the class and advances it *as* the good argument: valid or even sound (as in deductive arguments), cogent (as in inductive arguments), "the best" (as in the inference to the best explanation), the least objectionable (as in dialectical disputations), etc. In this sense, arguments are indeed like musical compositions (Simard-Smith & Moldovan, 2011).

Being so, where do arguments belong to vis-à-vis Fregean distinctions? As already argued, argument *forms* (modus ponens, etc.) – and even all the propositions used as reasons or conclusions – sit comfortably within "the third realm" of thoughts, as long as one is at ease with the very idea of the third realm to start with. But to combine these propositions into arguments via various inference forms so that they support each other is what humans do. This process is, in the mental world, parallel to the act of *judging* a proposition true or false: it is an active human choice to consider something true or false, and to pick this unique argument from among other alternatives. In the world of linguistic communication, argument is parallel to the act of *asserting* a proposition as true or false: it is an active human choice to pronounce something true or false, and to perform this unique argument from among other alternatives. In four words: arguments are speech acts.

Arguments are thus not mind-independent abstract objects or structures. Like most other speech acts, they *use* or *rely on* such abstract objects, notably propositions. They also *use* or *rely on* inference rules, much in the same way explanations do. But the support relation – a constitutive part of the concept of ARGUMENT as defined above – is first entertained in the mental act of reasoning and then advanced in the speech act of arguing. It's important to stress that this formulation leaves it open whether arguments are *invented* or merely *discovered*. Both the constructivist and realist metaphysics would work just fine, similarly to the way they work for assertions. On the realist view, our judging and asserting that something is the case is a mere statement of a mind-independent fact; for constructivists, it contributes to or even constitutes the fact. Yet in both cases an active human intervention is needed to assert and communicate the fact as true. And the same goes for arguments. One can imagine some mythological world where all the possible, and possibly infinite, combinations of reasons and of musical notes slumber, awaiting humans to come and bring them to life. But these possible combinations of reasons would not be arguments until humans advanced them as such – just as much as propositions cannot be assertions without being asserted and notes cannot be compositions without being composed.

The speech act approach can also shed light on the genesis of arguments many have struggled with (see esp. Goddu, 2009; Goodman, 2018; Simard-Smith & Moldovan, 2011). Doesn't adding the human intervention make all arguments simply subjective? But then wouldn't the arguments disappear if arguers stopped entertaining them? (Just as pain or pleasure disappear when we stop feeling them.) If so, isn't it better to return to the objective abstract realm and embrace its puzzling complexities? Distinctions found within Searle's social ontology (1995; 2010) seem to provide a simple solution. Arguments are *ontologically subjective*. As speech acts, they are products of human creativity, just like works of art, games, money, and social institutions. But once they get created and recognized in the social world, they become *epistemically objective*: everyone has the right and obligation to entertain them as existing independently of their perceptual capacities, beliefs, desires, and intentions. Anselm's ontological argument, once made by Anselm, lives with us alongside Bach's suites and the game of chess.

To conclude: Rather than being atemporal and mind-independent abstract entities, arguments are speech acts, not unlike assertions, promises, contracts, or declarations. Once made, they (can) enter the social world and keep living there as man-made, even if abstract entities.

5. Three weird consequences

Based on the conceptual argument of the previous sections, I dedicate this section to further discuss some of the consequences of the so-conceived concept of argument.

5.1. One concept of argument

Pace O'Keefe (1977), there is no need to conceptually distinguish arguments$_1$ and arguments$_2$: arguments made *just are* arguments had. As a result, there is no topic change between these (*pace* Goddu, Goodman, Parsons, etc.). Instead, there is one concept of argument that shares some of the key qualities often attributed to these supposedly different sub-concepts.

Above in section 2, I have sketched how O'Keefe's (and *OED*'s) two concepts collapse. Argument$_1$, argument *made*, inherently, even if sometimes implicitly, involves contrast and thus disagreement with other "diverging or opposite views" (*OED*, see above). It thereby encroaches on the territory of what traditionally belonged to argument$_2$, argument *had*. Whatever remains in that territory of argument$_2$ – people having a verbal fight, a "heated or angry exchange" (*OED*, see above) – but is not support- or justification-relevant simply belongs to another concept, for example that of a VERBAL FIGHT or ANGRY EXCHANGE.

Now, against the background of a consistently contrastivist view, this can be further elaborated. On this view, "a reason or set of reasons given in support of an idea, action or theory" (*OED's* def. (2), O'Keefe's argument$_1$, see above) cannot but exist against the contrast class of "diverging or opposite views" (*OED's* def. (1), O'Keefe's argument$_2$ see above), although not necessarily in the context of an actual "exchange", let alone one that is "heated or angry one". Here, an "exchange" is a word in question. On the most straightforward interpretation, exchange is an activity engaging two or more parties actively trading something for something. Communication scholars tend to overemphasize the actual exchange element, while philosophers focus exclusively on the sets of reasons supporting a conclusion. But contrary to communication scholars, for arguments to be advanced, exchange *need not* be actual, and contrary to most philosophers, reasons *need to* be presented in the context of "of diverging or opposite views". That's the mediating and unifying position this paper defends. The wrongly excluded middle between straightforwardly dialogical and logical concepts of arguments is a dialectical concept, very much in line with Blair's (1998) proposal. Blair has argued that while not all arguments are *dialogical*, in the sense of emerging in actual communicative exchanges, all arguments are *dialectical*, as they need to address possible objections and discharge the burden of proof against these objections, to be worthy of being arguments (see also Johnson, 2000; Dutilh Novaes, 2020). On the concept defended here, this is so since the support relation advanced in an argument inescapably operates against the contrast class of competing alternatives.

5.2. Argument has a function

Pace Goodwin (2007), argument does have a proper function – namely that of supporting conclusion with reasons – as determined in its definition presented above. While Goodwin does a tremendous job at disentangling and criticizing various senses and shades of functionalism, a speech act theorist can resort to the distinction between the illocutionary and perlocutionary goal of a speech act to avoid Goodwin's bashing (see Austin, 1962). On the current proposal (see also Bermejo-Luque, 2011), the illocutionary goal of arguing is to support a conclusion with reasons. Conclusion can be an assertive, but also any other type of a speech act, especially in the case of practical reasoning, where commissive and directive speech acts are typically concluded (as in: For all these reasons, I will step down as a prime minister next week, or, You should go and apologize, because...) (see Hitchcock, 2007; Lewiński, 2021a). An argument thus conceived is a linguistic mechanism to manifestly demonstrate our support for other speech acts we perform. Metaphorically speaking, advancing arguments is like saying: "I have money in my bank account to support my offer, here's the bank statement". In this sense, this proposal smoothly aligns with Jackson & Jacob's (1980; Jackson, 2019; Jacobs, 1989) account of argumentation as a disagreement-relevant expansion of (any) speech act and with Johnson's (2000) notion of "manifest rationality".

Arguments as illocutionary acts that use reasons to support conclusions can then be used for an incredible variety of perlocutionary goals. Some of these goals are very close to the function specified here. For epistemically oriented philosophers, one distinct perlocutionary goal of argument is that of "the achievement of knowledge or at least of justified belief" (Siegel & Biro, 1997, p. 278) or, at the *very* least, of "doxastic coordination" between arguers (Patterson, 2011). (These functions are obviously restricted to the cases of theoretical reasoning where conclusions are assertions.) But Goodwin is right that once we start formulating the function of argument and argumentation in terms of communicative aims or even consequences such as "rational dispute resolution", we quickly get to the uses of argument such as "to maintain relationships" or "to experiment with the line between acceptable and unacceptable behavior" (Goodwin, 2007, p.

76). Yet, the fact that I have just used my kitchen knife as a screwdriver and a paper holder, does not is any way undermine the fact that the (human-dependent, proper) function of a knife is to cut things.

5.3. There are no conductive arguments (for all arguments already are "conductive")

Pace Blair & Johnson (2011), and the incredibly rich discussion on the topic, there are no conductive arguments in Wellman's 3rd sense of arguments which include pro/con considerations; or, if one so wishes, all arguments are conductive (including formally deductive arguments), as contrasting considerations is simply what any argument does by definition. It's certainly a discussion for a standalone paper, but one important note is essential. A conductivist would argue that pro/con considerations are prior to any contrastive work any argument does: we first "conduct" (as in "bring together") various pro-and-con considerations for a conclusion, then we weigh them producing a balance-of-considerations score, and only then can we contrast our result with arguments for other conclusions (see Blair, 2016). A contrastivist would instead claim that the "conductive pro/con considerations" essentially depend on the prior contrastive options and, as such, are derivative of the contrast classes under consideration (see Dretske, 1970, 1972; Sinnott-Armstrong, 2008; Temkin, 2012). A car that costs $ 40.000 gets a pro-argument for "economic cost" consideration when contrasted with a $ 50.000 option but a con-argument when the other option costs $ 30.000. Values and weights assigned to reasons are thus contrast-dependent, rather than the other way round. I have defended this position for practical reasoning in (Lewiński, 2017), drawing on Temkin's (2012) distinction between the "Internal Aspects View" and the "Essentially Comparative View" on the nature of practical reasoning. Here, I went a step further and built it into the very definition of argument, whether practical or theoretical (epistemic).

6. Conclusions

Argument is a set of reasons $\{R1, R2, R3, ..., Rn\}$ advanced to support conclusion C1 rather than any other conclusion Cn. Arguments are thus speech acts (note "advancing") aimed at providing reasoned support to other speech acts (conclusions), against other contrastive option(s) (note "rather than"). By proposing this definition, I carve out a unified, speech-act-based, mono-functional, and contrastive concept of argument.

In most respects, there is hardly anything new about this. There are many theories of arguments as speech acts on the market (Bermejo-Luque, 2011; van Eemeren & Grootendorst, 1984; 2004; Goodwin, 2014; Hitchcock, 2007; Jackson & Jacobs, 1980; Jacobs, 1989; Walton, 1990; for an overview, see Lewiński, 2021b). My approach presented here is entirely indebted to them. But it is a barebones approach that preserves a minimal concept of argument as a speech act, without engaging in a broader discussion over the uses of argument in actual argumentative conversations. In this way, my approach is in many respects inspired by and consistent with the philosophical work of Hitchcock, Goddu, Simard-Smith & Moldovan, and Goodman. Yet, it leads to different, sometimes directly contradictory conclusions. For one, Simard-Smith & Moldovan "argue that it [argument - ML] does not have a sense that refers to a kind of speech act" (2011, p. 232). [12] There is certainly more contrastive argumentative work to be done to further revisit and re-engineer the concept of argument.

References

Aikin, S. (2021). Argumentative adversariality, contrastive reasons, and the winners-and-losers problem. *Topoi, 40*(5), 837-844.
Aikin, S., & Casey, J. (2022). Argumentation and the problem of agreement. *Synthese, 200*, 134, 1-23.
Austin, J. L. (1962). *How to do things with words*. Oxford: Clarendon Press.
Bermejo-Luque, L. (2011). *Giving reasons: A linguistic-pragmatic approach to argumentation theory*. Dordrecht: Springer.
Blair, J. A. (1998). The limits of the dialogue model of argument. *Argumentation, 12*(3), 325-339.
Blair, J. A. (2016). Advocacy vs. inquiry in small-group deliberations. In D. Mohammed & M. Lewiński (Eds.) (2016). *Argumentation and Reasoned Action: Proceedings of the 1st European Conference on Argumentation, Lisbon 2015. Vol. I* (pp. 53-68). London: College Publications.
Blair, J.A., & Johnson, R.H. (eds.). (2011). *Conductive argument: An overlooked type of defeasible reasoning*. London: College Publications.
Blaauw, M. (Ed.) (2012). *Contrastivism in philosophy*. New York: Routledge.
Chang, R. (2016). Comparativism: The grounds of rational choice. In E. Lord & B. McGuire (Eds.), *Weighing reasons* (pp. 213-240). Oxford: Oxford University Press.
Dretske, F. (1970). Epistemic operators. *Journal of Philosophy, 67*(24), 1007-1023.
Dretske, F. (1972). Contrastive statements. *Philosophical Review, 81*(4), 411-437.
Dummett, M. (1993). *Origins of analytical philosophy*. London: Duckworth.
Dutilh Novaes, C. (2020). *The dialogical roots of deduction: Historical, cognitive, and philosophical perspectives on reasoning*. Cambridge: Cambridge University Press.

---

[12] Further, as "abstract objects that are created by human intellectual activity" arguments are not acts (whether speech acts or mental acts) and, further, "are also not subject to the act/object ambiguity" (Simard-Smith & Moldovan, 2011, p. 232).

Eemeren, F. H. van, & Grootendorst, R. (1984). Speech acts in argumentative discussions: A theoretical model for the analysis of discussions directed towards solving conflicts of opinion. Dordrecht: Foris.

Eemeren, F. H., van & Grootendorst, R. (2004). *A systematic theory of argumentation: The pragma-dialectical approach*. Cambridge: Cambridge University Press.

Frege, G. (1892/1948). Über Sinn und Bedeutung, in *Zeitschrift für Philosophie und philosophische Kritik*, 100: 25–50; translated as 'On Sense and Reference' by M. Black in *The Philosophical Review, 57*(3), (1948), 209–230.

Frege, G. (1918/1956). Der Gedanke. Eine Logische Untersuchung, in *Beiträge zur Philosophie des deutschen Idealismus*, I (1918–1919): 58–77; translated as 'The Thought: A Logical Inquiry' by A. M. Quinton & M. Quinton in *Mind, 65*(259), (1956), 289–311.

Goddu, G. C. (2009). Refining Hitchcock's Definition of 'Argument'. In J. Ritola (Ed.), *Argument Cultures: Proceedings of OSSA 09, CD-ROM* (pp. 1-15). Windsor, ON: OSSA.

Goddu, G.C. (2011a). Is 'argument' subject to the product/process ambiguity? *Informal Logic 31*(2), 75–88.

Goddu, G.C. (2011b). Commentary on 'arguments as abstract objects' by Paul Simard-Smith and Andrei Moldovan. In *Proceedings of the 9th OSSA Conference: Argumentation: Cognition & Community*, University of Windsor.

Goodman, J. (2018). On defining 'argument'. *Argumentation, 32*(4), 589–602.

Goodwin, J. (2007). Argument has no function. *Informal Logic, 27*(1), 69–90.

Goodwin, J. (2014). Conceptions of speech acts in the theory and practice of argumentation: A case study of a debate about *advocating*. *Studies In Logic, Grammar and Rhetoric, 36*(49), 79-98.

Hample, D. (1985). A third perspective on argument. *Philosophy & Rhetoric, 18*(1), 1-22.

Hitchcock, D. (2007). Informal logic and the concept of argument. In D. Jaquette (Ed.), *Philosophy of Logic* (pp. 101-129). Amsterdam: Elsevier.

Jackson, S., & Jacobs, S. (1980). Structure of conversational argument: Pragmatic bases for the enthymeme. *Quarterly Journal of Speech, 66*(3), 251-265.

Jacobs, S. (1989). Speech acts and arguments. *Argumentation, 3*(4), 345-365.

Johnson, R. (2000). *Manifest Rationality*. Mahwah, NJ: Lawrence Erlbaum.

Lewiński, M. (2017). Practical argumentation as reasoned advocacy. *Informal Logic, 37*(2), 85-113.

Lewiński, M. (2019). Argumentative discussion: The rationality of what? *TOPOI: An International Review of Philosophy, 38*(4), 645-658.

Lewiński, M. (2021a). Conclusions of practical argument: A speech act analysis. *Organon F, 28*(2), 420-457.

Lewiński, M. (2021b). Speech act pluralism in argumentative polylogues. *Informal Logic, 41*(3), 421-451.

Lewiński, M., & Aakhus, M. (2023). argumentation in complex communication: Managing disagreement in a polylogue. Cambridge: Cambridge University Press.

O'Keefe, D. J. (1977). Two concepts of argument. *Journal of the American Forensic Association, 13*(3), 121-128.

Parsons, T. (1996). What is an argument? *The Journal of Philosophy, 93*(4), 164-185.

Patterson, S. W. (2011). Functionalism, normativity and the concept of argumentation *Informal Logic, 31*(1), 1-26.
Schaffer, J. (2004). From contextualism to contrastivism. *Philosophical Studies, 119*(1-2), 73-103.
Searle, J. R. (1995). *The construction of social reality.* London: Penguin Books.
Searle, J.R. (2010). Making the social world: The structure of human civilization. Oxford: Oxford University Press.
Siegel, H., & Biro, J. (1997). Epistemic normativity, argumentation, and fallacies. *Argumentation, 11*(3), 277–292.
Siegel, H., & Biro, J. (2008). Rationality, reasonableness, and critical rationalism: Problems with the pragma-dialectical view. *Argumentation, 22*(2), 191-203.
Simard-Smith, P., & Moldovan, A. (2011). Arguments as abstract objects. *Informal Logic, 31*(3), 230-261.
Sinnott-Armstrong, W. (2008). A contrastivist manifesto. *Social Epistemology, 22*(3), 257-270.
Snedegar, J. (2015). Contrastivism about reasons and ought. *Philosophy Compass, 10*(6), 379-388.
Svoboda, V. (2020). On defining 'argument': Comments on Goodman. *Argumentation, 34*(4), 537–542.
Temkin, L. S. (2012). Rethinking the Good: Moral ideals and the nature of practical reasoning. Oxford: Oxford University Press.
Walton, D. N. (1990). What is reasoning? What is an argument? *The Journal of Philosophy, 87*(8), 399-419.

LEWIŃSKI

COMMENTARY ON LEWIŃSKI, "THE CONCEPT OF ARGUMENT: REVISITED AND REENGINEERED"

HARVEY SIEGEL
University of Miami
hsiegel@miami.edu

This is a very rich and challenging paper; I congratulate Marcin Lewiński on it. It deserves more attention than I can give it here; I hope it will be taken up by a broad range of argumentation theorists. In what follows I focus on Lewiński's reengineered concept of argument.

It is a treat to see Frege discussed in a paper on argumentation, but I won't engage in Frege-interpretation here. Instead, I note the distinction between 'lumpers' and 'splitters'. The former are aptly characterized by the slogan associated with Occam's Razor: William of Ockham's 'No entities without necessity.' The latter are captured by Bishop Butler's 'Everything is what it is and not another thing.' Lewiński is a lumper: he offers a new account of 'argument' that collapses all extant legitimate senses of the term into his reengineered one. In what follows I will take the role of the splitter, and suggest that Lewiński's reengineered concept does not acknowledge different legitimate senses of 'argument' in use by both argumentation theorists and ordinary speakers of English.

Lewiński offers the following definition of 'argument': "a set of reasons $\{R_1, R_2, R_3,..., R_n\}$ advanced to support conclusion C_1 rather than any other conclusion C_n." The key element of this definition is that in order to count as an argument, a set of reasons must be *advanced* – that is, actually put forward, in speech or some other way, by actual arguers. For this reason, Lewiński's reengineered concept has it that "arguments are speech acts". I agree that this is one sense of 'argument'. But is this the only legitimate sense of the term? In particular, isn't the *abstract object* sense of the term also legitimate? I will suggest that the answers to these questions are No and Yes. But I first briefly consider the method Lewiński employs in order to arrive at his reengineered concept, that of *conceptual engineering*.

What is conceptual engineering? David Chalmers defines it as "the design, implementation, and evaluation of concepts." (Chalmers 2020) He offers several examples: 'supervenience', 'grounding', Carnap's 'intension', Frege's 'sense', Grice's 'implicature', Kripke's 'rigid designator', Fricker's 'epistemic injustice', Haslanger's 'woman', etc. These were all invented to help solve specific philosophical problems. He distinguishes between 'de novo' vs. re-engineered concepts; Lewiński's project is the latter.

Lewiński hopes to resolve extant problems in argumentation theory, in particular that of clarifying the central term 'argument', with his new, reengineered concept. But unlike 'supervenience' and other examples of conceptually (re)engineered concepts, 'argument' is not only a technical philosophical term. It is also used in ordinary discourse by people who are not argumentation scholars. Will the reengineered concept do either for them or those scholars? I'm doubtful, for the following reasons.

First, Lewiński writes that "the support relation – a constitutive part of the concept of ARGUMENT as defined above – is first entertained in the mental act of reasoning and then advanced in the speech act of argument." He is right that that relation is first *entertained* mentally. But why does its first entertainment matter to its epistemic/support status? Consider: 'The ground is wet; so it probably rained recently.' If no one entertained this little argument concerning yesterday's storm, didn't/doesn't the wet ground still provide support to the conclusion about recent rain? The support offered to the conclusion by the premise is not a function of its first entertainment in a mental act of reasoning or its advancement in a speech act, but rather by its ability to enhance the justificatory status of that conclusion.

Second, he writes: "One can imagine some mythological world where all the possible... combinations of reasons... slumber, awaiting humans to come and bring them to life. But these possible combinations of reasons would not be arguments until humans advanced them as such." This is Lewiński's main thesis: nothing is an argument until a human advances it as such. Here I have two worries. First, a minor point: why limit it to humans? When a mouse senses a cat and hides, it doesn't advance anything. Still, doesn't 'there seems to be a dangerous predator nearby' support – constitute a good reason for – 'I'd better hide!'? Or consider an advanced non-human intergalactic group of beings who have proofs of outstanding mathematical problems: don't their theorems and proofs constitute arguments, even though no humans advanced them? Placing humans at the center of the concept seems unduly human-centric.

More importantly: This is a thesis about the *causal origins* of arguments: to become an argument, a set of propositions need to be advanced; someone needs to advance it *as* an argument. Let's grant this. What follows? *Not* that its status as an argument depends over time on its having been advanced. Once it has been advanced, it's an argument, now and forever. That is Lewiński's stated view: "Rather than being atemporal and mind-independent abstract entities, arguments are speech acts, not unlike assertions, promises, contracts, or declarations. *Once made, they (can) enter the social world and keep living there as man-made, even if abstract entities.*" (my italics) But this seems to entail that they're not just speech acts. After all, speech acts aren't abstract entities. A man-made abstract entity is still an abstract entity. So what's the theoretical import of something's needing to have been advanced in order for it to be an argument?

In addition, he holds that only inference/argument *types* (e.g. *modus ponens*) constitute arguments; *tokens* do not: "Logicians' concept of argument…is…sustainable only for *types* of inference such as *modus ponens* but unsustainable for *tokens* of arguments containing concrete premises and conclusions." (emphases Lewiński's)

This seems wrong. Consider:
1. If p, then q
2. p
3. Therefore, q

This, on Lewiński's view, is an argument. But
4. If the ground is wet, it has probably rained
5. The ground is wet
6. Therefore, it has probably rained

is not. Why is the first of these an argument, but the second not? Lewiński spends several pages arguing for this, but his arguments seem to me uncompelling. As far as I understand it, his reason for thinking so is that the first is a schema that is not dependent on communication, whereas the second one is.

That claim hinges on his view of the centrality of *communication* to both philosophy and argumentation: "…at the very center of philosophical investigations lies the world of meaningful communication. Communication is not merely a display mechanism for the ideas we have and thoughts we grasp, but an indispensable tool to understand, express, and share thoughts." Let's grant that. Now recall the point made earlier: The causal origin of a particular communicative effort to provide support to a conclusion may be interesting and important, but it doesn't imply anything, in Lewiński's hands, other than that it has achieved its status as an argument. As he grants, it is then a (man-made) abstract entity. How then does this support his reengineered concept? He continues:

> …how did Anselm grasp or entertain his *Ontological Argument* for the existence of God, or how did John Searle entertain his *Chinese Room Argument* against the possibility of strong Artificial Intelligence? Historians of philosophy would tell us they did so exactly the way Goodman [2018, discussed by Lewiński] came to 'entertain' the arguments for his concept of argument, namely, by being part of a larger intellectual conversation of the time… Through various intellectual activities… including… direct spoken and written exchanges with colleagues – they eventually 'grasped' their respective arguments and wrote them down in the form known to us today.

Lewiński then considers a possible reply: "The way arguments might come into existence is a curious historical and empirical question, but that doesn't undermine the study of arguments *for their own sake*, as mind-independent abstract objects." Lewiński replies to this reply: "However, arguments are never quite for their own sake, the way the Pythagorean theorem can be."

So arguments about the nature of arguments, on Lewiński's view, are not 'for their own sake', and in this respect are unlike Pythagoras' theorem – or rather, the argument for that theorem. Why not? Because thinking so, he argues, involves both *category mistakes* and intractable *individuation problems*. I won't address these here; I hope that the several theorists who construe arguments as abstract objects that Lewiński criticizes will do so. Rather, I end with this question: Why aren't arguments that are advanced and recognized as arguments also worth studying *"for their own sake*, as mind-independent abstract objects" once they have been advanced and recognized, just as the argument for the Pythagorean theorem is so worth studying? The Pythagorean theorem may be, as Frege held, timelessly true, but the argument for it – that is, the premises that, together with the conclusion, constitute the argument that establishes that particular conclusion as timelessly true – is just like any other argument. All have causal origins; all allege epistemic support; all are evaluable in terms of that alleged support; all are both advanced, and in that way brought into existence as arguments, and are also abstract objects, even if man-made. So what's the difference?

References

Chalmers, David (2020): 'What is conceptual engineering and what should it be?' *Inquiry*, DOI: 10.1080/0020174X.2020.1817141.

ELICITING ARGUMENTATION IN ENGINEERING-SCIENCE EDUCATION. MAKING JUSTIFIED RESEARCH-AND-DESIGN DECISIONS IN CONCEPTUAL MODELLING

MARIANA OROZCO
University of Twente
m.orozco@utwente.nl

MIEKE BOON
University of Twente

Abstract

This work deals with research-and-design (ReDe) decisions in science-engineering education and how to elicit students' argumentation, as rationales behind their decisions. The skill of making ReDe decisions plays a prominent role in science-engineering education and further professional practice. This skill is relevant to students' ability to develop conceptual models of complex real-world problems, and is part of students' overall cognitive skills repertoire. It can be expected that instructors and peers in challenging each other to make judgements and assumptions explicit have the power to elicit argumentation. This argumentation takes place within normative language games, according to a mixed deliberative and inquiry-oriented dialogue. Still new insights into how science-engineering students in interaction become skilled in making ReDe decisions are needed if we aim to advance current knowledge on cognitive skills, and make a novel contribution to educational science and practice.

This paper reports a phenomenological investigation conducted in a biomedical science-engineering programme, where students develop the skill of making justified decisions by working collaboratively in the conceptual modelling of a healthcare problem. The results show how, and to what extent, argumentation is used to support ReDe decisions, or fails to do so. Additionally, the results reveal what kind of the decisions are taken, whether decisions are primarily pragmatic, logical or criterial, what triggers a decision, and what follows a decision. In all, it appears that epistemological considerations are at the root of ReDe decisions that students make, but also at the root of those decisions that students avoid making. In considering these empirical findings, various recommendations are made for science-engineering education.

This work makes a call for more argumentation in science-engineering education, both in a deliberate fashion and as a habit of mind, i.e., by promoting both the skill to use argumentative strategies appropriately and the disposition to do so. The conclusions are expected to have scientific relevance and practical implications for science-engineering education in general.

1. Introduction

Engineering education is concerned with forming professionals who are capable of facing societal and scientific challenges, today and in the future. In particular, academic engineers are expected to be able to conduct scientific research in order to solve pressing problems and advance new knowledge in their transdisciplinary fields (Meijers, Van Overveld, & Perrenet, 2003; Van den Beemt et al., 2020). While some of such professional challenges are known or predictable, other ones are not; addressing these challenges will require that professionals engage not only substantial domain knowledge but also a myriad of skills, e.g., individual, and collective cognitive skills. The present work focuses on *cognitive skills*, more particularly in biomedical *science-engineering education* (BME), were the students conduct research on given healthcare problems. In this context, the students' activity develops according to the conceptual modelling (CM) approach (Boon, 2020; Boon & Knuuttila, 2009; Knuuttila & Boon, 2011), i.e., the students study -or research- real-life problems to understand underlying phenomena at play, and envision -or design- coherent technological solutions.

Beyond frequently studied cognitive skills, such as critical and analytical thinking, the skill of *making justified decisions during CM* emerges prominently in our empirical investigations and in everyday educational practice. Our study, therefore, concentrates on students' making research-and-design (ReDe) decisions in interaction[1] during the process of CM. The skill of making such justified ReDe decisions is relevant to students' ability to develop models that meet epistemic and pragmatic criteria related to the science-engineering problem at hand. We propose that, although ReDe cognitive activities during CM involve incessant making decisions, it is only when students conceive their cognitive activities as 'making decisions', that they will be inclined to provide the

---

1 'In interaction' refers to collective reasoning skills that are elicited by a group tasks. In the empirical context of the present study, the students work collaboratively on project-based assignments, while interacting with others within small groups, between groups, and with their tutors.

corresponding justificatory arguments. If they do not do this spontaneously, their peers will challenge them to make their arguments (and assumptions) explicit. Note that this interaction takes place in a normative space of reasons (Brandom, 2000; Sellars, 1956), where the claimer is made responsible for their claim or decision, and the challenger plays an active role in *eliciting argumentation*.

Our work carries *a call for argumentation* in science-engineering education. We advocate promoting students' frequent and appropriate argumentation to provide rationales for ReDe decisions during their CM activity, where the grounds and the implications of such decisions are prompted through normative language games (Brandom, 1994, 2000) with peers and tutors challenging each other to take their epistemic responsibility (van Baalen & Boon, 2014). Furthermore, if we expect that the students learn to justify their ReDe decisions satisfactorily, first they have to understand that their decisions (a) motivate their modelling process and output, and (b) are informed by their epistemic values[2] (De Regt, 2009, 2020). In all, science-engineering education has an important role to play in promoting not only the skill to use argumentative strategies appropriately, but also the disposition to do so. Indeed, it has been suggested that "argumentative thinking requires not only *the skill* to apply argumentative strategies, such as supporting theories with evidence, but also *the will* to apply these strategies by considering argumentative thinking to be both reasonable and worthwhile" (Hefter et al., 2018, p. 325).

The relevance of students' ReDe decisions, including the argumentation necessary to defend them, is supported by the literature on (science-) engineering education for their contribution to the students' understanding of contents, and to a high level of skill development (Garmendia, Aginako, Garikano, & Solaberrieta, 2021; Hazelrigg, 1998). However, little is known about the students' ability and disposition to make and justify modelling decisions in interaction, as part of their overall skills development during their programmes. We claim that new insights into *how these modelling skills are deployed in science-engineering education* can contribute to advancing our knowledge on cognitive skills, and that this contribution constitutes a novelty to educational science and practice. Therefore, we set up an investigation addressing this central question.

We conducted a *phenomenological study* embedded in a biomedical science-engineering programme, where students develop the skill of making justified decisions by working collaboratively in the CM of a healthcare problem involving biomaterials. We operationalised our central question into two *research questions* that further guided our own research

---

2 It has been proposed that epistemic values often remain implicit. We take it to be true that making own epistemic values explicit (and identifying those in others' discourse) can conduce to more coherent and cogent argumentation.

decisions and activity: (RQ1) How do science-engineering students in interaction deploy their developing skills in making justified ReDe decisions? and (RQ2) How can science-engineering education promote the development of such skills?

2. The empirical context

The literature on cognitive skills is complex. It deals with the topic from different disciplines (e.g., philosophy, cognitive psychology, instructional design), and it addresses such skills in terms of conceptualisation (occasionally), teaching and learning (to some extent), or measuring (more extensively, and in particular contexts). We observe that this body of literature often refers to typically expected cognitive skills, such as critical thinking and analytical thinking, while it fails to acknowledge the relevance of some other skills that appear crucial to the professional roles for which science-engineering programmes educate, according to our empirical investigations and our everyday educational practice. This refers to two sets of skills, i.e., *'making research and design decisions'*, and 'raising genuine and critical questions'. The present paper deals with the former, i.e., the ReDe decisions students make in interaction[3], which we inscribe in the more encompassing concept of 'modelling decisions', as they are taken during the process of CM. Engineering instructors agree that the many opportunities for decision-making in interaction that problem- and project-based learning offer, contribute the students' practical understanding of contents and to a high level of skill development (Garmendia et al., 2021).

The case considered here, regards the first module of a biomedical science-engineering programme (BME). This programme deals with research, design, and development of (innovative) technological products and processes for the healthcare sector (e.g., rehabilitation robots, artificial organs, and medical imaging). This field integrates engineering with natural and life sciences, such as biology, nanotechnology, physics, electrical engineering, chemistry and mechanical engineering (University of Twente, 2020a). The students are expected to become *academic*

---

3 Although the students develop their reasoning skills mainly during the interactions with each other, this does not necessarily imply that they are further incapable of individual reasoning. The aim of the educational setting considered is that students learn to reason in a conducive fashion in future (unexpected) situations, be it individually and/or in collaboration with others. We have no reasons to suggest that the collaborative character of the learning process would hamper individual learning outcomes. We advocate that the development of cognitive skills in interaction can yield independent thinkers as well.

researchers and problem-solving engineers[4], with the ability to study complex clinical problems, transform multidisciplinary insights into concrete, feasible technical solutions and, ideally, re-think those technical solutions to theorise new insights and transform their prior knowledge in a spiral way. The BME bachelor programme is organised according to a project-based learning approach, while the first module is dedicated to *biomaterials*. Here, the students investigate what the required properties of a (bio)material should be for it to function in a human body, given a particular biomedical problem; based on their findings and a deep understanding of the problem to tackle, the students propose a defendable concept for a technological solution (University of Twente, 2020b). The physical and technical feasibility of the overall solution has to be estimated and justified considering aspects of e.g., biochemistry, anatomy, physiology, feasibility of the chirurgical intervention, and the well-being of the patient. The process is facilitated by a learning assistant who, next to the teacher, accompanies various project groups throughout the project, playing a formative and an evaluative role.

An important aspect of the project is CM (Boon, 2020; Boon & Knuuttila, 2009; Knuuttila & Boon, 2011), by which students learn to structure their approach to the assignment into a phase of problem framing and a phase of solution design. Here, the students *build models that conceptually relate* various known (and still unknown) elements in a plausible story-like explanation of the phenomena involved in the problem and in the solution. Although the students may initially perceive CM only as a heuristic technique, this perception tends to develop during the execution of the assignment.

The cognitive challenges posed by the modelling activity are deemed to promote students' *learning process* and development of their cognitive skills. The CM approach to scientific reasoning in practice (Boon, 2020; Boon & Knuuttila, 2009) provides guidance by, e.g., structuring ideas, raising awareness on the various elements of a coherent model, making the connectedness among those elements more explicit, and giving a vocabulary for individual and collective reasoning. In other terms, the *CM approach invites students reasoning*, both individually and in collaboration/interaction with each other. In particular, the skill of making

---

[4] The focus of the learning objectives of the course is on the students' reasoning and the justification of their decisions, rather than on the design of solutions. In other terms, the aim of the course is that the students learn to think as a scientific researcher in a design context. This implies, for example, that the evaluation criteria are split; on the one hand there are requirements for the biomaterials for the intended function (and the students need to figure out what these requirements are) and, on the other hand, there are learning objectives related to how the students conduct scientific research, construct conceptual models, and think critically and creatively. With this clarification, we wish to emphasise the reasoning and the justification of decisions, while de-emphasising the mere design activity.

justified ReDe decisions is promoted, we propose, by the confrontation with all kinds of choices (e.g., what pieces of knowledge are relevant to the model, or adequate for the problem of interest, and why certain simplifications are justified in view of the uses of the model). Those choices are not unrestricted; *epistemic and pragmatic criteria are crucial to the justification* of those choices and, eventually, justify the quality of the entire model for its intended purpose (Boon, 2020).

Furthermore, we advocate that the CM activity is much more than the production of models as 'mental representation of external reality' used to reason and make decisions (Jones, Ross, Lynam, Perez, & Leitch, 2011), or models as 'argumentative devices' (Aydinonat, Reijula, & Ylikoski, 2021). The appropriate use of CM in this programme -and beyond- constitutes an intended learning objective (ILO), for being considered a conducive way of reasoning, a desirable habit of mind and a transferable cognitive skill. Becoming skilled in CM is consistent with the pursued profile of the engineer-scientist (Boon, 2020).

3. Theoretical considerations

In general, researchers (-to-be) make decisions all the time, while these are only likely to be well thought-out and justifiable if researchers *"acknowledge these decisions, and recognise them as decisions"* (Braun & Clarke, 2006, p. 80), and further proceed to conducting their investigation accordingly. Ideally, such decisions should be logged and accompanied by an *argument-based justification,* as well as discussed in an "ongoing reflexive dialogue on the part of the researcher" (Braun & Clarke, 2006, p. 82). Another example of the crucial role of justificatory reasoning in ReDe can be found in decision-based engineering design (Hazelrigg, 1998): the recognition that engineering design (a particular case of modelling activity) as a decision-making process not only forces such design into a total systems context[5], but it also has consequences on engineering education. In the same line, existing studies on engineering-design learning experiences (Wendell, Andrews, & Paugh, 2019) emphasise the many decisions involved in the design activity suggesting that these are productive for learning[6]. CM frameworks (Aydinonat et al., 2021; Boon, 2020; Boon & Knuuttila, 2009) also support the idea that the (collective) cognitive activities and decisions taken in iterative phases of research and

---

[5] A total systems context aims at exhaustiveness, considering multiplicity of aspects rather than reducing complexity (e.g., design decisions account for a product's total life cycle).
[6] It can be noted that the literature often keeps design methodology (and education in it) separated from science-engineering research, whereas the CM approach aims to keep them integrated.

design constitute essential parts of the overall modelling. Beyond the goodness of the decision (e.g., in terms of its content[7] and timing), it is *important to understand* what (kind of) decisions are taken, on what bases they are taken (i.e., the dimensions of thinking), what follows a decision, what triggers a decision, and the difficulties in deciding.

In terms of the thinking that underpins the students' decisions, Ennis (1962), in his philosophical work on critical thinking, proposes that one or more dimensions are at play. Indeed, we propose that part of Ennis' work to conceptualise 'critical thinking' can be used to analyse the modelling decisions that science-engineering students make. This framework posits that "there are three basic analytical distinguishable dimensions" (Ennis, 1962, p. 84): a logical, a criterial, and a pragmatic dimension. Roughly, the *logical dimension* is about judging presumed relationships between meanings of words and statements; for example: knowing what follows from a statement by virtue of its meaning (in the field in which the statement is made), knowing what it is for something to be a member of a class, and knowing how to use logical operators. In its turn, the *criterial dimension*, is about knowing the criteria for judging statements (other than the logical criteria); for example: using criteria for judging whether an observation statement is reliable (i.e., a set of rules built up in a particular field for judging accuracy of observation), whether an inductive conclusion is warranted, the adequacy of a definition, and whether a statement made by a presumed authority is acceptable. Finally, the *pragmatic dimension* regards the role of the background purpose, and it refers to whether the decision (in the form of a statement) is good enough for that purpose. The background purpose plays a legitimate function in appraising the acceptability of statements, as it includes an acknowledgement of the necessity for the balancing of factors that precede the statement (whereby complete criteria cannot be established). Noteworthy, these dimensions resonate with epistemic and pragmatic criteria used in CM[8].

Wendell and colleagues (2019), having studied engineering design learning experiences, suggested several strategies as being *productive for learning* (e.g., the students must document their design iterations), while emphasising the many decisions involved in the design activity. Additionally, ethnographic studies focusing on interactive decision-making episodes, argue for "an explicit *retrospective focus* on the processes

---

[7] The goodness of the content refers to the quality of the model construction and resulting model. It includes epistemic and pragmatic criteria relevant to the specific problem.
[8] The logical dimension: logical consistency, internal coherency, empirical adequacy. The critical dimension: coherency with accepted knowledge. The pragmatic dimension: intelligibility (for model users) and relevant (to intended epistemic purpose).

and consequences of [group][9] decisions following projects as an avenue for fostering the development of design judgment that engineering graduates will take into their professional practice" (Campbell, Roth, & Jornet, 2019, p. 294). In its turn, the CM framework (Boon, 2020; Boon & Knuuttila, 2009), as well as other modelling frameworks (e.g., Justi & Gilbert, 2002) support the idea that the (collective) cognitive activities and decisions taken in iterative phases of research and design constitute *essential parts* of the overall modelling. From the preceding references, it seems that there is much agreement on the relevance of ReDe decisions and their justifications. Concurrently, it must be noted that 'mainstream engineering design education' not often includes scientific reasoning as goal, as it often targets product design focused on functions and their uses, rather than the *creation of processes or functionality that need some mechanistic understanding* in order to technologically conceive them. Therefore, decisions in CM are not to be equated to decisions in mainstream design, given that they require *different kinds of justifications*. In other terms, different frameworks refer to models with different epistemic purposes and, accordingly, different cognitive activities become more central (e.g., prediction and explanation, rather than design), while other activities remain in common (e.g., knowledge generation).

Before we can focus on the proper argumentative strategies in the context of science-engineering education, students' *frequent and conducive argumentation in support of ReDe decisions are necessary*, both in a deliberate fashion and as a habit of mind. First, argumentative thinking requires the students' will to do so, for considering it sensible, and meaningful (Hefter et al., 2018). Furthermore, students and instructors are supposed to acknowledge the need to provide rationales (to themselves and to others) and to request rationales from others, thus making themselves responsible for their claims and the implications thereof. This involves that they act as 'scorekeepers' (Brandom, 2000) in the construction of their 'normative space of reasons' (Sellars, 1956). In *inferentialist and normative pragmatist* terms (Brandom, 1994, 2000), students are expected to actively play the 'game of giving and asking for reasons' to enhance their social reasoning skills. Brandom's non-representationalist theoretical framework is considered conducive to resolving ongoing debates in educational theory[10] (Derry, 2013; Su & Bellmann, 2018).

If we consider the preceding theoretical ideas in the light of the context of study (as described in the previous section), the question raises: *What*

---

9 We replace the term 'team' by the term 'group', as the latter is more appropriate (Blankenship & Ruona, 2009) for the features and dynamics of the collaborations among students we investigate.

10 For example, what it means to understand concepts, and how to conceptualise the role of knowledge in educational processes.

kind of dialogue game do science-engineering students need to play?[11] In the description of the empirical context and the expectations vis-à-vis the students, we emphasised the reasoning and the justification of decisions, while we de-emphasised the design activity. Such cognitive focus has bearings on (1) what types of dialogue are mainly at play, (2) what is exactly at stake in the trade-off that the students go through, and (3) the kind of reasoning involved.

Firstly, the type of dialogue at play can be considered as *'mixed deliberative and inquiry-oriented argumentation'*, characterised by cooperation and/or collaboration rather than by competition (Walton & Krabbe, 1995). In seeking a common 'space of reasons' (Sellars, 1956), there is no negotiation or need to resolve in the sense that one speaker needs to persuade or convince the others. Rather, the arguments have no owner who wins or loses in a debate-like conversation. In this 'game' there is no winner or loser, rather the group of interlocutors move together into a broader and more complex common space. It can be objected, however, that our use of the term 'argumentation' to refer to a deliberative and inquiry-oriented dialogue, stretches beyond the more common-sense conception that restricts 'argumentation' to exchanges intended to resolve disagreements (Krabbe & van Laar, 2007). Secondly, if there is any trade-off, such trade-off is not a balancing of students' opinions or preferences, but a *collaborative appraisal* of the acceptability of pieces of knowledge for the intended purpose and particular case (including considerations of uncertainties, and risk taking). Thirdly, the *cognitive activity* involved concerns, e.g., explanatory reasoning, exploration, imagination, idealisation, conceptualisation, categorisation, and integration (thus going beyond negotiating, persuading, or epistemic sacrificing).

Such collective reasoning is *normative*, rather than probative. It is normative because the framing of the problem and the proposed solution are scientifically and ethically informed (here, the students need to acknowledge the new knowledge that is being generated during this reasoning process, including its kind, validity[12], and epistemic value). It is not probative because it does not seek to prove hypotheses, or to serve testing or trial such as compliance to fixed evaluation criteria for solutions to the given problem.

---

11 We thank Jan Albert van Laar for raising the question on the kinds of dialogue at play in our context of study, and for his thought-provoking commentary.
12 'Validity' is used here to denote epistemic adequacy (which considers the purpose of the knowledge at stake), and trustworthy (which considers the methods that gave rise to the piece of knowledge considered and is accompanied by an appraisal of its value and its limitations).

4. Methods

Our empirical educational research proceeded in a *natural setting* (Cohen, Manion, & Morrison, 2011), while the research design followed as much as feasible the recommendations from the methodological literature (Cohen et al., 2011; Small, 2021; White, 2017). We observed thirty project groups (consisting of 4 to 5 students) during a semester while they were working in interaction with their peers and tutors on a ReDe task. More specifically, we *observed* the students' making and justifying ReDe decisions during the modelling of a real-life problem and a plausible solution. For each group, we recorded several students-tutor *conversations* and collected various students' *documents* (i.e., logbooks, and preliminary and final project reports). All students were informed and consented to participate in this investigation. First, we considered all data systematically to grasp *contextual issues*, explore for generalities and particularities, and identify potential 'outliers' in any sense. Further, we prioritised all groups on convenience and theoretical bases, i.e., considering the amount and richness of the data collected (rather than selecting cases at random). Eventually, three cases were shortlisted for further *in-depth analysis* of their multiple data sources (i.e., all transcripts of conversations and documents), including within-case comparisons over time and between-case comparisons (Yin, 2012).

During the coding steps, which involved content-conversational analysis and thematic coding, we looked for patterns in the ReDe decisions the students made and explored for changes in those patterns over time. The analysis focused on the bare decisions as verbalised by the students, and on their argumentation. This work did not judge the goodness of the decisions, rather it sought to grasp what kind of decisions were taken, on what bases, what triggered a decision, the difficulties in deciding, what followed a decision, and whether decisions were accompanied by argumentation or not.

5. Results

5.1. Directly addressing our research questions

Concerning *RQ1*, our findings suggest that the content of students' decisions mostly concern the relevance and the trustworthiness of pieces of knowledge and pieces of information they found or received from others. These were often pragmatic considerations, rather than criterial or logical (Ennis, 1962). In general, students did not feel urged to justify their decisions, but when they did provide a justification, they tended to use logical language. It appears, however, that the very use of seemingly logical reasoning may lead to deception (albeit unintentionally).

Regarding *RQ2*, and based on our findings and conclusions, we formulated a number of recommendations (on which we elaborate in the concluding section) for science-engineering education at the instructional level. At this point, we anticipate that some of these recommendations are in line with research on the 'consistency argument' (Montanari, 2019), an epistemic and argumentative norm allowing students to recognise contradictions in the process of argumentation and familiarise with certain argumentative strategies. In general, we recommend continuing fostering the CM skill in connection to the argumentative skill, e.g., by eliciting rationales for why and how students include certain aspects (and not other ones) into the conceptual models of the problem and the solution. In such argumentation, students are expected, e.g., to use assessment criteria for validation of the conceptual models, such as (internal) consistency, coherence, and empirical adequacy (Boon & Knuuttila, 2009; Van Fraassen, 1980).

5.2. Across the cases

Firstly, the results show, what kind of decisions were at stake. Here, 'what kind' refers to the content of the decision, which we connect to the dimension of the underlying thinking. We found decisions: (a) to investigate something further for considering it relevant, useful, and not yet sufficiently known or understood, and, oppositely, (b) to weed out or let go something for considering it irrelevant or leading too far. Here, 'something' refers to e.g., a scientific article, a piece of advice from the peer feedback, or an own idea. It seems then that most decisions pertain to inclusion/exclusion into the model (rather than e.g., how to approach the project task). Remarkably, most (if not all) identified decisions seem to respond to pragmatic considerations, while there is virtually no evidence of criterial and logical considerations.

Secondly, the results show that only in exceptional cases the students felt the need to log and *justify the decisions made*. Moreover, when this happens, quite some logical language is used thus denoting logical thinking, next to the pragmatic. However, on several occasions, the students believed that they were providing an acceptable justification, while the explicit rationale was missing.

Thirdly, the results show *when decisions appear* (i.e., what motivates or drives a particular decision). We found decisions appearing when triggered by new knowledge, when challenged by others (e.g., peers and/or tutors), on deliberate revision of the work done so far. In general, the decisions were taken both proactively (e.g., when deliberately designing) and reactively (e.g., when an issue is encountered).

Fourthly, the results show *what follows a decision*, e.g., decisions could lead to new questions, hypotheses, new decisions, approaches to the project task. In particular, there is evidence that students sometimes *need to revise decisions*, as they encounter problems that call for iterative work (which cannot be fully avoided, as many insights only emerge in retrospect). This revision, however, was seldom observed: the tutors needed to remind the students about revising past decisions and daring change them if necessary.

In an exemplary (and exceptional) passage in a logbook, a particular decision and its context were written in a narrative fashion.

Illustration. *While searching information, [Student X] found that the anatomy and physiology of the oesophagus are important elements of the conceptual model. We discussed this in our meeting. Together with the tutors, we decided we would study the anatomy surrounding the oesophagus as well. For example, it is important to know that the trachea is next to it. It may be the case that the trachea needs the oesophagus as support, or the other way about. At this point, we decided that [Student Y] would study the surrounding anatomy and that [Student X] would look at the anatomy of the oesophagus in even more detail.*

This rich passage revealed: (i) what gave rise to a felt need to discuss in order to decide something, (ii) that the main cause of the decision was primarily a pragmatic consideration, (iii) that the students considered it necessary to discuss within the group before deciding, (iv) that they preferred to consult the tutor, (v) what the core content of the decision was, (vi) that the decision raised new hypotheses and questions, (vii) that the decision led to new decisions.

Furthermore, we observed instances of *impossibility to decide*. For example, in a research question –taken from a student's report– which refers to efficiency of the solution, the meaning of 'most efficient' is unclear and, therefore, it is also unclear how this question should be approached. This witnesses lack of criteria: the students did not break down 'efficiency' into criteria that: (a) shows what aspects they consider and their relative importance, (b) provide them guidance in their research and decisions, (c) are specific enough to assess the goodness of the proposed solution.

5.3. Between-case and within-case analyses

A comparative analysis between the cases studied reveals that only Case III often is explicit about its making decisions and quite often uses logical vocabulary to introduce a justification (either a proper justification or not). That this behaviour is not observable in the other cases (neither in their logbooks and reports, nor in their conversations), does not mean that it is not happening, i.e., all groups make decisions (a) either deliberately or not and (b) either overtly justified or not.

In searching for any patterns of *decisions over time* we performed a within-case analysis. For Case III, we observed more explicit attention to decisions (using the logbooks as evidence) in the last third of the project duration. For Cases I and II, no evidence is available (neither confirming, nor disconfirming).

6. Discussion and conclusion

In our analysis, we were interested in grasping -particularly connected to argumentation- on what bases ReDe decisions are supported, as well as what kind of the decisions are taken, what triggers a decision, and what follows a decision.

6.1. How students deploy their ReDe decision skills

Our first research question sought description of how students use their developing skills in making and justifying ReDe decisions. The results revealed that students made decisions both *proactively and reactively*, due to different motives that originated either inside or outside their work groups (e.g., own curiosity, or an external challenge). This evidence on the multi-temporal character of decisions is in line with prior calls for focus on decisions at various stages (Braun & Clarke, 2006; Campbell et al., 2019), as introduced in our theoretical considerations section. Our findings also suggest that the content of the decisions students made mostly pertained to the significance (for the particular purpose considered) and the reliability (in broad terms) of pieces of knowledge and pieces of information that they found or received from others. These were often *pragmatic considerations* -rather than criterial or logical (Ennis, 1962)- intended to outline and organise their own work. Presumably, the lack of evidence of logical considerations resides in them not being verbalised and, therefore, we cannot simply claim lack of logical reasoning. On the other hand, the lack of evidence for criterial reasoning may actually point to the students' unattendance to criteria (e.g., the module's ILOs and the rubrics in the project's assessment instrument). We can think of several plausible explanations for this unattendance and, in the next subsection, we suggest corresponding ways to tackle the problem.

Most of the times, students did *not feel urged to justify* their decisions (on writing or in discussions); when they did provide a justification, they tended to use logical language. It seems, however, that the very use of *seemingly logical* reasoning may lead to deception (albeit unintentionally), as the students sometimes provided unacceptable arguments that lack any rationale but succeed in eventually dazzling the inattentive tutor.

Interestingly, students' decisions appeared to have *consequences and implications* ranging from further thinking activity to some hands-on activity. Such activity reflected most often modelling steps towards the design of an engineering solution (Boon & Knuuttila, 2009; Wendell et al., 2019). However, the students groups often seemed to depend on the tutors' coaching skills to acknowledge such implications (i.e., to make them visible or tangible). This was particularly the case, when a decision needed revision or called for iterative work that threated to spoil (part of) the work carried out so far.

One limitation of this investigation is that the logbooks (our most important data source) insufficiently witness the point at which students draw a conclusion (and the arguments behind such conclusion) and decide to act accordingly. In general, students seem not to be skilled at *verbalising their reasoning* (neither in speech, nor in narrative writing); other researchers have also indicated as a limitation that students failed to report their reasoning as they are often 'not used' to writing down their thoughts (Treagust, Mthembu, & Chandrasegaran, 2014). Consequently, we only have limited evidence for patterns of decisions over time (to respond to the question about the developing skill): the apparent increase in explicit decisions-justifications towards the end of the project work cannot be generalised to all cases, nor beyond. When such increase was observed, plausibly the students felt triggered by the need to converge to a report and by the need to respond to the peer feedback. If this was the case, we could state that both the *project reporting* task and the task to *respond to peer feedback* benefited the development of thinking skills in terms of (a) more deliberate and informed decisions and (b) more argumentation as justification. This conclusion aligns with the literature about the usefulness to think in terms of 'design decision' in research (Braun & Clarke, 2006).

Finally, the results suggest that in some instances, there was an *impossibility to decide*. When the three dimensions of thinking are missing (i.e., there are no criteria, no logics, and no pragmatic considerations): How can the students make decisions? As discussed before, pragmatic decisions appeared to be the most extended, while logical decisions may have been more extended than the evidence reveals, due to their expected implicitness. Criterial decisions, on their turn, remained surprisingly absent and this raises a new hypothesis: criteria are confrontative and students tend to avoid confrontation. We propose that, in general, criteria (any requirements for biomaterials or the modules' ILOs) may be unknown or unclear in the beginning but become progressively tangible and, eventually, must be used for an evaluation (any evaluation of a proposed solution or of the own learning). Often it is not straightforward how to measure such criteria or the available methods are unsatisfactory; eventually, we must deal with lack of certainty (e.g., make assumptions, estimate validity, take risks, and hedge our claims). *Uncertainty* puts us all (also these students) in uncomfortable situations that we prefer to avoid, to the extent that criteria are neglected, and decisions arrested. This is an epistemological issue that requires attention in (science-engineering) education.

Further research on the justification of ReDe decisions (that would count on rich data in terms of frequent and extensive argumentative justifications) could employ 'argumentation profiles' (Macagno, 2022) to analyse argumentative strategies by taking into account the types of argument used, and their quality.

6.2. How science-engineering education can promote ReDe decision skills

Our second research question intended to prompt recommendations for science-engineering education to promote the development of skills in making justified modelling decisions. From our theoretical considerations, it follows that engineering education needs to promote thinking in terms of making decisions both in research and in design work (Hazelrigg, 1998). As explained before, if we expect that students' ReDe decisions are satisfactorily justified, the students need to acknowledge that these decisions motivate their modelling activity in the first place, while science-engineering education needs to actively *promote the 'skill and will' of argumentative thinking* (Hefter et al., 2018). We advocate for more frequent and appropriate argumentation, i.e., rationales for ReDe decisions elicited during the CM activity through normative language games (Brandom, 1994, 2000) in which peers and tutors challenge each other to take their epistemic responsibility (van Baalen & Boon, 2014) for the grounds and the implications of their claims and decisions. In short, the present contribution is *a call for argumentation in science-engineering education*. Moreover, we propose that the interaction among students (and those with the tutor) functions as a language game that elicits argumentation in a normative environment, thus promoting the development of the interlocutors' argumentative skills, particularly -but not only- in connection to CM.

Our findings and conclusions allow us to make a number of *recommendations*: see Table I. The instructors (both teachers and tutors) need to be or become capable of implementing the following strategies, in terms of understanding the theoretical backgrounds and having the required professional skills.

Table I. Recommendations

Number	Content
1	Do not take students' ability to make proactive/reactive decisions for granted, nor assume that all students are equally skilled or develop the skill at the same pace.
2	At times, analyse a good/bad decision with the students, e.g., (i) whether it is primarily criterial, logical and/or pragmatic, (ii) what originates it, (iii) what follows it, (iv) how is it justified or can be.
3	Challenge the students to (further) justify their decisions. Do not accept modelling decisions lacking a rationale or an acceptable rationale (on the instructor consideration). This can be done by questioning.
4	Engage with the students in their (logical) reasoning to strive for sound justifications.
5	Identify and diagnose any need to assist the students, providing them guidance in the aspect of their ReDe decision-making they struggle with (e.g., by confronting them with vagueness and poor operationalisation, lack of acceptable justification, insufficient acknowledgement of the implications and further action).
6	Encourage the students to reflect, preferably on writing, in a narrative fashion.
7	Remind the students that they should review past decisions and dare change them if necessary.
8	Remind the students to attend to criteria such as the module's ILOs and the rubrics in the project's assessment instruments, and to use such criteria to guide their learning (i.e., in a formative rather than in an evaluative fashion). In particular, emphasise the epistemic and pragmatic criteria captured by CM, as offered to students through lectures, workshops and syllabus.
9	Often use opportunities to make logical criteria explicit and discuss them. Discuss epistemological problems overtly as they emerge (e.g., problems of stability, relativity, validity, uncertainty, and plausibility).

6.3. Final conclusion

This work situates in the field of cognitive social argumentation in higher education; it deals with ReDe decisions in science-engineering education and how to elicit students' argumentation, as rationales behind their decisions. It aims to contribute new insights into how science-engineering students in interaction become skilled in making ReDe decisions, and into how science-engineering education can promote such development. Our findings result from a rigorous in-depth analytical method that relied on multiple comparisons, iterations, and a dialogue among various data sources, while building on a conducive theoretical considerations.

Despite the limitations of the investigation, we advanced new insights into what (kind of) decisions are taken, on what bases they are taken (i.e., according to what dimensions of thinking), what triggers them, what follows from them, and some difficulties in deciding. Also, we advocated that the apparent development of the modelling decision-making skill (or at least the change over a relatively short time), is triggered by the instructional design of the module and, more particularly, the design of the project task through CM.

In all, it appears that epistemological considerations are at the root of modelling decisions that students make (that are connected to the relevance and trustworthiness of knowledge), but also at the root of those decisions that students avoid making (that are connected to the stability and certainty of knowledge). Such root considerations situate in the realm of students' epistemological believes (Bromme, Pieschl, & Stahl, 2010; Hofer & Pintrich, 2012) and epistemic values (De Regt, 2009, 2020). Based on our conclusions, we suggested what is at stake in modelling decisions and made recommendations for practitioners to enhance the students' skill development. A condition is that instructors be or become capable of implementing the recommended strategies. As advanced in the introduction to this paper, our work makes a call for more argumentation in science-engineering education, both in a deliberate fashion and as a habit of mind, i.e., by promoting both the skill to use argumentative strategies appropriately, and the disposition to do so. This includes students acknowledging that their modelling decisions are informed by their epistemic values (De Regt, 2009, 2020), and taking their epistemic responsibility (van Baalen & Boon, 2014).

Our conclusions are expected to have both scientific relevance and practical implications for science-engineering education in general. The cognitive skills here investigated are key both in the context of biomedical engineering education and in all other science-engineering programmes, as long as their aim is to form professionals who are able to conduct scientific research related to challenging problems and to advance new knowledge in their fields. Finally, we advocate that our focus on modelling decision-making, as one key element of a broader set of cognitive skills, constitutes a novel contribution to educational science and practice.

References

Aydinonat, N. E., Reijula, S., & Ylikoski, P. (2021). Argumentative landscapes: the function of models in social epistemology. *Synthese*. doi:10.1007/s11229-020-02661-9

Blankenship, S., & Ruona, W. E. A. (2009). Exploring Knowledge Sharing in Social Structures: Potential Contributions to an Overall Knowledge Management Strategy. *Advances in Developing Human Resources*, 11(3), 290-306.

Boon, M. (2020). Scientific methodology in the engineering sciences. In D. P. Michelfelder & N. Doorn (Eds.), *The Routledge Handbook of the Philosophy of Engineering* (1 ed.). New York: Routledge.

Boon, M., & Knuuttila, T. (2009). Models as Epistemic Tools in Engineering Sciences: A Pragmatic Approach. In A. Meijers (Ed.), *Philosophy of technology and engineering sciences. Handbook of the philosophy of science* (Vol. 9, pp. 687-720). North-Holland: Elsevier.

Brandom, R. (1994). Making it explicit : reasoning, representing, and discursive commitment. Cambridge, Mass.: Harvard University Press.

Brandom, R. (2000). Articulating reasons: An Introduction to Inferentialism. In *Articulating Reasons*: Harvard University Press.

Braun, V., & Clarke, V. (2006). Using thematic analysis in psychology. *Qualitative Research in Psychology*, 3(2), 77-101. doi:10.1191/1478088706qp063oa

Bromme, R., Pieschl, S., & Stahl, E. (2010). Epistemological beliefs are standards for adaptive learning: a functional theory about epistemological beliefs and metacognition. *Metacognition Learning Metacognition and Learning*, 5(1), 7-26.

Campbell, C., Roth, W. M., & Jornet, A. (2019). Collaborative design decision-making as social process. *European Journal of Engineering Education*, 44(3), 294-311. doi:10.1080/03043797.2018.1465028

Cohen, L., Manion, L., & Morrison, K. R. B. (2011). *Research methods in education* (7th ed.). New York: Routledge.

De Regt, H. W. (2009). The epistemic value of understanding. *Philosophy of Science*, 76(5), 585-597.

De Regt, H. W. (2020). Understanding, values, and the aims of science. *Philosophy of Science*, 87(5), 921-932.

Derry, J. (2013). Can Inferentialism Contribute to Social Epistemology? *Journal of Philosophy of Education*, 47(2), 222-235. doi:10.1111/1467-9752.12032

Ennis, R. H. (1962). *A Concept of Critical Thinking*: Harvard Educational Review.

Garmendia, M., Aginako, Z., Garikano, X., & Solaberrieta, E. (2021). Engineering instructor perception of problem- and project-based learning: learning, success factors and difficulties. *Journal of Technology and Science Education*, 11(2), 315-330. doi:10.3926/jotse.1044

Hazelrigg, G. A. (1998). A Framework for Decision-Based Engineering Design. *Journal of Mechanical Design*, 120(4), 653-658. doi:10.1115/1.2829328

Hefter, M. H., Renkl, A., Riess, W., Schmid, S., Fries, S., & Berthold, K. (2018). Training Interventions to Foster Skill and Will of Argumentative Thinking. *Journal of Experimental Education*, 86(3), 325-343. doi:10.1080/00220973.2017.1363689

Hofer, B. K., & Pintrich, P. R. (2012). Personal epistemology: *The psychology of beliefs about knowledge and knowing*: Routledge.

Jones, N. A., Ross, H., Lynam, T., Perez, P., & Leitch, A. (2011). Mental Models: An Interdisciplinary Synthesis of Theory and Methods. *Ecology and Society*, 16(1).

Justi, R., & Gilbert, J. K. (2002). Models and modelling in chemical education. In J. K. Gilbert, O. De Jong, R. Justi, D. Treagust, & J. H. Van Driel (Eds.), *Chemical Education: Towards Research-Based Practice* (pp. 317-337). Dordrecht: Kluwer Academic Publishers.

Knuuttila, T., & Boon, M. (2011). How do models give us knowledge? The case of Carnot's ideal heat engine. *European Journal for Philosophy of Science*, 1(3), 309. doi:10.1007/s13194-011-0029-3

Krabbe, E. C. W., & van Laar, J. A. (2007). About Old and New Dialectic: Dialogues, Fallacies, and Strategies. *Informal Logic*, 27(1), 27-58. doi:10.22329/il.v27i1.463

Macagno, F. (2022). Argumentation Profiles: A Tool for Analyzing Argumentative Strategies. *Informal Logic*, 42(1), 83-138.

Meijers, A., Van Overveld, C., & Perrenet, J. (2003). *Academische criteria voor bachelor en master curricula*: Technische Universiteit Eindhoven.

Montanari, E. (2019). Educating Students to Consistency via Argumentation. *Informal Logic*, 39(3), 263-286. doi:10.22329/il.v39i3.5100

Sellars, W. (1956). Empiricism and the Philosophy of Mind. *Minnesota studies in the philosophy of science*, 1(19), 253-329.

Small, M. L. (2021). What is "Qualitative" in Qualitative Research? Why the Answer Does not Matter but the Question is Important. *Qualitative Sociology*. doi:10.1007/s11133-021-09501-3

Su, H., & Bellmann, J. (2018). Inferentialism at Work: The Significance of Social Epistemology in Theorising Education. *Journal of Philosophy of Education*, 52(2), 230-245. doi:10.1111/1467-9752.12292

Treagust, D. F., Mthembu, Z., & Chandrasegaran, A. L. (2014). Evaluation of the Predict-Observe-Explain Instructional Strategy to Enhance Students' Understanding of Redox Reactions. In I. Devetak & S. A. Glažar (Eds.), *Learning with Understanding in the Chemistry Classroom* (pp. 265-286). Dordrecht: Springer Netherlands.

University of Twente. (2020a). Biomedical Engineering (Master programme). Retrieved from https://www.utwente.nl/en/education/master/programmes/biomedical-engineering/

University of Twente. (2020b). De maakbare mens: Construeren met moleculen. [The makeable human: Building with molecules]. Retrieved from https://www.utwente.nl/onderwijs/bachelor/opleidingen/biomedische-technologie/studieprogramma/studiejaar-1/

van Baalen, S. J., & Boon, M. (2014). An epistemological shift: from evidence-based medicine to epistemological responsibility. *Journal of Evaluation in Clinical Practice*, 21(3), 433-439.

Van den Beemt, A., MacLeod, M., Van der Veen, J., Van de Ven, A., van Baalen, S., Klaassen, R., & Boon, M. (2020). Interdisciplinary engineering education: A review of vision, teaching, and support. *Journal of Engineering Education*, 109(3), 508-555.

Van Fraassen, B. C. (1980). *The scientific image*. Oxford: Clarendon Press.

Walton, D., & Krabbe, E. C. (1995). *Commitment in dialogue: Basic concepts of interpersonal reasoning*. Albany, NY: State University of New York Press.

Wendell, K. B., Andrews, C. J., & Paugh, P. (2019). Supporting knowledge construction in elementary engineering design. *Science Education*, 103(4), 952-978. doi:10.1002/sce.21518

White, P. (2017). *Developing research questions* (2 ed.). London: Palgrave.

Yin, R. K. (2012). Case study methods. In APA handbook of research methods in psychology, Vol 2: Research designs: Quantitative, qualitative, neuropsychological, and biological. (pp. 141-155). Washington, DC, US: American Psychological Association.

COMMENTARY ON "ELICITING ARGUMENTATION IN ENGINEERING-SCIENCE EDUCATION: MAKING JUSTIFIED RESEARCH-AND-DESIGN DECISIONS IN CONCEPTUAL MODELLING," BY MARIANA OROZCO AND MIEKE BOON

JAN ALBERT VAN LAAR
University of Groningen
j.a.van.laar@rug.nl

1. Commentary

In response, I have one extended comment, that I regard as fully supportive of the paper. The authors contend that in order to educate researchers working in situations where good design solutions need to be developed, students need to learn, in small groups, to make all kinds of decisions when doing research in the service of moving from problem to design solution, and these decisions need to be justified in the sense that they are supported by argumentation. On the basis of research into the ways students do actually make such decisions, and the extent to which these decisions are justified with arguments, they advance a number of recommendations on how to improve the skills and attitudes required for developing well-argued research-and-design decisions.

One central idea is that students need to engage in games where they challenge decisions, ask for reasons, and thus elicit argumentation, and where decisions are in response justified, argued for, and well-considered. What kind of conversational game should this be?

Krabbe and Walton (1995) distinguish six main types of dialogue, and in addition elaborate on what they call 'mixed types'. A *persuasion dialogue* starts from a disagreement, and aims at its resolution. It proceeds by persuasive reasoning (Krabbe and van Laar 2007). A *deliberation dialogue* starts from a practical question on what course of action to adopt. The participants do not need to start from a disagreement over the answer. The main goal is to arrive at a shared and correct response, and the interpersonal reasoning serves a probative function. An *inquiry dialogue* is similar to a deliberation dialogue, except that the initial question is not practical, but has a cognitive orientation. Finally, a *negotiation dialogue* starts from a difference of interest (or a difference of opinion that is 'commodified' so as to enable a negotiation – like in politics), aims at a compromise solution, and proceeds mainly on the basis of an exchange of offer and counteroffer. Persuasion and negotiation dialogue start from some difference, whereas such competition is not intrinsic to deliberation and inquiry dialogue. What kind of dialogue game do the engineering students need to play when doing their research?

Plausibly, it will be a mixed type of dialogue. The initial situation is an open question about a (research-and-design) decision that, in the end, will be helpful to a satisfactory design solution for the engineering problem. The students themselves need to find out what need to be the evaluation criteria for the final design solution. Many of these criteria can be qualified as pragmatic, in the sense that they concern the intended application of the design solution. An example of such a criterion could be that the most suitable material has been selected for the innovation. Another that the design solution gets released in time (as stipulated by a lecturer in educational settings, or in real life by a contract). I expect that from such criteria for the final design solution, more specific evaluation criteria for the various in-between research-and-design decisions can be derived. For example, if the group needs to decide to spend their time only on an inquiry into the features of materials #1 and #2 or whether also to include material #3, then the quality of that decision must, in some way or other, connect to the final criteria, and the students need to find or determine the best balance between finishing in time and selecting the best material. But in the educational setting, it seems equally justified to regard them as epistemic criteria, as they determine the quality of the design solution, and in a derived way also of the interim decisions.

The main goal of the mixed dialogue is to find or make a decision that the members of the group can subscribe to, and that fits the (derived) evaluation criteria. Is this outcome practical, or does it have a cognitive orientation? It seems to be a matter of mere framing. If the students make a decision that facilitates the development of a proper proposal for a design solution, then the main structure of the dialogue is that of a deliberation. Whereas if they are trying to find or discover a decision that facilitates the development of design solution that best fits the criteria, then it is more an inquiry dialogue. Not much hinges on the distinction, for in both cases the collective reasoning is probative, and needs to move and propel them from the initial situation to the decision. One could call the resulting reasoning 'argumentation', although this would diverge from a nowadays popular use of the term that restricts it to situations where reasoning is used to overcome disagreements, rather than extending available knowledge (Krabbe and van Laar 2007). Questions that trigger students to engage in high-quality probative reasoning are: "What additional information do we need to arrive at a robust decision? Can this candidate decision be derived from the information at our disposal, our aims, and the decision-making methods we have adopted?"

In the mixed dialogue, we can also expect situations that trigger persuasive reasoning. First, a disagreement could arise between the group members. One student may believe that they should examine all three materials (#1, #2 and #3), whereas another student believes that including also material #3 takes too much time whereas the pay-off will probably be limited. They then may want to engage in a persuasion dialogue, and resolve this disagreement. Second, they may want to determine the quality of some mid-term decision by playing the devil's advocate, just to find out whether the possible decision withstands future critical questioning. In such cases, a persuasion dialogue is embedded in the overall structure of the deliberation (or: inquiry) dialogue, and instrumental to avoiding bad or arriving at better decisions. Kinds of questions that enhance the quality of the persuasive reasoning are: "How can you convince me that this decision is the best one? How can you respond to that objection?"

A further feature of the plausible structure of the mixed dialogue has to do with the complexity of real-life problems: nature, technology and society typically block solutions that have all perfections. Any feasible design solution is at best a trade-off, and must, for example, strike a balance between time constraints and good materials selection, or between financial feasibility and sustainability, and some of the decisions will prefigure such trade-offs. Ideally, students arrive at a full-hearted consensus about what decision furthers such a good balance or optimal trade-off. But the diversity in a group may impede any such agreement to be realized in time: disciplinary, cultural, political or ideological differences may be deeply felt and the dissent may, for all practical purposes, simply be too deep. In such settings it could be wise to examine the possibility of a compromise solution, especially when a failure to agree will be too costly (failing the course, or breaching the contract). A compromise solution resembles a trade-off, but differs, because the parties do not full-heartedly endorse the outcome, and each needs to make, what she conceives as, an epistemic sacrifice. Again, different questions are helpful: "Why would I, given my perspective on this issue, accept your proposal? What can you give me in return? Do you recognize that my offer is your best interest?" The reasoning triggered is directive as they steer the interlocutor to accept an offer or mitigate her demands.

The argumentation elicitation approach is rightly labelled pragmatic, in the special sense that an epistemic problem is dealt with as a dialogical problem.

The game of asking for and giving reasons admits of different versions, and the reasons elicited may serve different functions. Next to reasoning that is persuasive, probative and directive, there is also reasoning that explains, or motivates or accounts for a decision. Students need to be proficient in each of these argumentative games, and to be capable of shifting from one type to another (cf. Walton and Krabbe 1995). Sometimes they need to warned not to engage in specific dialogues: "Never start bargaining!' or "Never yield to competitive debating!" Two conclusions to draw from this: educational dialogue itself needs design (see for an educational application that enables the design of online discussions: van Laar 2022), and students need to become skilled players in them (van Laar 2021).

References

Walton, Douglas N., & Erik C. W. Krabbe (1995). Commitment in Dialogue. Basic Concepts of Interpersonal reasoning. Albany, NY: State University of New York Press.
Krabbe, Erik C. W. & Jan Albert van Laar (2007). About Old and New Dialectic: Dialogues, Fallacies, and Strategies. Informal logic, 27, 27 – 58.
van Laar, Jan Albert (2021). Improving Argumentative Skills in Education: Three Online Discussion Tools. In: Cattani, Attani & Bruno Mastroianni (eds.), Competing, Cooperating, Deciding: Towards a Model of Deliberative Debate (pp. 87-98). Firenze: Firenze University Press.
Van Laar, Jan Albert (2022). Deliberative Debate. Software, accessible at: https://deliberativedebate.nl.

THE INTERACTION BETWEEN ARGUMENTATIVE NORMS: THE CASE OF MUNĀẒARA

RAHMI ORUÇ
*Ibn Haldun University, Comparative Literature,
ArguMunazara Research Center*
rahmioruc@ihu.edu.tr

Abstract

In this paper, I argue for argumentative holism against argumentative reductionism by elaborating on the interdependence between norms in Ādāb al-Baḥth wa-l Munāẓara, a seven centuries-old argumentation theory, and practice. Argumentative reductionism is a possible yet undesirable outcome of argumentative perspectivism. Whereas argumentative perspectivism states that argumentation can be examined in logical, rhetorical, or dialectical frameworks with the adjoining norms they require, reductionism highlights one set of norms over others. I first elaborate on the attempts and successful applications of argumentative holism in the case of pragma-dialectics, Johson's dialectical tier of arguments, and de Grefte's modal safety norm. I then examine the objections that these proposals face and add my objections if applicable. Building upon these proposals and my objections, I move into Munāẓara. I show argumentation is set to be achieved by the interdependence between product-based, procedural, and agential norms. Product-based logical norms state that: 1) unless incontrovertible, a premise needs defense, 2) an argument can not be flawed, and 3) the standpoint should not admit another counter-argument. The process-based norms state that a strict turn-taking procedure is to be established for argumentation, where -reminiscing Johnson's dialectical tier- an inquiry is deemed adequate if the protagonist addresses the legitimate critical moves of the antagonist. Having discussed the product-process ambiguity in Munāẓara, I move into the interdependence of agential and procedural norms to values of argumentation made explicit in the sequencing of the antagonist's critical moves. Values such as coalescent cooperation or performative efficiency are embodied in a certain sequence of the procedure, and they demand a virtuous arguer for realizing them. Finally, as an example of interdependence between argumentative norms, I show how derailments from the procedure can also be characterized as argumentative vices. In Munāẓara, a move amounts to hastiness if the antagonist begins with a counter-argument rather than objecting to the premise. It amounts to arrogance when an incontrovertible premise is objected to. It is subjugation when a claim is held true without an argument. Finally, it amounts to usurpation when the onus of proof is shifted. In conclusion, I discuss the wider implications of argumentative holism.

1. Introduction

This paper discusses the interaction between argumentative norms from the perspective of a once widespread, now largely forgotten Arabic/Muslim argumentation theory and practice: Ādāb al-Baḥth wa-l Munāẓara. Following Wenzel (1992), I first detail argumentative perspectivism, a methodological framework for studying argumentation, by dividing it into different accounts (e.g., rhetorical, dialectical, and logical). Then I elaborate on applications of what I call argumentative holism in the case of pragma-dialectics (van Eemeren & Grootendorst 2004; van Eemeren 2009), Johnson's dialectical tier (2003), and de Grefte's (2022) modal safety norm as a bridge between product and process-based norms. Argumentative perspectivism is a principle that holds different perspectives on argumentation are important in their focus on rhetorical principles, dialectical methods, or logical products, as with their focus, we can zoom in and out on the intricacies of argumentation. However, we should also caution that argumentative perspectivism might lead to argumentative reductionism. I define argumentative reductionism as the prioritization of one perspective to argumentation over another or the complete denial of other perspectives.

I will work from the premises of argumentative holism, arguing that argumentative norms are not mutually exclusive. Adding on to that, however, I will claim that argumentative norms are not only mutually inclusive; they, in fact, are interdependent. The norms I consider are logical (product-based), procedural (process-based), and agential (agent-based). To demonstrate the interdependence between argumentative norms in the Munāẓara literature, I will first show logical norms determine the soundness of an argument. Then I will show how the Munāẓara procedure translates these product-based norms to design a strict turn-taking procedure, so much so that there is an ambiguity between the product and the process of argumentation. I will continue by showing how procedural norms and agential norms inherit an ambiguity as they derive from a common source, i.e., the values of argumentation, to the extent that it is not possible to demarcate and prioritize either set of norms over another. Finally, as a case study, I will elaborate on how the derailments from the procedure are also characterized as character failures in the Munāẓara literature.

While we have a plethora of argumentative theories that take different perspectives on argumentation, there have also been successful attempts to achieve argumentative holism by diminishing the gaps between different theories and the related norms: 1- Pragma-dialectical theory of argumentation has first defended an ambiguity between arguments as product and arguments as process showcasing how fallacies are not simply derivative of product-based norms. In its extended version (van Eemeren 2009), where the notion of strategic maneuvering is introduced, pragma-dialectics included the rhetorical goal of effectiveness with the critical dimension of the joint pursuit of reasonableness. 2- Johnson (2003; 2012) has introduced a dialectical tier of arguments where a good argument is one that also anticipates and addresses the possible objections the protagonist might encounter. 3-Recently, de Grefte (2022) introduced the modal criterion of safety as a bridge between epistemic and dialectical norms, arguing that contrary to widely shared assumptions, these norms do not override each other.

Notwithstanding the three attempts -and possibly a wide range of others- the endeavor to provide a unified theoretical framework that will include these norms are not wholeheartedly welcomed by other scholars: 1- In response to the pragma-dialectical account of argumentative ambiguity, objectivist (Biro and Siegel 2006) and epistemological approaches to argumentation (Lumer 2005; 2010) have raised objections. 2-Jonhson's dialectical tier has been attacked from a number of angles, demanding what objections there are and how can a protagonist account for them. (Johnson 2003). 3- The virtue approach to argumentation views the preceding theories as "act-based", defending an agent-based approach to argumentative norms (Aberdein 2010). These objections signify a partly justified hesitation to renounce the division of labor in argumentation theory and thus argumentative perspectivism. However, as I will show, the justified hesitation runs the risk of turning into argumentative reductionism. I will gloss over the objections to the attempts of argumentative holism. Evaluating these objections, I will either confirm or reject them. In consequence, I will list my own objections, if applicable, with the aim to indicate barriers to argumentative holism and Munāẓara's possible contributions.

I will first enumerate attempts to move beyond argumentative perspectivism in argumentation scholarship and the objections & rebuttals these attempts have faced (section 1), then I will elaborate on Ādāb al-Baḥth wa-l Munāẓara and the interdependence between argumentative norms (section 2). Section 1 will help me elaborate on Ādāb al-Baḥth wa-l Munāẓara, the literal translation of which is the manner of inquiry and argumentation (Belhaj 2016; El-Rouayheb 2015; Faytre 2018; Arif 2020) to which a more technical rendering would be virtuous conduct for the monological and dialogical argumentation.

In section 1, I will first deal with pragma-dialectics and its hypothesis of product-process ambiguity. The hypothesis leads to a characterization of fallacies as derailments from the critical discussion procedure. Pragma-dialectics is criticized for its omission of product-based norms&goals (Lumer 2005; 2010) and the obscure place of logic in the critical discussion procedure (Blair 2010). Then I will continue with Johnson's (2012) theory of the dialectical tier of arguments, where he aims to enlarge the focus of product-based theories of argumentation to the process by arguing that a good argument is one that also anticipates and addresses the possible objections it might get. Johson's dialectical tier is criticized for not spelling out what are the anticipated objections and which objections will need to be addressed (Johnson 2003). Finally, I will discuss de Grefte (2022) and his proposal to add modal safety as a norm for epistemic arguments. Modal safety will be the norm in deciding which objections should be accounted for during the argumentation process.

In section 2, I will move on to Ādāb al-Baḥth wa-l Munāẓara. There I will first discuss argumentative perspectivism in the Muslim world before the emergence of Munāẓara. Then I will show how with the emergence of Munāẓara, product-based, process-based, and agent-based norms of argumentation have been included in the study of argumentation. Agreeing with pragma-dialectics and Johnson, I will show how Munāẓara holds that there is a product and process ambiguity. However, unlike Johson's dialectical tier, Munāẓara differentiates between an argument and the dialectical inquiry process (al-Jaunpūrī 2006, 12; Gelenbevi 1934, 32) where the antagonist has in her arsenal three kinds of critical moves: Objection, Refutation, and Counter-Argument (al-Samarqandī 1934, 126). While we can still have a traditional account of arguments as a claim and reason pair, we can argue that a good argument is one that has gone through the dialectical inquiry delineated in the argumentation procedure. Having established the product-process ambiguity, I then claim that there is also an ambiguity between the procedure and the agent. I argue that there are values of argumentation such as coalescent-cooperation and performative-efficiency embedded in a certain procedure demanding reliable and responsible agents to realize them. The values argumentation as the common source for procedural and agential norms of argumentation showcase that it is not possible to prioritize one set of norms over the other. The interdependence between the procedure and the agent is also reflected in Munāẓara's characterization of derailments from the Munāẓara procedure after argumentative vices such as arrogance (al-Āmidī 1900, 58), hastiness, and subjugation (Ahmet Cevdet 1998, 112).

2. Perspectivism and its Discontents in the Contemporary Argumentation Scholarship

Argumentative perspectivism holds that there are different perspectives to argumentation in accordance with varying norms relating either to the product, process, and procedure (Wenzel 1990, 9). Accordingly, there do not exist logical, rhetorical, or dialectical arguments. Argumentation is one and only, and it is manifested in logical products, rhetorical persuasion processes, or methods for critical decisions. Each manifestation has a different set of norms (Godden 2016). Over recent years in the normative domain of arguments, a new perspective has emerged, i.e., the virtue approach to argumentation (Paglieri 2015). The virtue approach holds that the above-mentioned three perspectives are "act based", whereas argumentation should be understood as "agent-based". The agent-based perspective holds that the agent has a conceptual priority over the act, be them the product, the process, or the procedure. Accordingly, the goodness of an argument is measured vis-a-vis the virtues of the agent (Cohen 2005, 1).

Although argumentative perspectivism underlines that each perspective is only part of the story and that these perspectives work as a framework for better understanding different facets of argumentation, no unified theory of argumentation has been developed. Moreover, let alone a unified theory of argumentation, there is an argumentative reductionism buttressed by the history of philosophy (van Eemeren 2009, 130), holding that argumentation is primarily a process, procedure, product, or agent-based endeavor.

2.1. Pragma-dialectics and Argumentative Norms

Since its Renaissance in 1958, taking place in Europe and America, argumentation theories have fought with formal logic, and its standards posed as forms to be adhered to, i.e., the validity of argument form, and the incontrovertibility of argument content (Lewiński and Mohammed 2016). Toulmin has developed the Toulmin Model/Scheme, where new elements are introduced, such as backing, warrant, and qualifier. For him, the goal was to show that argumentation anticipates a field-invariant procedure and that the soundness of an argument is measured context-dependently. In the same line, Perelman argued that argumentative reasonableness should not be reduced to the norms of formal logic (Rigotti and Greco 2019, 131). Pragma-dialectics is one such theory conceived as a pseudo-geometrical ideal procedure in which certain rules are posited for critical discussion.

Following Hamblin, pragma-dialecticians argued against the standard treatment of fallacies where the interactional context is not considered (van Eemeren & Grootendorst 2004, 158). In response, inspired by the Dutch linguistic convention where the word "argumentation" can simultaneously mean the product and the process, it posits a product-process ambiguity (van Eemeren et al., 2019, 3). In this approach, the soundness of an argument is measured by whether the appropriate and intersubjectively agreed-upon scheme is employed and whether the critical questions pertaining to the argument scheme have been satisfied. If a move deviates from the ideal discussion procedure and therefore does not serve the common purpose, it is characterized as a fallacy.

Pragma-dialectic is criticized for overshadowing product-based norms&goals with process-based norms&goals (Lumer 2010). It disregards the possibility of argumentation as an epistemic endeavor (Biro and Siegel 2006). However, the idea of product-process ambiguity is justified, and the data from Munāẓara will also supplement its justification as we will see that there is an ambiguity between inquiry and argumentation. Moreover, as Johson will argue, without a dialectical tier, it is not possible to have adequate treatment of an argument. However, pragma-dialectical adherence to the critical rationalist philosophy of reasonableness leads them to argumentative egalitarianism, where the antagonist has the right to question any incontrovertible premises. Nevertheless, as we will see with de Grefte, modal safety as the norm for epistemological argumentation characterizes irrelevant cases of argumentative egalitarianism as irrelevant both from a product-based and procedure-based perspective.

2.2. Johnson and the Dialectical Tier of Arguments

Informal logic as a movement aims to mend the incongruity between the formality of logical texts and the diversity of real-life argumentative encounters (Johnson and Blair 2006). As much as in pragma-dialectics, the norms of formal deductive logic are deemed inconclusive at best. Accordingly, what makes an argument good will be its acceptability, relevance, and sufficiency. The premises of an argument should be acceptable, and the argument itself should be relevant to the standpoint. However, this is not enough. The argument should supply sufficient support for the standpoint (Johnson and Blair 2006, xiii). These criteria show that the rules of deductive logic do not bring out a good argument.

Over the years, Johnson (2012, 2003) has defended a "dialectical tier" of arguments as the natural result of the informal logical conception of good arguments. He argues that next to the illative core, i.e., the standard elements of an argument, there should be a dialectical tier in which the goodness of an argument will be measured by how successfully an arguer will anticipate and address the possible objections that might be raised. A dialectical tier to arguments is necessary for various reasons as the formal deductive logic is silent in the face of premises whose truth is not known; arguments that are not valid enough, and finally, a sound argument might have an equally sound counter-argument (Johnson 2003, 178).

Johnson argues that pragmatically arguments require a dialectical tier as an arguer must satisfy dialectical adequacy, meaning that she must perform her dialectical obligations. This brings the question of what the objections are and what principles are there for dialectical adequacy. For the latter, he holds that an argument should accurately deal with objections by adequately and appropriately responding to them. When it comes to objections, he acknowledges that in its history of two millennia, a proper study of objections is not extant. In my analysis of Munāẓara tradition, I will list three types of objection available to the antagonist. Further, I will argue that it might not be necessary to enlarge the scope of what we traditionally call an argument. Drawing on the Munāẓara procedure, I will argue that we can differentiate between an argument and an inquiry, where an inquiry is the result of a dialectical encounter in which objections are raised and responded to by the parties.

2.3. de Grefte and Modal Criterion for Epistemic Arguments

In his recently published paper (2022), de Grefte introduces the modal norm of safety for epistemological argumentation. His aim is to show that we do not need to assume either the product-based or process-based account of argumentation must be necessarily true. He acknowledges that argumentation is a context-dependent activity. Whereas a politician in a public political debate setting might interpret truth very liberally, this is not the case for scientific argumentation. Different contexts require different norms. Accordingly, he argues that the epistemological approach to argumentation –a subvariant of product-based account– should work within the safety principle. As the epistemological approach argues that the function of argumentation is to arrive at acceptability (truth, high probability, and verisimilitude), it needs to acknowledge that there are cases where justified true belief is not enough for knowledge. Knowledge requires more than truth. The safety principle holds that to produce knowledge, the risk of false-belief formation must be eliminated (de Grefte 2022, 408).

We do not need to delve into the intricacies of the safety principle with its nearby worlds and their similarity to our actual world. In the case of epistemological argumentation, the antagonist raises motivated or unmotivated challenges to the protagonist to eliminate the risk of false belief formation. However, what if the antagonist adheres to the principle of dialectical egalitarianism, which states that she has the unconditional right to raise a challenge and that the protagonist is bound to offer reasons to ward off the challenge? This would mean that it would be almost impossible to stop the antagonist from raising objections, leading to the inevitable conclusion that epistemic argumentation or product-based norms will always fall behind dialectical norms. de Grefte argues that if the challenge raised by the antagonist pertains to a far-off possibility such as that of the spell of a Cartesian evil demon, the protagonist does not need to respond. Some objections fall behind the domain of relevance, and therefore knowledge production does not hinge on the epistemically irrelevant behavior of the antagonist (de Grefte 2022, 403).

de Grefte allows us to save incontrovertible premises from the critical-rationalist philosophy of pragma-dialectics. However, for this paper's purposes, I reject argumentative pluralism of norms, I.e., the idea that different contexts will require different norms. Instead, as related to my investigation of argumentative holism, I will argue that all argumentative contexts require the interdependence of norms.

3. Ādāb al-Baḥth wa-l Munāẓara

In the remainder of the paper, I make use of the implications of Section 1 when I elaborate on Ādāb al Baḥth wa-l Munāẓara and its application of argumentative holism through the interdependence of norms. I will show that, as pragma-dialectics argued, there is an ambiguity between the product and the process. This is explicated by a linguistic and technical convention: The term Baḥth translates as inquiry, and it refers to the justification of a claim through arguments. The term Munāẓara is translated as argumentation. These two terms can be used interchangeably in most instances. The close relationship between the two terms, corresponding to monological and dialogical argumentation, is conspicuous in the product-based and process-based norms as well.

The logical norm states that a proposition is to be defended until arriving at an incontrovertible starting point. This norm obliges the protagonist to defend a proposition in the face of an objection, which is the right and duty of the antagonist. In the same manner, the logical norm states that an argument can not be flawed or fallacious. The protagonist is bound to satisfy this requirement, whereas the antagonist has the right and duty to refute the argument. Finally, product-based norms hold that a good argument withstands a counter-argument. Therefore, the antagonist has the duty and the right to produce counter-arguments, whereas the protagonist must respond to counter-arguments (al-Samarqandī 1934, 125-6).

The product-process ambiguity that is defended by pragma-dialectic shows itself in the interchangeability of the terms Baḥth and Munāẓara (al-Jaunpūrī 2006, 12; Gelenbevi 1934, 32). Here we also find a practical application of Johnson's dialectical tier of arguments. Remember, he argued that a good argument is one that anticipates possible critical moves it will face. In the case of Munāẓara, these critical moves are objection, refutation, and counter-argument (al-Samarqandī 1934, 125-6). The objection addresses the premise, and its illocutionary force is asking for defense (al-Jaunpūrī 2006, 76). The refutation addresses the argument overall, and its illocutionary force demands another deficiency-free argument from the protagonist (al-Āmidī 1900, 61). The counter-argument either addresses the claim or the conclusion of an argument. Its illocutionary force is the denial of the claim and undertaking its negation (al-Āmidī 1900, 60). Importantly, by making a counter-argument, the antagonist wants from the protagonist her role – she requests to be the protagonist.

The interaction between argumentative norms is not restricted to the ambiguity between the product and the process. Again, building upon a linguistic convention, Munāẓara scholars posit an ambiguity between the act and the agent, more specifically with the procedure and the agent. It is argued that Munāẓara as a term is equivocal: It both refers to the procedure while also referring to attributes of an agent (al-Āmidī 1900, 8). Just as the act of quarreling signifies a quarreler, the act of proper arguer denotes an arguer. This presupposed interdependence between the agent and the act reflects itself in two aspects of Munāẓara: 1- Sequencing of the antagonist's critical moves, and 2- Derailments from the procedure. When it comes to sequencing, I argue that prescriptions that have emerged over the history of Munāẓara show concern for different ways of cooperation between the antagonist and the protagonist, making explicit different values such as coalescence and efficiency. These values are both procedural and agential: The sequencing embeds certain values in the procedure (say, performing the role of the antagonist with utmost cooperation), and it demands an agent reliable and responsible enough to realize these values.

3.1. Muslim/Arabic Account of Argumentative Perspectivism

We had seen that Wenzel argues that different perspectives on argumentation do not posit there are rhetorical, dialectical, or logical arguments *per se*. Each perspective zeroes in on different aspects of argumentation, be its product, procedure, or process, corresponding to logic, dialectic, and rhetoric, respectively. This is a stark difference from the context theory of argumentation, according to which there are five kinds of syllogism (Black 1990), namely demonstrative, dialectical, rhetorical, sophistical, and poetical (Coşkun 2011, 78). Context theory is a late development in the Greek and Arabic Aristotelian tradition for the study of *Organon*, where rhetoric and poetics are subsumed under the category of syllogism.

Aristotelian tradition in the form of context theory has become a huge hit in the Muslim world. However, over time dialectic and rhetoric lose their appeal. This is fundamentally to do with the metaphysical underpinnings of demonstration. That is, the demonstration is not merely a tool to be employed by all the sciences; it also has the alleged potential to perfect our sublunar minds. In the introduction of his famous *al-Risāla al-Shamsiyya*, al-Qazwīnī elaborates why he has written a manual on logic:

> Whereas, agreeably to the opinion of all men of mind and virtue, the sciences, more particularly the incontrovertible sciences, are the highest pursuits in life, and whereas the professors thereof are the most noble among human beings, their minds being sooner prepared to be absorbed into the angelic minds, and farther, whereas it is impossible to comprehend the subtleties of sciences and to preserve the acme of their varieties except by the assistance of the science, which is called Logic. (al-Qazwīnī 2007, 2)

It was believed that when the human mind is perfected, it would have the ability to conjunct with the Active Intellect to the extent that it would become a replica of the world. So much so that merely by contemplating itself, the human mind would be able to receive knowledge without establishing a causal relationship between objects.

Other than the purported benefit of demonstration, Aristotelian dialectic and rhetoric do not find ground as in the Muslim world there were already established disciplines dealing with those subjects: jadal (dialectic), and balagha (rhetoric). Jadal was primarily of two kinds: theological and juridical (al-Baġdādī, 2019). In theological jadal dealing with the domain of epistemic arguments, the aim was to arrive at truth by employing incontrovertible starting. In the juridical jadal corresponding to the practical arguments, however, the aim was to arrive at the preponderance of belief in matters where certainty is untenable as the Qur'an, and other religious corpora do not spell out an answer. Balagha, on the other hand, dealt with speaking in accordance with the requirements of the context one is in (al-Taftāzānī, 1891). In balagha, the presupposition was that one's deeds, literary or not, are derivative of one's ādāb (the manifestation of a character and virtuous conduct).

Ādāb al-Baḥth wa-l Munāẓara emerged at the end of the 13th century and quickly dominated the study of argumentation (El-Rouayheb 2016). The jadal faded into the background, and Munāẓara became a stable Muslim college, the madrasa. The students would first study logic, and then they would move into Munāẓara. The founder of Munāẓara had envisioned Munāẓara as an appendage to the study of logic (al-Samarqandī 2014, 500). Munāẓara established a strict-turn-taking procedure for virtuous conduct where the goal was the manifestation of truth or the preponderance of belief. Contrary to jadal, where there was a division of labor between epistemic and practical arguments, Munāẓara is a context-independent theory of argumentation. One interesting thing to be noted here is that there was not a clash between rhetoric, dialectic, and logic. Each had its own space, and although Munāẓara shares the term "Ādāb" with rhetoric, –to my knowledge of primary and secondary literature on these disciplines– there were no attempts to unify/diverge argumentation and rhetoric.

3.2. Product (Baḥth, Inquiry) - Process (Munāẓara, Argumentation) Ambiguity

An 18th-century Ottoman scholar Gelenbevī (1934, 32) defines Baḥth and Munāẓara as the exchange of defenses (mudāfʿa) for the manifestation of truth. In this definition, it is not understood whether inquiry and argumentation are two different things or not. In response to this obscurity, Gelenbevī's Kurdish commentator explains that the connective *and* in "Baḥth and Munāẓara" is for explication (1934, 32). This renders Gelenbevī's words as such: "Know that Baḥth, -that is Munāẓara- is the exchange of defenses for the manifestation of truth." In this definition, inquiry and argumentation are used interchangeably.

In the technical terminology of Muslim argumentation scholarship, Baḥth has three meanings: 1-Predication as in the example of "Rahmi is a Ph.D. student". 2-Inference as in the example of "Rahmi is a Ph.D. student because he has devoted quite a few of his years to complete his Ph.D.". 3-Argumentation occurring between the protagonist and the antagonist (al-Jaunpūrī 2006, 12). Jaunpūrī states that what is meant by Baḥth in the context of the science of Munāẓara is either the inference or the argumentation. This is because, in an argumentation, parties will deal with inferences and the kinds of objections they receive. Argumentation will serve as a platform where an argument about a certain issue will be exhausted by the parties. The antagonist will employ the available, legitimate kinds of critical moves in her arsenal: objection, refutation, and counter-argument. The protagonist will be obliged to respond to these objections if they undermine her claim or arguments. The dialogical encounter between parties will constitute the inquiry. The Munāẓara procedure, with its turn-taking rules and prescription for virtuous conduct, will regulate how the overall machinery is to be operationalized by either party in the argumentative setting.

The ambiguity between the product and process is further elaborated by Sājaqlīzāda (1872), a 17th-century Ottoman scholar when he discusses changing the subject during argumentation. He discusses legitimate and illegitimate cases of when one has the right to change the subject. For instance, in the face of an objection, the protagonist's duty is to defend her premise. Therefore, she can not argue that the objection is unfounded. Arguing that the objection is unfounded is only possible when the protagonist has no other means to secure her premise. In other cases, it means changing the subject and thus opening a new inquiry. If it is the case that the premise can only be secured by objecting to the objection itself, then this will not amount to opening a new line of inquiry, the parties will still be in the same subject. Sājaqlīzāda's elaboration shows that an inquiry is a joint endeavor between two parties where dialectical obligations and adequacy pertaining to an issue are satisfied (al-Sājaqlīzāda 1872, pp. 50-61).

3.3. Procedure and Agent Ambiguity

In Munāẓara, the ambiguity between argumentative norms is not confined to the product and the process. In the same line, an ambiguity between the procedure and the agent is also posited. Accordingly, the term Munāẓara can be interchangeably employed 1- for the procedure of argumentation where dialectical obligations and adequacy are satisfied, and 2- for the attribute of the arguing the agent bears when she complies with Munāẓara rules (al-Amidi 1900, 8). Here the idea is as much as quarreling refers to the acts that are not in line with the procedure of argumentation, acts that remain in the Munāẓara procedure make their agents an arguer.

The abstract idea that the procedure and the agent's attributes are the same becomes concrete in two cases, one relating to the sequencing of the antagonist's critical moves and the other, the derailments of the Munāẓara procedure. I will first detail how sequencing shows procedural and agential norms are interdependent and then move into how Munāẓara scholars label derailments from the procedure as vices of character.

3.3.1 Sequencing of Antagonist's Critical Moves

In the previous subsection, we saw that there are three kinds of critical objections available to the antagonist: Objection, refutation, and counter-argument. I say "kinds of critical moves" as there are different forms of objection, refutation, and counter-argument. What makes these moves of the same kind is their focus and illocutionary force (cf. Krabbe and van Laar 2011; van Laar and Krabbe 2013; van Laar 2001). The objection focuses on the premise, the refutation on the argument, and the counter-argument on the claim. By performing an objection, the antagonist demands an argument in support of the premise from the antagonist. The refutation of the antagonist demands a new argument for the claim that is free from any deficiency. The counter-argument, on the other hand, declares to the protagonist that her claim is untenable, as there is an opposing argument that negates the initial claim.

Whereas the interdependence of product-based and process-based norms determine the legitimate moves for an antagonist, it does not determine which of the three critical moves is to be employed first when all three are legitimate. That is, what should an antagonist do first when the premise needs defending, the argument is deficient, and the claim admits a counter-argument to its negation? As far as I am aware, three prescriptions for the sequencing of the antagonist's moves have emerged. In a paper, two of my colleagues and I (Oruç et al., *forthcoming*) have dealt with all three. However, in this subsection, I will first detail two of them.

Sequence I: The first prescription comes from the founder of Munāzara, al-Samarqandī (1934, 125-7). When detailing the moves available to the antagonist, he first mentions the objection, then the refutation, and finally the counter-argument. It is al-Jurjānī and Jurjānī's commentator, al-Jaunpuri (2006), who justifies Sequence I. Accordingly, the objection should come first as it is free from the risk of usurpation (shifting the burden of proof). In the case of refutation and counter-argument, the burden of proof shifts in varying degrees. The refutation requires the antagonist to not only claim that there is a flaw or fallacy in the argument but also show it with a *shahid* (witness, evidence). Nevertheless, the protagonist has the right to dismiss the antagonist's line of thought and produce another argument for the same claim, free from any deficiency. In the case of counter-argument, the burden of proof shifts completely to the antagonist; therefore, she becomes the new protagonist. In the case of objection, however, there is no risk of usurpation. The objection is a request for further arguments for the contested premise. The objection, therefore, widens the zone of disagreement and allows the protagonist to perform her role (al-Jaunpūrī 2006, 76-77).

Having done so with the objection, al-Jurjānī recommends continuing with the refutation rather than the counter-argument. The counter-argument addresses the claim with a declaration that the protagonist's argument is false because there is another argument for the negation of the claim. In that sense, it is the critical move with the strongest illocutionary force, as it demands from the protagonist her own role, i.e., being the protagonist. In the case of refutation, as we have seen above, the protagonist is not required to hand over her own role as the protagonist. She can continue her role by producing another argument. This is why refutation is preferred over the counter-argument with the consideration that the move with the strongest illocutionary force should be left to the end.

Sequence I, then, values opening spaces for the protagonist to let her perform her role. al-Jurjānī states the antagonist's primary role is to become informed and to question. In this sequence, a quiet, coalescent version of cooperation is prescribed. However, when we remember that the term Munāẓara is equivocal as it both refers to the procedure and the attributes of the agent, we see that Sequence I demand its performer to value coalescent argumentation, even in the cases where she might have fierce counter-arguments. She should actually help the protagonist and turn helping the protagonist into a disposition to perform argumentation in accordance with virtuous conduct, i.e., ādāb. It is, therefore, impossible to argue which comes first; the embodiment of the value in the procedure or the conceptual priority of the arguer as the one who realizes the value?

Sequence II: Sājaqlīzāda prescribes beginning with the objection, then with the counter-argument, and finally ending with the refutation (al-Āmidī 1900, 60). While arguing for starting with the objection, Sājaqlīzāda has a different rationale in mind. He does not view the objection as opening a communicative space where the protagonist will cheerfully perform her role. Sājaqlīzāda does not preach extreme other-caring. The antagonist should start with the objection as it is the easiest one: The objection is not tasking the antagonist. Only after having done so with the objection does Sājaqlīzāda recommend the counter-argument. For him, the duty of the antagonist is to contest the claim, so when she gets the chance, she directly jumps into the counter-argument. Sājaqlīzāda might have prescribed the antagonist to continue with the refutation, as we have seen in sequence I; however, refutation does not attack the claim but the argument. As it is possible that a new, deficient-free argument for the same claim will emerge, all the efforts the antagonist has undertaken might have been fruitless. It is tasking and not that rewarding (al-Āmidī 1900, 60-1).

Sequence II, then, does not value coalescent cooperation but performative efficiency. That is, by performing the legitimate moves, she is already cooperating with the protagonist, but when it comes to their sequencing, the consideration is to perform being an antagonist as efficiently as possible. Therefore, first, the most effortless move is chosen, and it is followed by the most cost-effective one. In this sequence, as well, we see that certain values are embedded in the procedure and are demanded from the antagonist to realize them. Munāẓara scholars prescribe another sequence with considerations leading to certain values serving as the justificatory force. It would be possible to consider different sequences as well. For instance, one might argue that the antagonist should start her critical reaction with the counter-argument, then the refutation, and finally the objection. Meaning that the most powerful move in terms of force is prioritized, then less powerful, and finally least powerful. In such a sequence, which is the exact opposite of sequence I, where the goal is coalescent cooperation, the underlying value would be adversality. By performing the legitimate moves, the antagonist would still be cooperating, but with utmost adversality.

3.3.2 Derailment from the Procedure as Character Failures

Sequencing of the antagonist's critical moves gives cues to the interdependence between agential and procedural norms blurred through values of argumentation embedded in a procedure, demanding a virtuous arguer for their realization. One could see how these values, when realized by the agent, will necessitate certain virtues and dispositions such as open-mindedness or honesty. Conforming to the procedure and, therefore, the prescribed values will earn one the title of a virtuous arguer, with the condition that these values have turned into certain virtues by habituation. On the opposite line, derailing from the procedure is also related to agency, but this time in a negative sense. As the term, Munāẓara is employed equivocally as both the procedure, and as the attribute of the agent (al- Āmidī 1900, 8), derailments from the procedure signify a negative attribute of the agent. The paradigm cases of this interdependence are epitomized by: Arrogance (al-Āmidī, 1900, 58), hastiness & subjugation (Ahmet Cevdet 1998, 112), and usurpation (Jaunpūrī 2006, 81).

Arrogance can be identified in an argumentative setting in several ways (Sājaqlīzāda 1872, 69), but here I will confine myself to objecting to incontrovertible premises. We have seen at the beginning of this section that product-based norms state that a premise needs further support if it is not incontrovertible. The premise might be of the many kinds of incontrovertibility: It can be axiomatic, observational, experimented, or mass-transmitted for instance (al-Qazwīnī, 2007, 35). The premise that the sum is greater than the part is an example of axiomatic incontrovertible premises. Internal states of mind and body, such as feeling cold or hungry, are examples of observational premises. Experimented premises are the outcome of constant testing, as in the case of a certain medicine curing an illness. Mass-transmitted premises are like the existence of Tokyo for someone who has never been there.

The kinds of incontrovertible premises enjoy varying degrees of incontrovertibility. A premise might be either monologically or intersubjectively certain. The example of a monologically certain premise, i.e., a premise that is certain for either of the parties, will not be necessarily intersubjectively certain (al-Āmidī 1900, 60). Observational and experimented premises are certain for the party that claims them. However, axiomatic and mass-transmitted premises are incontrovertible intersubjectively, i.e., they are certain for both parties (al-Āmidī 1900, 60). If a party objects to intersubjectively shared premises such as the axiomatic and mass-transmitted one, she will be derailed from the argumentation procedure, showcasing the habit of a quarreler who does not serve the manifestation of truth or preponderance of belief.

The second case where derailment from the procedure signifies an agent with an attribute of quarreling is hastiness (Ahmet Cevdet 1998, 122). If a party, performing the role of the antagonist, begins her critical engagement with the counter-argument right away, rather than the refutation and the objection, this signifies hastiness. The third case of interdependence between the procedure and the agent is subjugation (Ahmet Cevdet, 1998, 122). The product-based norms of logic command that a claim needs sufficient support, free from the possible charge of the objection, refutation, and counter-argument. However, if the protagonist and the antagonist insist on a claim (be it a standpoint or a charge of refutation), they will be subjugators, illegitimately derailing from the discussion procedure. The fourth paradigm case where the derailments from the procedure amount to argumentative vice is usurpation (al-Shankīṭī, 200, 60).

One thing to note here is that derailments from the procedure do not directly translate into habitual character defects. It might be the case that Munāẓara scholars imagined a water-tight correlation between the derailments and argumentative vices. However, contextual research is necessary to label someone, or some move as vicious. Imagine a heated debate on state-controlled TV between members of the government and the opposition. The opposition could usurp the burden of proof, but this might be in response to a systematic silencing campaign coordinated by the TV and the government by exploiting their institutional leverage over the opposition members.

Munāẓara literature has expanded with almost 300 independent contributions over the centuries, most of which are still in manuscript form awaiting discovery (Çelik 2022). The present subsection has employed a limited number of contributions to the literature that are published either in the lithograph form or via modern printing. More research is needed to excavate the literature and relate it to contemporary discussions. The present data shows that Munāẓara scholars clashed when it comes to certain derailments from the procedure. Just to give two examples: al-Jurjānī states that objecting to incontrovertible premises might be illegitimate, but it does not directly translate into the vice of character. He argues that it is the Muslim Aristotelian philosophers who hold that it is an easily tractable defect of character (Jaunpūrī 2006, 59). In the same line, Gelenbevī argues that if the usurpation is relevant to the discussion, then it contributes to the common goal, let alone being a vice (Gelenbevī, 1934, 57-58).

4. Conclusion

In this paper, I argued against argumentative reductionism, which states a certain perspective on argumentation, whether the act or agent, is enough to explain it. By elaborating on several contemporary contributions, I displayed examples of argumentative holism, which aim to enlarge the normative scope of argumentation. Pragma-dialectics posits a product-process ambiguity underlining the limits of product-based norms to account for fallacies. Further, by extending their analytical framework to strategic maneuvering, pragma-dialectics successfully exemplifies an interdependence between the procedure and the process, as "bad" cases of strategic maneuvering will be fallacious. Johnson, too, furthers the product and process ambiguity by positing a dialectical tier of arguments where a good argument is one that satisfies dialectical obligations adequately. Finally, I discussed de Grefte's modal safety as a norm for epistemological argumentation in the service of knowledge. Having elaborated on these developments and the objections they face; I detailed how I am going to make use of them in discussing Munāẓara and the interaction between argumentative norms.

I remained loyal to the pragma-dialectical idea of product-process ambiguity, but I agreed with the criticism that pragma-dialectics foreground product-based norms and disregard any thick epistemological goals. To further discuss the product-process ambiguity, I revisited Johnson's theory of the dialectical tier. I responded to Johnson's call to investigate the anticipated objections that need to be adequately dealt with by bringing in the three kinds of critical moves at the arsenal of the antagonist. Finally, notwithstanding the significance of modal safety, I disagreed with de Grefte in his call for normative pluralism, not because it is a bad idea, but for an inquiry into the interdependence of argumentative norms. In the way I imagine, argumentative holism in response to argumentative reductionism necessarily leads to a plurality of norms, not because there will be different contexts, but because of intricate cases where each norm will posit urgency vis-a-vis others.

Having done the review of contemporary argumentation scholarship with its applications of argumentative holism and the objections thereof, where I also located my agreements or rejections relating to specifics of each contribution, I moved into Munāẓara. I first discussed what might be viewed as the corresponding division of labor we witness in contemporary scholarship between rhetoric and argumentation. Therefore, I did not delve into the process-based norms. However, I argued that in Munāẓara, product-based, procedure-based, and agential norms are interdependent. The product-based norms affect the procedure of argumentation. An argument is good 1) when its premises are supported with acceptable arguments, 2) when the argument for a claim is deficiency-free, and 3) when the claim can withstand counter-arguments. In relation to these norms, the Munāẓara procedure determines the legitimate critical moves and duties of each party while arguing. It obliges&empowers the antagonist with three critical moves: Objection, refutation, and counter-argument. If a claim can withstand these objections in a dialectical inquiry process whose procedure is determined with minute-turn taking in an argumentative engagement, the inquiry relating to a claim is adequately exhausted. Therefore, there exists an ambiguity between the product (which includes what Johnson calls the dialectical tier with its obligations and adequacy) and the process.

After the exploration of product-process ambiguity in a new light by considering pragma-dialectics and Johnson, I then moved into the ambiguity between procedural and agential norms. To exemplify their interdependence at play, I discussed the sequencing of the antagonist's critical moves and different proposals thereof. Taking inspiration from the equivocal usage of the term Munāẓara as both the procedure and the attributes of an agent, I argued that a certain sequence prescription embodies values such as coalescent cooperation or performative efficiency that require an agent with the necessary character traits to realize them.

The interdependence between character traits and the argumentation procedure is not limited to virtuous conduct but to the derailment of the procedure, which signifies vicious behavior as well. I discussed arrogance, hastiness, subjugation, and usurpation as vicious derailments from the procedure. These derailments might earn their performers the attribute of a vicious arguer if the contextual data shows that either party has habituated these moves by performing them over and over. One final note here, I am aware that a direct translation of an argumentative move into a state of mind is not only misleading but also dangerous. However, a different way of furthering this line of research might be investigating the participants in a series of argumentation(s) over a stretch of time. One can not but admire Proust when he observes that a person behaving extremely viciously might also be the beacon of virtue when it comes to another set of behaviors. Virtues and vices are situational. And yet, this does not diminish the fact that an agent will habituate character failures in a certain context. American prime-time news commentary tradition with its liberals and conservatives are conspicuous examples of such character behaviours, and argumentation scholarship should address this phenomenon.

References

Aberdein, A. (2010). Virtue in argument. *Argumentation, 24*(2), 165–179.
Ahmet Cevdet. (1998). Miyâr-ı Sedâd and Âdâb-ı Sedâd in Mantık Metinleri 2. Ravza.
al-Āmidī, 'Abd al-Wahhāb. (1900). *Abd al-Wahhāb alā-l Waladiyya*. Dersaada.
al-Baġdādī, A. al-Qāhir. (2019). *'Iyār al-Naẓar Fī 'Ilm al-Jadal*. Asfār.
al-Jaunpūrī, 'Abd al-Rashīd. (2006). *Sharḥ al Rashīdīya*. Maktabat al-Īmān.
al-Samarqandī, S. al-D. M. (1934). Risāla fī Ādāb al-Baḥth. In *Al-Badr al-'Illāt*. Maṭba'at al-Sa'āda.
al-Samarqandī, S. al-D. M. (2014). *Qistās al-Afkār fī Tahqīq al-Asrār* (P. Necmettin, Ed.). Türkiye Yazma Eserler Kurumu Yayınları.
al-Taftāzānī, I. U. (1891). *Al-Mutawwal ala al-Talkhis*. Dar al-Tiba al-Amira.
al-Qazwīnī. (2007). *Al-Risala al-Shamsiyya* (W. Hodges, Ed.; A. Sprenger, Trans.).
al-Shankīṭī, M. al-Amīn. (2005). *Ādāb al-Baḥth wa al-Munāẓarah*. Dar al-'ilm al-favaid.
Arif, S. (2020). The Art of Debate in Islam: Textual Analysis and Translation of Ṭaşköprüzade's Ādāb al-Baḥth wa al-Munāẓarah. *Jurnal Akidah & Pemikiran Islam, 22*(1), 187–216.
Belhaj, A. (2016). Ādāb al-Baḥth wa-al-Munāẓara: The Neglected Art of Disputation in Later Medieval Islam. *Arabic Sciences and Philosophy, 26*(2), 291–307.
Biro, J., & Siegel, H. (2006). In defense of the objective epistemic approach to argumentation. *Informal Logic, 26*(1), 91–101.
Black, D. L. (1990). *Logic and Aristotle's "Rhetoric" and "Poetics" in medieval Arabic philosophy*. Springer.
Blair, J. A. (2010). Logic In The Pragma-Dialectical Theory. *ISSA Proceedings*.
Çelik, M. (2022). Âdâbu'l-Baḥs ve'l-Munâẓaranın Kaynağı ve Eserleri (14-20. Yy.). *Tahkik İslami İlimler Araştırma ve Neşir Dergisi, 5*(2), 207–249.
Cohen, D. H. (2005). Arguments that backfire. *OSSA6*.
Coşkun, A. (2011). Ibn Sina Felsefesinde Beş Sanat. *Islami Ilimler Dergisi, 6*(2).
de Grefte, J. (2022). A Modal Criterion for Epistemic Argumentation. *Informal Logic, 42*(2), 389–415.
El-Rouayheb, K. (2006). Opening the Gate of Verification: The Forgotten Arab-Islamic Florescence of the 17th Century. *International Journal of Middle East Studies, 38*(2), 263–281. JSTOR.
El-Rouayheb, K. (2015). *Islamic intellectual history in the seventeenth century: Scholarly currents in the Ottoman Empire and the Maghreb*. Cambridge University Press.
Faytre, L. J. C. (2018). "Münazara" and the internal dimension of argumentation ethics: A translation and commentary on Ahmed Cevdet's Adab-i Sedad in the light of sufism and western argumentation theories. [Master's Thesis]. Ibn Haldun University.
Gelenbevī, I. (1934). *Gelenbevī 'alā al-Ādāb ma' al-Ḥāshiyat*. Matbaat al-Sa'āda.
Godden, D. (2016). On the priority of agent-based argumentative norms. *Topoi, 35*(2), 345–357.
Johnson, R. H. (2003). The dialectical tier revisited. In *Anyone who has a view* (pp. 41–53). Springer.

Johnson, R. H. (2006). The ambiguous relationship between pragma-dialectics and logic. P. Houtlosser, A. van Rees, Considering Pragma-Dialectics. A Festschrift for Frans H. van Eemeren on the Occasion of His, 60, 135–148.

Johnson, R. H. (2012). *Manifest rationality: A pragmatic theory of argument.* Routledge.

Johnson, R. H., & Blair, J. A. (2006). *Logical self-defense.* Idea.

Krabbe, E. C., & van Laar, J. A. (2011). The ways of criticism. *Argumentation,* 25(2), 199–227.

Lewiński, M., & Mohammed, D. (2016). Argumentation theory. In *The International Encyclopedia of Communication Theory and Philosophy* (pp. 1–15). Wiley Online Library.

Lumer, C. (2005). The Epistemological Theory of Argument–How and Why? *Informal Logic,* 25(3).

Muḥammad ibn Abī Bakr, al-S. (1872). *Taqrīr al-qavanīn.* (n.d).

Oruç, R., Üzelgün, M. A., & Sadek, K. (forthcoming). Sequencing critical moves for an ethical argumentation practice: Munāẓara and the interdependence of procedure and agent. *Informal Logic.*

Paglieri, F. (2015). Bogency and goodacies: On argument quality in virtue argumentation theory. *Informal Logic,* 35(1), 65–87.

Rigotti, E., & Greco, S. (2019). Inference in argumentation: A topics-based approach to argument schemes.

van Eemeren, F., & Grootendorst, R. (2004). *A systematic theory of argumentation: The pragma-dialectical approach.* Cambridge University Press.

van Eemeren, F. H. (2009). Argumentation theory after the New Rhetoric. *L'analisi Linguistica e Letteraria,* 17(1), 119–148.

van Eemeren, F. H., Grootendorst, R., & Kruiger, T. (2019). *Handbook of argumentation theory: A critical survey of classical backgrounds and modern studies* (Vol. 7). Walter de Gruyter GmbH & Co KG.

van Laar, J. A. (2014). Criticism in need of clarification. *Argumentation,* 28(4), 401–423.

van Laar, J. A., & Krabbe, E. C. (2013). The burden of criticism: Consequences of taking a critical stance. *Argumentation,* 27(2), 201–224.

Wenzel, J. W. (1992). Perspectives on argument. In *Readings in argumentation.* De Gruyter

ARGUMENTATIVE VIRTUE AND DIALECTICAL OBLIGATIONS

WENQI OUYANG
Department of Philosophy, Sun Yat-sen University, China
ouywq@mail2.sysu.edu.cn

Abstract
The notion of dialectical obligation in the pragmatic informal logic theory has been controversial since Ralph Johnson has proposed it decades years ago(2000), and he fails to argue convincingly why the arguer will have to undertake these obligations. Facing varieties of challenges, Johnson makes a compromise to say that in dialectical tier we may need to consider both the dialectical obligations and the dialectical virtues (2007), but he didn't explain it in any further detail. So, what exactly is the relationship between dialectical obligations and dialectical virtues? This paper aims to address the question from particularly the perspective of virtue argumentation theory, which has been developed in the last decade by scholars like Cohen and Aberdein. Virtue argumentation theory has stressed the role of arguer's virtue, and theorized it in a systematic way in argument analysis and evaluation. It provides us with fresh new framework to reconsider Johnson's seminal idea in a more promising way. Therefore, I will first discuss the correspondence of obligation and virtue in the content to see whether the fulfillment of dialectical obligations can be fully explained by the arguer's possession of some argumentative virtues. Then, since the dialectical obligation still faces some unresolved problems, I will see whether the argumentative virtue could offer a better solution to the long-standing controversy over dialectical obligation. In the end, it will be contended from these two aspects that we can interpret Johnson's theory of dialectical obligations in terms of the arguer's argumentative virtue.

1. Introduction

Ralph Johnson has proposed a pragmatic informal logic theory in which an argument is conceived as possessing "a dialectical tier in which the arguer discharges his dialectical obligations." (Johnson, 2000). Although Johnson has made great effort to clarify different types of obligations and the ways for fulfilling them, he fails to argue convincingly why the arguer should be required to undertake these obligations. As a result, the notion of dialectical obligation has been subject to much controversy since its inception (van Rees, 2002; Alder, 2004). In one of his commentaries on Johnson's work, Finocchiaro (2007) made a distinction between dialectical adequacy and dialectical excellence, and then he pointed out that the arguer's efforts in discharging her dialectical obligations could be better understood "as a virtue, but not an obligation or duty". As his reply, Johnson (2007) made a compromise to say that in the dialectical tier we may need to consider both the dialectical obligations and dialectical virtues. However, neither of Finocchiaro and Johnson had explained their own view in any further detail. So, what exactly is the relationship between the dialectical obligations and dialectical virtues?

This paper aims to address the above question by a careful comparison of Johnson's view and the virtue argumentation theory, with the hope that in the end we can see more clearly how far we can interpret Johnson's theory of dialectical obligations in terms of the arguer's argumentative virtues. So, I'm going to analyze the question from two aspects. From the two theories themselves, I will first discuss whether the fulfilling of Johnson's dialectical obligations can be fully explained by the arguer's possessing of some argumentative virtues. Then, given that the dialectical obligation still faces some unresolved problems, I will try to offer a better solution to the long-standing controversy over Johnson's notion of dialectical obligation from the perspective of argumentative virtue, in order to use the virtue argumentation theory to explain the dialectical obligations to the greatest extent.

2. Johnson's argumentation theory and Virtue argumentation theory

With a general understanding of Johnson's theory, let's move to introduce the background of Virtue argumentation theory and do some preparations before making it help with interpreting dialectical obligations. In the last decade, virtue argumentation theory has been developed by scholars like Cohen and Aberdein. They have stressed the role of arguer's virtue in the act of arguing, and theorized it in a systematic way in argument analysis and evaluation. In particular, virtue argumentation theorists contend that the goodness of an argument should be explained to some extent through the characters of the arguer, thus they focus on exploring the relevant intellectual virtues that can make a good arguer. For example, Cohen (2005) has put forward a concept of "ideal arguer", and discussed in several occasions about different types of arguer's virtues such as open-mindedness and being reflective (Cohen, 2007, 2013). Based on Cohen's work, Aberdein (2010, 2016) further develops preliminary lists of the argumentative virtues and of their corresponding vices, and he also used the vice-list to reinterpret the traditional logical fallacies. In this connection, we can see that the emergence of virtue argumentation theory has indeed enriched our understanding of arguer's virtues, and thus provided us with a fresh new framework to reconsider Johnson's seminal idea about dialectical obligations in a more promising way.

Before making a comparison between dialectical obligation and argumentative virtue, we first give a description of both contents. First of all, dialectical obligation raises a new normative requirement based on the innovative conception of argument proposed by Johnson, and the way to fulfill it also provides the content of dialectical tier, which is generally considered to include the following issues: anticipating and responding to objections; anticipating and responding to other criticisms; dealing with alternative positions; anticipating the consequences and implications (Johnson, 1987). But in the practice of argumentation, not all of the above dialectical materials need to be responded, so what is the arguer really required to deal with? Johnson further clarified arguer's dialectical obligation by the specification problem and the criteria of dialectical excellence. The answer to the former was a list containing three kinds of objections at first, then developed to two divided objections according to the different phases in the practice of argumentation. Therefore, this paper will consider the content of dialectical obligation based on the above analysis and will discuss objections in particular since the other three issues of dialectical tier can refer to the discussion on objections in the same way. Secondly, with regard to argumentative virtue, we will mainly use the preliminary list of argumentative virtue drawn up by Aberdein (Table I), and try to give a more specific explanation of each virtue in the corresponding process. Finally, the approach in which we match the obligation and virtue in content must be taken into consideration. There is no doubt that the demands of obligation are generally seen to be stronger than virtue. That is, if a virtuous arguer has fulfilled some dialectical obligations, then he is supposed to possess some or more argumentative virtues. But in fact, it cannot necessarily be inferred from the premise that the arguer has some argumentative virtue to the conclusion that he must has fulfilled some certain dialectical obligations. Accordingly, we choose to start from the dialectical obligation to reflect that fulfilling it implies some kinds of argumentative virtue, so as to examine the correspondence between the two in terms of contents and further consider about the restriction of the argumentative virtue on the dialectical obligations.

Table I. A tentative typology of argumentational virtue (Aberdein, 2010)
(1) Willingness to engage in argumentation
    (a) Being communicative
    (b) Faith in reason
    (c) Intellectual courage
        (i) Sense of duty
(2) Willingness to listen to others
    (a) Intellectual empathy

        (i) Insight into persons
        (ii) Insight into problems
        (iii) Insight into theories
    (b) Fairmindedness
        (i) Justice
        (ii) Fairness in evaluating the arguments of others
        (iii) Open-mindedness in collecting and appraising evidence
    (c) Recognition of reliable authority
    (d) Recognition of salient facts
        (i) Sensitivity to detail
(3) Willingness to modify one's own position
    (a) Common sense
    (b) Intellectual candour
    (c) Intellectual humility
    (d) Intellectual integrity
        (i) Honour
        (ii) Responsibility
        (iii) Sincerity
(4) Willingness to question the obvious
    (a) Appropriate respect for public opinion
    (b) Autonomy
    (c) Intellectual perseverance
        (i) Diligence
        (ii) Care
        (iii) Thoroughness

3. Obligation and Virtue: correspondence

According to the nature of dialectical obligations, it is undoubtedly conditional rather than categorical for arguers, that is, the arguer is required to have such obligation only if he is willing to be a participant in the process of argument. In other words, if an arguer tries to fulfill his dialectical obligation, then it implies that he has the virtue of "willingness to engage in argumentation" (1). Let us demonstrate more concretely how the sub-items of this virtue manifest in fulfilling dialectical obligations. Firstly, if the arguer does not try to communicate with the audience, he will not be able to understand them and give a response. Therefore, it is necessary for the arguer to be communicative (a); Secondly, the arguer chooses to use arguments to persuade and respond to the audience without appealing to violence or other irrational approaches, which indicates arguer's faith in reason (b); Finally, in the face of objections, the arguer dares to respond instead of evading difficulties, which manifests his sense of duty (c)(i). At the same time, he does not respond blindly or fanatically to any objections but responds appropriately, showing his intellectual courage(c).

In order to clarify what the arguer's dialectical obligation is, Johnson made the first attempt to answer The Specification Problem: "The arguer must deal with The Standard Objections. In addition, the arguer is obliged to deal with any objections that the arguer knows the audience will expect that he or she deal with (if they are not included in TSO) and also those objections the arguer believes his or her position can handle (even if not included in TSO). (Johnson, 2000)." Obviously, Johnson came up with a list of three categories of objections. We will examine what kind of argumentative virtue corresponds separately to the dialectical obligation to respond to each category of objections.

First of all, as for the standard objections, Johnson refers to that class of salient objections typically or frequently found in the neighborhood of the issue (ibid.). The arguer here grasps the intuition about salientness, which applies the virtue of maximum relevance, that is, to pay special attention to the most crucial part. Furthermore, we believe that the arguer who fulfills this obligation must possess the argumentative virtue of "willingness to question the obvious" (4). Because for the arguer, the establishment of his own argument is supported by reliable reasons and at the moment of raising the argument, he already thinks that he has given sufficient reasons to make the argument obviously valid. So trying to anticipate the standard objection in such a situation requires the arguer to take the audience's point of view to re-question the argument that he has taken for granted. From the definition of the standard objection at this time, it can be seen that it overlaps with the public opinion to a certain extent, so it also reflects the arguer's "appropriate respect for public opinion" (a); In the process, the arguer needs to identify significant objections independently according to his own judgement, and to question and to respond actively, reflecting "autonomy" (b) of the arguer, which means being rationally independent of others; While "intellectual perseverance" (c) is manifested as follows: under the case where rational reasons can be obtained to support his argument, the arguer needs to keep constantly responding, defending his argument persistently until he can't deal with it.

In the next part, as for responding to any objections that the arguer knows the audience will expect that he or she deal with, it requires the arguer to attach importance to the audience, So there is no doubt that it corresponds to the virtue of "willingness to listen to others" (2). Then, we are going to make a more detailed and comprehensive examination of it. One can interpret the arguer's understanding of the audience's needs and grasp of the audience's questions as embodying his "intellectual empathy" (a), only when the arguer has an insight into persons and their problems can he reconstruct the audience's opinions or positions in a true and complete manner according to their requirements; After understanding the real ides of the audience, the arguer needs to evaluate the audience's opinions or positions at first, and the virtue of "fairmindedness" (b) is essential during this process, which means the arguer should maintain a fair attitude, neither favor himself nor others, always be open-minded when collecting and evaluating evidence, and insist in considering objectively whether the opinions of others are necessary and worthy of responding to; Moreover, since the audience may pay attention to some details that the arguer does not notice, the arguer is supposed to be sensitive to details in order to make a reasonable assessment of the audience's opinion, that is, to be able to "recognize the salient facts" (d).

In the end, the arguer also needs to respond to objections he believes his position can handle. For the arguer, if he knows how to deal with an objection but fails to respond to it, it is clear that he is dominated by the vices of intellectual cowardice and deriliction of duty. Contrasted with the list of argumentational vice given by Aberdein, this is the absence of virtue "willingness to engage in argumentation" (1) (Aberdein, 2016). For this kind of objection, Johnson only identifies it as a necessary, but not a sufficient condition, because there are still objections that the arguer is aware of but cannot respond to. So he considers two possible situations. Suppose it were a lethal objection which is strong enough to seriously jeopardize the argument, then "the arguer ought to withdraw the argument and think the matter over" (Johnson, 2000, pp.331). This definitely requires the virtue of "willingness to modify one's own position" (3), including "intellectual humility" (c): that is, when the arguer is faced with a strong objection that he cannot deal with, it is not a wise choice to defend blindly, but to try to revise and improve the original argument and position; If the arguer regards the objection as one that does not undermine the argument, then Johnson believes that the rational course would be for the arguer to include the objection in the dialectical tier and admit that it is not known how to handle it (ibid.). It also manifests the "willingness to modify one's own position" (3), while it implies more in "intellectual candour" (b): that is, regardless of the ability to respond, the arguer should honestly present to the audience the valuable objection he knows.

It can be seen that the list including the three kinds of objections given above is somewhat controversial. With the increasing discussion of The Specification Problem, Johnson also made a clearer analysis of dialectical obligation to deal with Govier's regress problem. He distinguishes between two types of objections by dividing the argumentative process into two different phases (Johnson, 2001): In the construction of the argument, the arguer's dialectical obligation is to respond to The Standard Objections. After taking into account both the cognitive limitations of the arguer and the expectations of the audience, it is actually an attempt to combine the last two categories of objections in the above list into the standard objection, whereby the definition of the standard objection expands more specific into three dimensions: strength, proximity, and salience (Johnson, 2007). Therefore, to examine the virtue corresponding to the fulfillment of the obligation to respond to the standard objections in this phrase, we can refer to the complete correspondence about the preliminary list above; In the revising phase after the argument has been proposed, the arguer faces the task of responding to actual objections, which Johnson further identifies as Serious Objections. Johnson believes that the objections faced by the arguer and their obligations are quite different in these two phases. Then, Will the two phases also focus on different requirements for the arguer's virtue? In my opinion, in the phase where the arguer is constructing his argument, he first needs to possess the virtue of "willingness to engage in argumentation" (1), which is the premise for the practice to proceed; During the process, anticipating those conceivable or known objections is also essential, which reflects more about the virtue "willingness to question the obvious" (4) compared to the other two virtues. Then, in the revising phase, the consideration of actual objections implies the inclination towards the audience; therefore, it focuses more on the virtue of "willingness to listen to others" (2). Whereas trying to respond to serious objections also requires the arguer to possess the virtue of "willingness to modify one's own position" (3). All above are considerations of the content correspondence between the dialectical obligation and the argumentative virtue in the context of The Specification Problem. We will turn next to the case of dialectical excellence.

The Specification Problem answers "How are those dialectical obligations to be identified and specified", while the criteria of dialectical excellence answers "what is required for an argument to discharge these obligations" (Johnson, 2003). Johnson proposed the "3A" standards: Accuracy, that is, the arguer must accurately and faithfully state the objection; Adequacy, that is, the arguer must make an adequate response; Appropriateness, that is, the objections anticipated must be appropriate. Since the criteria of adequacy are understood more regarding the sufficiency of the illative core, we will focus primarily on the criteria of accuracy and appropriateness in consideration of virtue correspondence. The 'accuracy' requirement means that the arguer should avoid the straw-man fallacy, ensuring that the response is to the position that the audience holds. Then referring to Aberdein's work of the 'Gang of Eighteen' and their distinctive corresponding vices, it can be seen that the straw-man fallacy is considered to be that the arguer is intellectual dishonest (3-)(b), while the audience is over-reliance on reason (1+)(b) and partiality to others in evaluating their arguments, (2+)(b)(ii). From this, we can conclude that if the arguer state the objection accurately, then he must be "intellectual candour"(b), which is the sub-item of "willingness to modify one's own position" (3). The requirements of accuracy here, it can be divided into virtues and skills. Accordingly, grasping the real position accurately in skill does not mean that you are willing to point it out by virtue, while the virtue of being willing to point it out accurately is always limited by the arguer's cognitive abilities to get the real objection, so that you end up not being able to express the real one accurately as well. Therefore, the standard still needs the requirement of argumentative virtue - "intellectual candour"(3)(b), that is, the arguer will not deliberately distort the real thoughts of others for persuasion, and can honestly point out the real objection to his own position. In addition, 'appropriateness' means that if the arguer has omitted/failed to deal with an objection which, it can reasonably be claimed, he ought to have dealt with, then the arguer's response, even if it satisfies the accuracy and adequacy requirements, is not rationally satisfying. (Johnson, 2007). This also implies that the arguer needs to be "intellectual candour"(3)(b), to honestly understand and state the objections he faces, rather than ignore well-known and essential views, and have intellectual courage (1) (c) to participate in the argument.

It should be noted that what we consider in this paper is the preference for a certain argumentative virtue when fulfilling a particular dialectical obligation in an argument, which does not mean that other virtues should be excluded simultaneously. But if we'd like to state that all argumentative virtues work comprehensively as a whole, how should we deal with the conflicting virtues followed by the fulfillment of some obligation? And whether having some certain virtues will constrain the fulfillment of obligations in the actual process of argumentation?[1]

4. Virtue: a new way to respond to challenges

The theory of dialectical obligations has caused many controversies since it was put forward. The criticisms it faces mainly include The Regress Problem, "too demanding" question, and the challenge of "dialectical virtue". Faced with the difficulties mentioned above, Johnson tried his best to fulfill his dialectical obligations and made partial responses respectively. However, after careful examination of these defenses, it is not difficult to find some obvious deficiencies and problems, so the objection has not been completely resolved. At present, the virtue approach of argumentation theory is flourishing. Based on what this paper has tried to explain and accommodate the content of dialectical obligation with virtue argumentation theory, we will further attempt to respond to the issues that dialectical obligation failed to address from the perspective of virtue.

---

1 There are indeed some conflicting situations we need to consider: 1. If rational persuasion is understood as moderate rationality, does the requirement to manifest rationality rely excessively on rationality, leading to the vice, contrary to the virtue of faith in reason (1)? 2. The arguer is required to respond to The Serious Objections in the revising phase, although he exerts the virtue of maximal relevance (focus primarily on what is most significant), does it violate the virtue of "sensitivity to detail" in "recognition of salient facts" (2)(d)? 3. We have acknowledged that the arguer cannot deal with all possible objections. Does it imply that he violates the virtues of the dialectical responsiveness (address all potentially sound criticism) but applies the virtue of maximal relevance? 4. Whether the argumentative virtue of "willingness to modify one's own position" (3) will constrain the arguer's handling of alternative positions to a certain extent?

4.1. Challenge1: Is dialectical tier necessary for the definition of argument?

4.1.1 The objections from Govier and Finocchiaro

Johnson's theory contains substantial revision to the concept of argument. The relationship between the illative core and the dialectical tier has been extremely controversial. The main objection, however, is that the dialectical tier is unnecessary, or even that the definition of argument should not include the dialectical tier at all.

In Govier's view, an argument only needs the illative core because the requirement of dialectical tier leads to an infinite regress. Furthermore, according to Johnson, every argument is incomplete without a dialectical tier (Johnson, 1996). That is, the arguer has an obligation to buttress the main argument with supplementary arguments responding to alternative positions and objections, thereby supporting the original argument. However, Govier argues that supplementary arguments are also arguments, thus they will continue to require supplementary arguments to address alternatives and objections. As a result, this line of reasoning will constantly need to be supplemented, and Johnson's view seems to imply an unacceptable infinite regress. We won't agree that the arguer is obliged to make an infinite number of arguments to provide a good reason for a conclusion. So on this interpretation, the dialectical tier would not be a tier, instead it would be a staircase that mounts forever (Govier, 1999).

Finocchiaro proposes a natural way to moderate Johnson's double requirement by disjoining the two conditions, which gets a weaker but still dialectical conception as following: an argument is an attempt to persuade someone that a conclusion is true by giving reasons in support of it or defending it from objections (Finocchiaro, 2003). In this definition, Finocchiaro separates Johnson's dual demands through a more inclusive disjunction, and naturally reconciles the two. While disjunction allows the argument to contain both illative and dialectical tiers, it also implies the cases where the argument no longer needs the dialectical tier. Thus, the problem that the dialectical tier is not needed once again challenges Johnson's conception of argument.

4.1.2 Johnson's response and the problems

Regarding The Regress Problem, Johnson gives a three-prong strategy as follows. First of all, the reason why such doubts arise, Johnson thinks, is mainly due to their superficial understanding of his concept of "dialectical tier". He clarifies that not every argument requires a dialectical tier, but rather that the paradigm case of argument should display this structure (Johnson, 2003). Specifically, not all the arguments require the dialectical tier but only the paradigm case of argument, and since supplementary arguments are not necessarily the paradigm cases of argument, infinite regress does not necessarily arise either. Second, in the traditional concept of argument, which only refers to the illative core, one does not usually think that reasons are infinitely regressive in support of a conclusion, and our rationality allows us to admit that it will eventually stop somewhere. And such the same way of reasoning that prevents an infinite regress in the illative core, Johnson argues, can also be deployed to prevent the exfoliation of the dialectical tier (ibid.). Finally, Johnson connected the specific content of the dialectical tier with the broader issue of dialectical adequacy. By answering two relevant sub-questions, he initially formulated the appropriate criteria of dialectical adequacy (ibid.), thus avoiding The Regress Problem. Because just as the RSA standard needs to be satisfied in the illative core, the dialectical adequacy criteria in the dialectical tier will tell the arguer when to stop, and the argument will no longer need to be endlessly added. However, as for Finocchiaro's early thoughtful criticisms (Finocchiaro, 2003), Johnson did not make any response.

Reviewing the above responses, we find that the following problems exist. 1. Although Johnson limits the requirement of dialectical tier to the paradigm case of argument, thus clarifying that the requirement does not necessarily lead to the dilemma of infinite regress, the fact that it does not necessarily occur does not mean that it won't occur, so the case of infinite regress is not excluded from the concept of argument. Obviously, if the traditional concept of argument contained only the illative core, there would be no concern about this problem at all, and adding the requirement of a dialectical tier would only increase the risk of infinite regress. In addition, the concept of "paradigm argument" is ambiguous, and Johnson does not explicitly give the criteria to distinguish it from other arguments. 2. It is inappropriate to defend the dialectical tier by using the same line of reasoning that prevents an infinite regress in the illative core where we only look for the reason that provides support for a certain conclusion, and it can be also regarded as the product displayed. However, in the dialectical tier, based on responding to the objections, we often provide more than one reason. The two differ in content and nature, and in the way they look for reasons and responses, so the points from which they are considered to cause the regress cannot be the same. In other words, the responses to objections are not the same as the reasons for supporting the original argument, but are more complex because of the dynamic process of supplementing new arguments. 3. The dialectical adequacy standard itself has been questioned since it was put forward. The relevant standard formulated by Johnson implies that the judgment of adequacy will be more or less with the subjective idea of the arguer or the audience. As a result, the related concepts like the standard objection are difficult to determine without qualitative criteria. They are constantly changing and adjusting, so The Regress Problem cannot be completely avoided.

4.1.3 Can virtue deal with them?

Suppose the content of dialectical tier is explained or even accommodated by the virtue argumentation theory. In that case, we will try to deal with the difficulties mentioned above from the perspective of virtue. Whether it is Govier's regress problem or Finocchiaro's new disjunctive definition, it is a discussion of the relationship between illative core and dialectical tier. The controversy is actually whether the content of the dialectical tier can be regarded as a part of the argument product, or rather, whether the requirement of dialectical tier is necessary for the concept of argument. According to Johnson's definition, the fulfillment of arguer's dialectical obligation is only limited to the dialectical tier. However, when we explain and cover these contents with the virtue argumentation theory, we will find that the arguer is not required to have the obligation which is only in dialectical tier. Rather, the requirement develops into the normative of virtue that run through the entire argumentation process. Based on this shift, the relationship between illative core and dialectical tier will no longer matter, and we should focus on the conduct of the arguer that infiltrates in the argumentation process in a virtuous manner and ultimately manifests itself in the product of the argument. There is no doubt that Tindale has already pointed out that arguers often have the possible and actual responses of others in mind when they formulate their premises, and in that sense, a particular premise in the illative core may itself have a dialectical tier (Tindale, 2002). Now we can explain this behavior of the arguer by argumentative virtue, so that the distinction between illative core and dialectical tier dissolves completely, thus addressing the critic's fear about the separation of the two tiers and providing support for the necessity of dialectical tier.

4.2. Challenge2: Does dialectical obligation require too much?

4.2.1 The objections from Tindale and Adler

In addition, Johnson's definition of argument has been criticized as having an overly exclusive perspective, which is too restrictive and excludes all sorts of specimens that would get counted as an argument by more inclusive definitions. As well as concerns that Johnson's theory was explicitly designed to cover written arguments, as contrasted with verbal argument, Tindale also worried that his theory privileged the arguments of philosophers. Clearly, everyday casual arguers may not be able to fit within such rigorous and professional practice of specialists, or that something would be lost if they were made to do so (ibid.). Thus, Tindale is concerned that Johnson's approach will put argument out of reach for the everyday casual arguer, that is, those casual arguments would be excluded for not including the dialectical tier. Investigating its essence, Tindale pointed out that in practice, the requirement of the arguer's dialectical obligation is not entailed by the standard view, that is, the requirement for arguers to discharge their dialectical obligations is beyond the capacities of casual arguers. At the same time, there are indeed scholars who have given evidence that ordinary arguers have difficulty in fulfilling their dialectical obligations (Perkins et al., 1983; Kuhn, 1991).

Another criticism related to this comes from Adler. From an epistemological point of view, he believes that Johnson's dialectical tier is too demanding, and the costs on time and resources necessary to satisfy it render it unfeasible. Moreover, the efforts toward fulfilling it would diminish the vitality of argumentation and inquiry (Adler, 2004). Furthermore, since argumentation is a social-epistemic activity, Adler states that the valuable aim of responding to objections, which Johnson's dialectical tier is meant to satisfy, can be achieved in better ways.

4.2.2 Johnson's response and the problems

So, does the dialectical obligation require too much or even go beyond the capacity of some arguers? Confronted by Tindale's skepticism, Johnson argues that the difficulty of casual arguers in accomplishing this task is not a reason to downscale the requirement, but rather a reason to give them better instruction (Johnson, 2002). At the same time, there are also two additional considerations presented by Johnson to suggest that the dialectical obligation is not actually beyond the ken of the ordinary arguer. First, we indeed have plenty of such examples in practice where ordinary arguers seem to be attempting to discharge their dialectical obligations by anticipating objections. Second, Johnson argues that the fallacy of straw person occurs when arguers go astray in discharging their dialectical obligations. Thus, in ordinary arguments, the occurrence of this fallacy indicates that casual arguers are aware of and attempts to dispatch their dialectical obligations (ibid.). As for Adler's objection, Johnson did not make any specific response.

It can be seen that Johnson's responses are still not reliable: obviously, no matter how hard one tries, one cannot accomplish tasks beyond one's ability. As a result, people generally do not regard behavior beyond their ability as a guide. In addition, Johnson's two additional reasons for clarifying that the dialectical obligation is within the abilities of ordinary arguers are also problematic. Firstly, rising from the occurrence of some individual cases to a general requirement is not allowed, and normative claims for all cannot be derived from individual descriptive facts alone, so the observation of such examples is not an appropriate sufficient reason for the claim that everyone should be able to fulfill the dialectical obligation. Secondly, the fallacy of straw person usually occurs in response to others rather than in the process of proposing an argument, so the audience most likely commits it. While we can conclude that the arguer has committed the fallacy on the basis of his failure to respond to the real objection, we cannot thereby limit the occurrence of this fallacy only to the fact necessarily that the arguer goes astray in discharging dialectical obligations. Furthermore, the occurrence of this fallacy always allows us only to see that the arguers have failed to accomplish this task, that is, they still lack the ability to fulfill their dialectical obligations. To be clear, going through all the outcomes that ultimately fail just to show that they have an implicit intention to attempt to do so is not a good strategy, but to justify that they are truly capable of doing it.

4.2.3 What if we turn to Virtue?

It is not difficult to find that, as far as the challenges faced by the dialectical obligations are concerned, the toughest one is undoubtedly the criticism that its requirement is too demanding to be realized in the practice of argument. On the one hand, it is due to the lack of the arguer's own cognitive ability. On the other hand, it is also attributed to the high cost of resources required to address the obligation. However, when this corresponding requirement of obligation is changed into the norm of argumentative virtue, everyday casual arguments will no longer be strictly excluded from the concept of argument, although virtue is not an ability that everyone already possesses, even if the ordinary arguers, we won't say that virtue is beyond their capabilities. Honestly, virtue is weaker than obligation in terms of the normative strength. As for dialectical obligation, it is a strict requirement for the arguer to have to do, and the huge gap between it and the actual ability of casual arguers is so apparent, which is bound to be worrying. In Govier's words, the requirement of obligation implies that we are able to do so (Govier, 1999), whereas virtue does not. Rather than the norm of obligation, we might think of virtue as a capacity that can be cultivated by providing some kind of instruction to the arguers and their behavior. Since virtue is innate and acquired, it is reasonable to believe that it is easier and more likely to be realized by more arguers (including ordinary arguers) than obligations. Of course, the requirement of making the arguers possess virtue and manifest it in the process of argumentation does not require a great deal of time and intellectual expenditure in the practice of argument, because the cultivation of virtue is a long-term process, and it pays more attention to the accumulation in daily life.

4.3. Challenge3: Dialectical obligation or dialectical virtue?

4.3.1 The rise of "Dialectical virtue"

According to Johnson's theory, Finocchiaro developed two new requirements for arguers: "strengthening objections" and "defending alternatives". Just as the idea of dialectical obligation can be traced back to Mill's argument in On Liberty, Finocchiaro provides the similar evidence that these two new requirements can be regarded as dialectical obligations as well as "anticipate and respond to objections". However, he was hesitant to say that an arguer has an obligation to do the above two practices. In terms of these two new issues, if and to the extent that arguers do these things, that adds extra value to their arguments, at least on some occasions, but he was not sure that not doing them is in itself a flaw or fault (Finocchiaro, 2007). Based on the rejection of the "new obligations", Finocchiaro continued to be convinced that what Johnson was proposing could not be regarded as the dialectical obligation either. In this way, he clearly distinguishes between dialectical obligation and dialectical virtue, then infers that Johnson understood the concept of excellence as synonymous with adequacy (Johnson, 2007a) because he used "adequacy" in a broad sense. Obviously, if the other more basic principles pertaining to adequacy are satisfied, we cannot say that not following the above two requirements does make one fall short of dialectical adequacy, but only short of dialectical excellence. Therefore, Finocchiaro's proposal to replace dialectical obligation with dialectical virtue completely breaks with Johnson's attempt to elevate the former to the universalist ideal of the argumentation theoretical norms.

4.3.2 Johnson's response

As to whether the arguer should "strengthen objections", Johnson gives an explanation by analogy with formulating a missing premise: the choice of putting one in the missing premise actually depends on the purpose of one's engagement in the argument. And the requirements for arguers to present objections in their strongest possible light when anticipating them is similar to asking for supplements to the needed premises. Regarding the practice of "defending alternatives", Johnson did not make any substantive response. In dealing with dialectical issues, Johnson has always considered in terms of obligations, although he has also drawn contrasts between obligations and interests as determinants (Johnson, 2003). However, in the face of Finocchiaro's challenge, Johnson has indeed been torn between a virtue-based approach to dialectical matters and an obligation-based approach. While Johnson's intuition is that there are both dialectical obligations and dialectical virtues, both has been not yet well understood (Johnson, 2007b). He also acknowledges that the two new demands proposed by Finocchiaro should indeed be regarded as virtues. However, he insists that the dialectical obligations should still be regarded as obligations. In fact, this is based on Johnson's assertion that his theory simply cannot derive the requirements proposed by Finocchiaro, that is, the two intuitively stronger new issues cannot be justified by Johnson's theory reasonably.

4.3.3 Can Dialectical virtue be justified also by "manifest rationality"?

Johnson's justification of his dialectical obligation ultimately appeals to the requirement of "manifest rationality". The key, therefore, is whether the two "new obligations" proposed by Finocchiaro can equally be necessarily derived from the requirement of "manifest rationality". If Johnson's theory can properly justify the content of the "new obligations", as it justifies the dialectical obligation it originally proposed, then, when we agree with Finocchiaro's point of view, acknowledging that the "new obligations" should indeed be reduced to the requirement of argumentative virtues, it is of course also reasonable to conclude that all other elements of the dialectical obligation under Johnson's theory should also be treated as argumentative virtues, including anticipating objections.

Let us first return to the justification of the dialectical obligation. In order to distinguish argumentation from other rational practices (especially rhetoric), Johnson added "manifest rationality" as a new feature to argumentation practice from the perspective of pragmatics, further clarifying the boundary between logic and rhetoric. Based on this, we must view the practice of argumentation as an exercise in manifest rationality (Johnson, 2000). Specifically, the practice of argumentation is not only rational, it also needs to appear to be rational (Johnson, 1996). And so do the participants in the practice of argumentation, which meas they need to be seen as rational and must therefore manifest rationality in the process of argumentation. It also provides a rationale for why rhetoricians are not obliged while arguers are obliged to deal with objections and criticisms, whose responses to dialectical materials realize this rationality in precisely the way people can see.

Does Johnson's demand to "manifest rationality" lead to Finocchiaro's "new obligations"? Let us consider the new issue of "strengthening objections", that is, when an arguer is confronted with an objection and knows how to strengthen it, according to the requirement of manifest rationality, whether the arguer has an obligation to strengthen it before responding. The question requires us to revisit the part of Johnson's theory that links the strength of objections to the arguers: it is clear that the definition of The Standard Objections involves both, which cannot be bypassed by answering The Specification Problem in the discussion of the dialectical adequacy. Furthermore, after dividing the process of argumentation practice into two phases, Johnson comprehensively considers the cognitive limitations of the arguer and the expectations of the audience, and defines The Standard Objection more specifically into three dimensions: strength, proximity, and salience (Johnson, 2007a). Among them, "strength" refers to: the stronger the objection is, the more it needs to be responded to. Undoubtedly, the person who knows the argument best and is most likely to give the strongest objection to an argument is the arguer himself. It seems that the measure of the strength criterion is more in the hands of the arguer than the audience. Thus, under the requirement of manifest rationality, an arguer, when confronted with an objection, is obliged to present it in the strongest possible way to make it a standard objection before responding.

As with the dialectical obligations Johnson originally justified, the new obligation to strengthen objections can also be justified by the requirement of "manifest rationality". Thus, as in Finocchiaro's point of view, Johnson's dialectical obligation can indeed be reduced to the dialectical virtue. Although neither of them go into more details about the dialectical virtue, with the development of the virtue argumentation theory recently, this finding will be more helpful for the argumentative virtue interpreting Johnson's theory of dialectical obligations.

5. Conclusion

As far as the results of the first aspect are concerned, there are two conclusions can be drawn through the analysis. For some dialectical obligations, (i) there are possibly more than one candidate virtue that can provide an explanation for their fulfillment, and (ii) there are also some argumentative virtues the possession of which would actually prevent the arguer from discharging them. As far as the first question is concerned, it is contrarily a good thing for interpreting Johnson's theory via virtues, which also demonstrates the inclusiveness of virtue argumentation theory. And for the second question, we can consider the suggestion about choice of virtue given by Stevens in the analysis of the case that an ideal arguer might have to possess virtues that at first sight look contrary to each other. And she proposes that the arguer should use practical wisdom to decide which virtues should guide her in which argumentative situation (Stevens, 2016). Similarly, this requirement can help us avoid the virtue from preventing the fulfillment of obligations in different situations. Therefore, it can be said that all the efforts in discharging dialectical obligations are generally explainable in light of the arguer's possessing of certain relevant argumentative virtues.

Through the analysis of the second aspect, it can be seen that three main challenges to dialectical obligations could get better resolved through virtue argumentation theory. And we do have sufficient reasons to admit that the dialectical virtue can also be justified by "manifest rationality", which further supports our attempt to interpret Johnson's theory of dialectical obligations from the perspective of a virtue theory of argument. Therefore, we can connect the discussion of associated dialectical obligations with agent-centered virtue theories of argumentation, and provide a completely new explanation to Johnson's theory.[2]

Reference

Aberdein, A. (2010). Virtue in argument. *Argumentation, 24(2)*, 165-179.
Aberdein, A. (2016). The vices of argument. *Topoi, 35(2)*, 413–422.
Adler, J. E. (2004). Shedding Dialectical Tiers: A Social-Epistemic View. *Argumentation, 18*, 279–293.
Cohen, D. (2005). Arguments that backfire. In Hitchcock.D, Farr.D (Eds.), *The use of argument.* OSSA, Hamilton, 58-65.
Cohen, D. (2007). Virtue epistemology and critical inquiry: Open-mindedness and a sense of proportion as critical virtues. Association for Informal Logic and Critical Thinking.
Cohen, D. (2013). Virtue, in context. *Informal Logic, 33(4)*, 471-485.
Finocchiaro, M. A. (2003). Dialectics, evaluation and argument. *Informal Logic, 23*, 19-49.
Finocchiaro, M. A. (2007). Commentary on Ralph H. Johnson: "Anticipating objections as a way of coping with dissensus." In H.V. Hansen, et. al. (Eds.), *Dissensus and the Search for Common Ground,* CD-ROM (pp. 1-6). Windsor, ON: OSSA.
Govier, T. (1999). *The Philosophy of Argument.* Newport News, VA: Vale Press.
Johnson, Ralph H. & J. Anthony Blair. (1987). Argumentation as Dialectical. *Argumentation, 1*, 41-56.
Johnson, R. H. (2000). *Manifest Rationality: A Pragmatic Theory of Argument.* Mahwah, NJ: Lawrence Erlbaum Associates.
Johnson, R. H. (2001). Clear Thinking in Troubled Times: An Integrated Pragma-Dialectical Analysis. *Informal Logic, 21(2)*, 17-30.
Johnson R. H. (2002). Manifest Rationality Reconsidered: Reply to my Fellow Symposiasts. *Argumentation 16*, 311–331.
Johnson R. H. (2003). The dialectical tier revisited. In F.H. van Eemeren, J.A. Blair, C.A. Willard & A.F. Snoeck Henkemans (Eds.), *Anyone Who has a View: Theoretical Contributions to the Study of Argumentation* (pp. 41-54). Dordrecht: Kluwer Academic Publishers.

---

2 I want to express my deep appreciation to all parties of the ECA 2022 and the chair Prof. Fabio Paglieri for providing me the opportunity to present this research work at the conference online. This work was supported by the Major Program of National Social Science Foundation of China (Grant No. 19ZDA042).

Johnson, R. H. (2007a). Anticipating objections as a way of coping with dissensus. In H.V. Hansen, et. al. (Eds.), *Dissensus and the Search for Common Ground*, CD-ROM (pp. 1-16). Windsor, ON: OSSA.

Johnson, R. H. (2007b). Reply to my commentator. In H.V. Hansen, et. al. (Eds.), *Dissensus and the Search for Common Ground*, CD-ROM (pp. 1-4). Windsor, ON: OSSA.

Kuhn, D. (1991). The skills of argument. Cambridge: Cambridge University Press.

Perkins, D. N., Allen, R. and Hafner, J. (1983). Difficulties in everyday reasoning. In J. Maxwell (ed.), Thinking: an interdisciplinary report. Philadelphia: The Franklin Institute Press.

Stevens, K. (2016). The virtuous arguer: one person, four roles. *Topoi 35*, 375-383.

Tindale, C. W (2002). A Concept Divided: Ralph Johnson's Definition of Argument. *Argumentation 16*, 299–309.

ARGUMENTATIVE VIRTUES: BACK TO BASICS

FABIO PAGLIERI
Istituto di Scienze e Tecnologie della Cognizione, Consiglio Nazionale delle Ricerche (ISTC-CNR), Roma, Italy
fabio.paglieri@istc.cnr.it

Abstract

This paper discusses three fundamental questions for virtue argumentation theory that, to the best of my knowledge, do not yet have satisfactory answers, even though some of them have been around since the very inception of this approach to argumentation. First, are argumentative virtues specific of argumentation, i.e. are they bound to manifest exclusively or more frequently during argumentative engagements, as opposed to other situations? Second, are argumentative virtues causally tied to the quality of an argument, rather than to some other desirable outcome of the social situation where the argument occurs? Third, what are exactly these argumentative virtues, and how could (and should) they inform our educational practices in critical thinking instruction?

1. Introduction: does virtue argumentation theory need a jiminy cricket?

In my previous engagements with virtue argumentation theory (VAT, from now on), I have always took the stance of a curious spectator (Paglieri, 2015): this was not intended to distance myself from the approach, nor to shield my remarks from criticism, but simply to acknowledge that I was discussing someone else's work, instead of trying to directly contribute to the development of what I perceived as a novel and promising approach to argumentation. The godfathers of VAT, Daniel Cohen and Andrew Aberdein, were kind enough to describe my forays in their area of expertise as "studiously even-handed" (Aberdein & Cohen, 2016), and perhaps my effort in that regard also contributed to enshrine my name in one of the entries of the Arguer's Lexicon.[1] In that much beloved online source of enlightened fun for argumentation scholars, a "paglieri" (noun) is defined as "a trusted adviser to the main spokesperson in an argument between

---
[1] See https://web.colby.edu/arguerslexicon/

rival factions; someone who can be counted on to do the dirty work in conflict 'resolution'." The scare quotes around 'resolution' emphasizes the tongue-in-cheek reference to the role of the "consigliere" in Italian mafia, yet it remains an apt definition of what I tried to accomplish in my past dealings with VAT.

This paper continues in that tradition, yet my position is shifting: nowadays I am more like a concerned bystander, with a touch of the angelic devil's advocate (Stevens & Cohen, 2021). As it will become apparent in the following pages, it seems to me that VAT, while treading new interesting grounds, is also neglecting to clean house on some basic issues that stands at the core of its theoretical edifice. This, I fear, may prove to be a recipe for disaster in the long run, thus I urge a course correction. With that, I am now taking the self-appointed role of the jiminy cricket in Pinocchio's tale, with respect to VAT: I am pestering virtue theorists on what they should do differently, if they want to succeed in their endeavors. This is at risk, of course, of being both annoying and foolish: not only because my concerns might be misplaced and ultimately result in a mere loss of time for all involved, but also because the jiminy cricket met a very brutal and untimely end in the story. However, I trust VAT scholars will prove to be more argumentatively virtuous than Pinocchio and kindly refrain from smashing me with a hammer...

We shall see how it goes. The fundamental questions I suggest VAT theorists should engage with more directly are three: I will simply list them here, and then the rest of the paper will be dedicated to discussing them, one at a time, before summarizing my conclusions and offering some suggestions on how to proceed from here.

First question: are argumentative virtues specific of argumentation, i.e. are they bound to manifest exclusively or more frequently during argumentative engagements, as opposed to other situations?

Second question: are argumentative virtues causally tied to the quality of an argument, rather than to some other desirable outcome of the social situation where the argument occurs?

Third question: what are exactly these argumentative virtues, and how could (and should) they inform our educational practices in critical thinking instruction?

2. Are argumentative virtues specific of argumentation?

This question was posed, in very much the same terms, by Geoff Goddu at the first European Conference on Argumentation in Lisbon, back in 2015. He presented a paper entitled "Are there any argumentation specific virtues?", with the following abstract:

"The purpose of this paper is to explore the question of whether there are any argumentation specific virtues or whether the virtues that theorists such as Aberdein and Cohen point out are merely generic virtues that have roles in other intellectual activities besides argumentation. If there are argumentation specific virtues, are they interesting or significant? If there are none, is that a problem for virtue argumentation theory?"

By the time the paper got published in the conference proceedings (Goddu, 2016), its title had morphed into "What (the hell) is virtue argumentation?": this shift suggests that Goddu did not find any satisfactory answer to his original question, during the conference or after it. Indeed, I believe that question still remains unanswered today, which is the reason why it is worth reconsidering it once again.

Asking whether argumentative virtues are specific of argumentation is tantamount to test whether they are bound to manifest exclusively or more frequently during argumentative engagements, as opposed to other situations. To do that, we must start from a set of likely candidates to the role of argumentative virtues. Currently, the best option for that is still the list compiled by Andrew Aberdein: to the best of my knowledge, this was first presented in 2010 as a "tentative typology of argumentational virtue", yet the same taxonomy is found in all subsequent papers by Aberdein on this topic (2016, 2021). This suggests that either such proposal is no longer to be considered "tentative", or that not enough attention has been given by VAT scholars to the task of refining a principled taxonomy of argumentative virtues. Either way, the following seems to be the current consensus on what should count as argumentative virtues (Aberdein, 2010):

(1) Willingness to engage in argumentation
    (a) being communicative
    (b) faith in reason
    (c) intellectual courage
        (i) sense of duty
(2) Willingness to listen to others
    (a) intellectual empathy
        (i) insight into persons
        (ii) insight into problems
        (iii) insight into theories
    (b) fairmindedness
        (i) justice
        (ii) fairness in evaluating the arguments of others
        (iii) open-mindedness in collecting and appraising evidence
    (c) recognition of reliable authority
    (d) recognition of salient facts
        (i) sensitivity to detail

(3) Willingness to modify one's own position
   (a) common sense
   (b) intellectual candour
   (c) intellectual humility
   (d) intellectual integrity
      (i) honour
      (ii) responsibility
      (iii) sincerity
(4) Willingness to question the obvious
   (a) appropriate respect for public opinion
   (b) autonomy
   (c) intellectual perseverance
      (i) diligence
      (ii) care
      (iii) thoroughness

Following Goddu, specific argumentative virtues should *not* be "generic virtues that have roles in other intellectual activities besides argumentation". With that criterion in mind and looking at Aberdein's list, specificity does not seem to characterize these entries: some are clear instances of *fully generic* virtues, i.e. courage, sense of duty, justice, honour, responsibility, autonomy, diligence, care, thoroughness; others are well-known examples of *epistemic* virtues, i.e. faith in reason, insight into problems and theories, open-mindedness in collecting and appraising evidence, recognition of reliable authority and salient facts, sensitivity to details, common sense, intellectual perseverance; and the rest appear to be *social* virtues (that is, virtues that make any social interaction better, regardless of its argumentative nature), i.e. being communicative, willingness to listen, insight into persons, fairness in evaluating the arguments of others, intellectual candour and humility, appropriate respect for public opinion.

To be fair to Aberdein, genericity plagues all proposals for argumentative virtues, not just his list. The usual suspects of VAT, that is, the most frequently studied argumentative virtues (things like open-mindedness, mutual respect, intellectual honesty), are all obvious examples of pretty generic virtues, and recent proposals to expand the standard list conform to the same trend, with suggestions to include courage (Aberdein, 2021), patience (Phillips, 2021), and other-regarding attitudes in general (Oliveira de Sousa, 2020; Stevens & Cohen, 2021), none of which would pass Goddu's specificity test. Indeed, Aberdein himself demonstrates awareness of this issue, yet he leaves it "as an open question whether argumentational virtues must be distinct from virtues associated with other activities, such as epistemic or ethical virtues. It may well be that some virtues require localization to argumentation, whereas others are best treated uniformly across argumentation, epistemology, ethics, and perhaps other areas" (2021, p. 1206).

Even though some (perhaps most) VAT scholars may be inclined to consider specificity as not really problematic for their approach, there have been attempts to provide a non-trivial definition of what would make a virtue "argumentative". Aberdein, in particular, links this to the identification of what these virtues are expected to track: if ethical virtues track the good and epistemological virtues track truth, then what do argumentative virtues track? Aberdein finds the answer to that question in *truth propagation*, as follows:

> "We might say that the virtues of argument *propagate* truth: where virtuous knowers are disposed to act in a way that leads to the acquisition of true beliefs, virtuous arguers are disposed to spread true beliefs around. The outcome of an argument between virtuous arguers would be a wider distribution of true beliefs (or a reduction in false beliefs). [T]ruth-preserving, but non-virtuous, arguments (…) would not achieve that outcome, even if their premises were true and mutually accepted, since they would fail to add their true conclusions to a interlocutor's store of beliefs. This is either because they are unconvincing or, as with circular arguments, because they require the conclusion to already be accepted. Conversely, virtuous arguments, whether truth-preserving or not, should provide interlocutors and audience with new true beliefs, or at least reasons to increase the confidence with which they hold to existing true beliefs. This account assumes a view of the goal of argumentation. There are many different goals that we may pursue through argument; not just familiar ones such as persuasion or resolution, but less obvious outcomes such as understanding, self-knowledge, and respect (…). Truth propagation is not itself a goal, unless randomly propagating arbitrary truths is worthwhile in itself, but all of these goals are consistent with the propagation of truth. There could be some circumstances in which it was unimportant, eristic dialogues perhaps, but these would seem to be atypical" (Aberdein, 2010, pp. 173-174).

I shall call this proposal *the spread approach to argumentative virtues*, since it assigns them the key function of "spreading true beliefs around", as Aberdein puts it. It is an interesting suggestion, yet it seems to run into some problems. Firstly, we live in a world where knowledge is eminently tied with testimony and communication (Dutilh Novaes, 2020): as a result, the virtues involved in acquiring true beliefs (which would count as epistemological for Aberdein) and those required to spread true beliefs are bound to overlap to a massive degree, and this dulls the proposed distinction between epistemic and argumentative virtues. Secondly, and relatedly, the spread approach characterizes argumentative virtues as belonging primarily, or even solely, to argument producers, not to argument interpreters – since all virtues exhibited by the latter would be, by definition, epistemic rather than argumentative. This seems just wrong, considering that VAT has always presented argumentation as giving the same dignity and relevance to both production and interpretation of arguments, often emphasizing how arguers play both roles and how the virtues they exhibit (or not) in each of them are intertwined: as Cohen put it, "the practice of giving reasons requires that we practice hearing reasons" (2019). Finally, even accepting the spread account of argumentative virtues, this would still fail to grant their specificity, because the list of virtues would remain the same and, therefore, quite generic, as discussed.

Overall, the state-of-the-art in VAT does not seem to offer a positive response to Goddu's specificity test: thus, until better evidence to the contrary is presented, we are forced to tentatively conclude that argumentative virtues are not, after all, exclusively specific of argumentation. This does not necessarily mean that VAT is inconsistent or conceptually ill-conceived, yet it does beg the question: what do we exactly mean, when we talk of argumentative virtues? The best answer, I believe, is to concede that *a virtue is considered argumentative insofar as it systematically helps improving the quality of argumentation*, regardless of whether the same virtue is relevant also in other human endeavors. Such an answer accepts the genericity of argumentative virtues, yet it insists on the usefulness of keeping the moniker, as a shorthand for "virtues that make argumentation systematically better". This, however, brings us to the second challenge that I want to discuss in this paper: i.e., what does it mean to "make argumentation better", and how do we proceed to establish that certain virtues achieve such an outcome "systematically"?

3. How are argumentative virtues tied to argumentation quality?

Even if we accept the non-specificity of argumentative virtues, we are still left with the task of determining a robust connection between possession of one of these alleged virtues and production of positive argumentative outcomes: in the absence of such a link, speaking of "argumentative virtue" would make little sense.

This undertaking requires first establishing what counts as "argument quality": if we take the somewhat old-fashioned view that validity constitutes the cornerstone of argument quality, then the prevailing opinion (Bowell & Kingsbury, 2015; Bondy, 2015; Godden, 2016) is that argumentative virtues cannot guarantee validity of the resulting arguments, here intended as premises-conclusion structures – pace Aberdein (2014). However, as I have argued elsewhere (Paglieri, 2015), this should not worry virtue theorists, since their point is precisely that argument quality should *not* be reduced to validity, or even rhetorical success and dialectical closure (Cohen & Miller, 2016). According to VAT, a significant portion of what makes an argument good goes beyond the quality of the reasons exchanged, and virtue theory helps argumentation theory to account for these layers of quality (Cohen, 2019).

In fact, some VAT scholars (Daniel Cohen, most notably) have been very vocal in claiming that this broader view on argument quality requires making argumentative virtues *constitutive* of it: in other words, a good argument should be *defined* as one in which the arguers have argued virtuously, instead of insisting that something counts as an argumentative virtue only if it delivers good arguments. Incidentally, this would seem to dissolve the problem of establishing a link between virtues and outcomes, since now that connection would be achieved by the very definition of argument quality.

However elegant, this solution is only apparent, since we still need to ensure that what falls under the broad purvey of "arguing virtuously" is *plausible*, and the only sensible way of understanding plausibility in this context is in terms of consistency with our prima facie intuitions on argument quality (here understood in the extended sense favored by VAT scholars). This is what allows, for instance, to consider open-mindedness as a perfectly suited candidate as argumentative virtue, whereas abusiveness and prevarication are immediately dismissed as manifestly non-virtuous. Why? Because it is intuitively clear that being abusive and prevaricating your fellow argues does not result, typically, in good argumentative exchanges, therefore it cannot plausibly be considered as argumentatively virtuous. This shows that the problem of establishing a connection between virtues and outcome quality does not go away by considering the former constitutive of the latter: to avoid vicious circularity, such a strategy still needs to articulate clear criteria on what makes certain virtues systematically deliver laudable results.

One of the most direct options available for that purpose would be to show how a certain argumentative virtue (or, alternatively, a set of them) constitutes a necessary and sufficient condition for (some aspect of) argument quality. Unfortunately, this is a non-starter for VAT, and for virtue theory in general: virtues are notoriously sensitive to context, thus it is very hard to articulate their relevance for argument quality in terms of necessary and sufficient conditions. As Katharina Stevens noted, "what seems to be a virtue in one argumentative situation could very well be called a vice in another" (2016, p. 375). Moreover, even when a certain trait or attitude acts as a virtue in an argumentative exchange, its virtuous character may still be related to something that does not pertain argumentation per se (however broadly understood), but rather the more general social situation.

For example, let us consider courage, that Aberdein recently proposed as an argumentative virtue (2021), and let us see what role it plays in a very famous (albeit probably spurious) historical example: Galileo's defiant statement "And yet it moves!" ("Eppur si muove!"), allegedly uttered in the face of his inquisitors, just after being forced to recant his astronomical theories. This act is widely regarded as an example of courage, alongside other virtues, such as intellectual honesty and respect for facts (both of which are also typically listed as "argumentative"). Yet what makes Galileo's deed noble does not seem to pertain the quality of his arguments, nor the broader argumentative exchange he was engaged in with his inquisitors: he had already publicly recanted his theories at that point, and, even assuming that his courageous last statement might have impressed the Inquisitions' judges, it did not have any such effect in practice and certainly did not sway the final outcome of the legal proceedings. This does not make Galileo's act any less virtuous, yet it clarifies that, whatever virtue he displayed, it was not an argumentative one in that context, since it did not affect the quality of argumentation in any obvious way. In fact, courage would have been exemplified equally well if Galileo had protested his fate without any pretense of argument, e.g. crying "Go to hell, you sanctimonious fools!", against his inquisitors.

What about cases where the result of acting virtuously is indeed argumentative, yet the connection does still not seem appropriate to define the notion of argumentative virtue? To find a suitable example, let us fast-forward four centuries and move to the other side of the globe, landing in Taiwan during the early stages of the Covid-19 pandemic: at that time, the Taiwanese government (as well as the governments of other countries worldwide) had to deal with irrational stockpiling of toilet papers by panicked citizens, fueled by the unfounded rumor that the raw materials and production lines used for sanitary masks were the same necessary to produce toilet papers, hence the need to stockpile on the latter before there was a shortage (which, of course, ended up causing such a shortage, in a classic instance of self-fulfilling prophecy). The premise of this bizarre line of reasoning was pure fabrication: the manufacture of facemasks did not in fact use the same materials or production lines as toilet papers. Yet that bit of fake news had spread like a wildfire worldwide, hence the Taiwanese government was faced with the problem of setting the record straight in ways that would be quickly heard by as many citizens as possible. It needed, in short, to make truth viral.

Luckily, they could count on Audrey Tang, Taiwan's "Digital Minister" and civic hacktivist, who applied to the problem the aptly named strategy of "humor over rumor" (their label, not mine): it consists in using humorous memes to spread quickly and widely the correct information about a controversial issue. Since the jury is still out on whether humor-based or logic-based correction of fake news is more effective (according to extant evidence, it depends on context; Vraga et al., 2019), the Taiwanese pragmatic approach was to employ both in synergy, instead of considering them as alternative options.

In the case of the toilet paper panic, they issued a cartoon of the then Prime Minister, Su Tseng-chang (Fig. 1), standing in front of a table reporting all the correct information on how sanitary masks are produced (nothing to do with toilet paper, as mentioned): the image was an animated

GIF, so that the Premier could be seen vigorously shaking his bottom (or, as they say today, "twerking") under the punchline of the meme, "Remember: we all have only one butt!". As expected, the GIF went viral in a matter of hours, soon becoming much more visible than the rumor that it was designed to counteract: after having a laugh about their twerking Prime Minister, Taiwanese were able to get their facts straight about sanitary mask production, so the toilet paper stockpiling stopped almost overnight. This kind of success vindicates yet another motto of the Taiwanese government strategy for communicating with their citizens: "fast, fair, and fun" (for a more detailed reconstruction of this fascinating case study, see Paglieri, 2022; Tischer, 2022).

Figure 1. How the Taiwanese administration used humor to curtail disinformation

This story nicely exemplifies a virtuous act of arguing: it promptly rectified mistaken beliefs on a relevant issue and thus prevented negative consequences, without patronizing the public and promoting instead critical engagement with accurate information. Moreover, using humor to make the message viral was crucial to guarantee argumentative success: virality is a necessary precondition of such virtuous campaigns, and more generally of any attempt to effectively curtail widespread disinformation. Yet virality does not affect argumentative quality, just circulation: it makes the communicative act good, not the argument – again, no matter how broadly one defines "argument". Whereas the expert use of irony in crafting the message (sharing a laugh at the expense of the highest authority in the country, instead of making fun of the silly beliefs of the citizens) did act as an argumentative virtue, insofar as it facilitated the uptake of the correct information conveyed by the message, its viral character did not make it better qua argument, even though it was essential to the argumentative success of the campaign. Another indication that necessary and sufficient conditions cannot give us the kind of connection we are looking for between alleged argumentative virtues and argumentation quality.

These examples comply with the requirement that, to give VAT a fair shot at delivering on argumentative virtues, we need to consider argumentation and its quality in broad terms, conceiving it primarily as an act, and a complex one at that. Yet I argue that this is not sufficient either: it is not enough to look for virtues that make the act of arguing good; we need virtues that make it *argumentatively good*, i.e. *good as an argument*, not as a speech act in general. This does not necessarily mean endorsing a narrow definition of argument: in fact, we can make it as broad and accommodating as we see fit (within reason), and still require argumentative virtues to be relevant for argumentation specifically, not just for the overall social situation. This is where courage (in Galileo's case) and virality (in the Taiwanese example) fail the test as argumentative virtues: they improve social interaction, without making it any better qua argumentation. In contrast, a virtue should count as argumentative only if it makes an act *argumentatively good*, however we define argumentation. Unfortunately, necessary and sufficient conditions cannot deliver on that, as discussed.

As an alternative, it might be useful to frame the problem in probabilistic terms (without necessarily aspiring to put a precise number on those probabilities): on such a view, an argumentative virtue would be a trait or disposition that, when possessed or manifested, makes it more likely that a good argument will ensue, all other things being equal. For instance, arguing courageously would count as virtuous if it increases the chances of producing good arguments, on some definition of argument and quality – and the same for any other proposed argumentative virtue. The property of being virtuous, i.e. increasing the likelihood of good quality argumentation, could be made dependent upon context as needed, thereby accommodating the well-known context-sensitivity of virtues. Granted, the usual challenges would still remain: most notably, how to define argument and its quality, and to what extent the broad agenda proposed by VAT can be justified without stretching the definition of "argument" too much or without confusing different senses of the term (for in-depth discussion of these issues, see Siegel, 2024 in this volume). In addition, a probabilistic approach would have to specify what count as a sufficient increase in the likelihood of argumentation quality, so that a certain trait or disposition may qualify as argumentatively virtuous: how more likely must a good argument be, for something to count as an argumentative virtue? Yet, once these conceptual issues are solved, and assuming they can be solved satisfactorily, there remains empirical work for VAT scholars to undertake: do proposed argumentative virtues work as expected in improving the chances of good argumentation, or not?

In other words, the probabilistic approach to argumentative virtues is *data hungry*, and this should prompt VAT scholars to occasionally leave the virtue armchair and engage in some empirical verification. Luckily, the means for such a verification abounds in argumentation studies: not only exemplary case studies, like the ones briefly discussed here, but also corpus analysis, laboratory experiments, field studies, surveys. All these methods can be leveraged to find answers to questions like: Does virtues X increases the likelihood of argumentation exhibiting quality Y? To what extent? Under what circumstances? Unfortunately, so far VAT scholars have been reluctant to leave the comfort zone of philosophical speculation, relying almost exclusively on isolated (and often fabricated) examples to support their intuitions of what should constitute an argumentative virtue. I suggest it is time to reconsider and abandon such reluctance, since the demand for empirical verification on argumentative virtues is not some unfair stress test for VAT, but rather the consequence of posing legitimate and even fundamental questions on its theoretical underpinnings: namely, how to define argumentative virtues, *precisely*, and how to ensure that such definition is *verifiable*?

It might be tempting to think that such verification is not needed, insofar as intuitions on argumentative virtues are clear and uncontroversial enough. This is a common mindset for philosophers, yet I caution VAT scholars against embracing it: experimental philosophy (for an overview, see Knobe et al., 2012) has provided good reasons for putting our philosophical intuitions and thought experiments to the empirical test, and we do have some early indication that what might plausibly seem "good" for argumentation could actually turn out to be systematically bad for it, depending on context. As a case in point, Mäs and Flache (2013) showed that allowing people to exchange arguments, as opposed to mere opinions (something that would certainly qualify as "virtuous" in the intuitions of any argumentation scholar), can *increase the level of polarization and entrenchment*, even in the absence of negative influence, but simply assuming some degree of homophily, i.e. a tendency to engage primarily with like-minded individuals. This result was replicated both with various computer simulations and in an empirical group study (N=96). It should not be seen as a nail in the coffin of VAT, but rather as a welcome, evidence-based refinement of its predictions: framed in a virtue theory perspective, it suggests that, when arguers do not display the virtues of open-mindedness and intellectual curiosity (that is, they exhibit instead high levels of homophily), exchanging arguments is not only ineffective, but even detrimental – regardless of whatever other virtues the arguers might embody. This is precisely the kind of interesting, counterintuitive results that taking seriously the empirical verification of argumentative virtues might deliver for VAT: hence, I renew my invitation to virtue theorists to seek collaborations with empirical scientists to verify at least some of the predictions made by their models.

4. How can we use VAT for critical thinking education?

Last but not least, VAT scholars should address more consistently, I argue, a practical challenge: how to make their theoretical framework useful for *educational purposes*. This is something that even critics of VAT are willing to consider valuable, after all, so virtue theorists should make sure they can deliver on that. As a case in point, this is how Patrick Bondy summarizes his (largely critical) stance on VAT:

> "I do not think that the virtue-theoretic approach can provide the theoretical underpinning for a general account of argument. But, having said that, I want to acknowledge that employing virtues of argument in our accounts can nevertheless be useful, especially in teaching critical thinking and introductory argument theory, even though the virtues will not be basic elements in our accounts. (...) More generally, teaching people to be virtuous arguers is good, even though the virtues are not constitutive of what makes arguments good" (2015, p. 464).

Bondy's point on the potential value of VAT for critical thinking instructions is a somewhat familiar refrain, even among scholars that would not describe themselves as 'virtue theorists'. The following is what Sharon Bailin and Mark Battersby, two well-known experts on critical thinking education, had to say on the matter:

> "There is widespread agreement that fostering the virtues of critical thinking is central to a rational community and a democratic society. (...) Argumentation theorists tend to have a real interest in education and have devoted a great deal of attention to the content of courses in critical thinking. Insufficient attention has been paid, however, to the kind of educational outcomes that we hope to achieve through critical thinking instruction and to the pedagogical practices that might best achieve these outcomes. Our contention is that conceiving of our enterprise in terms of initiating students into the practice of inquiry in its various forms and organizing our teaching to achieve this is the most effective way to foster the virtues of inquiry" (2016, pp. 372-373).

In this passage, argumentative virtues are discussed more as a positive consequence guaranteed by the approach to critical thinking education proposed by the authors, rather than as a tool needed to make that approach work: "initiating students into the practice of inquiry", as they put it, is argued to be effective in making sure they will develop the desired virtues.

Here I want to consider whether VAT may play a more central role in critical thinking instruction: not just as a checklist on which various approaches are supposed to deliver, but rather as the fundamental source of inspiration in designing virtue-based educational practices. If we want argumentative virtues to be more than a happy side effect of the practice of inquiry (as Bailin and Battersby seem to suggest), then it is reasonable to ask for *a well-structured list of virtues worth pursuing in our students*, under the assumption that this will make them better at arguing.

This brings us back to Aberdein's list of virtues (see section 2), as the best option currently available on the VAT market. That list, however, runs into some problems, when one attempts to use it for the sake of educational design: some of these concerns were recently discussed by Cassie Finley, who noted that "the essential features of dialectic and rhetoric are in tension with the defining characteristics of virtue argumentation theories", and thus suggested "a more viable route forward for virtue argumentation theorists, one that dissolves this tension through reframing their project as a virtue *dialogue* theory" (2023, p. 153). Here I will focus on a slightly different set of problems for VAT, yet I encourage virtue theorists to pay close attention to Finley's arguments for a shift toward virtue dialogue theory.

The first complication concerns *priorities*, which are conspicuously absent from Aberdein's catalogue: argumentative virtues are listed there without assigning them any priority, yet "where to start?" (and why) is a necessary question for any educational practice. The need for a principled order of priorities in teaching virtues is emphasized when considering virtuous attitudes that cannot realistically be practiced simultaneously. Take for instance Ian Kidd's reflections on the respective roles of intellectual humility and confidence in fostering virtuous argumentation:

> "Argumentation can contribute to the cultivation of the capacities constitutive of the virtue of intellectual humility, but only if it is conceived and practiced as an edifying discipline. Any such conception ought to be sensitive to psychological and social facts about the ways that anxiety, bias, confidence and other phenomena affect our capacity to engage in shared intellectual practice" (2016, p. 401).

In summarizing Kidd's contribution on this topic, Aberdein and Cohen note that "humility, then, is a sine qua non for certain kinds of argumentative success; but confidence is a sine qua non for others" (2016, p. 341). Yet this does not provide educators with any guidance on what to do with these alleged argumentative virtues: should they have their students practicing both? Or only one? Assuming they cannot be exerted simultaneously (or at least not easily, and rarely with positive results), then should we start with intellectual humility and then slowly transition to confidence, or the other way around? Or none of the above?

A second, related problem has to do with the *conflicts* that might arise among different virtues: indeed, as already mentioned, it is quite normal for the same trait or disposition to count as a virtue in a context and as a vice in another. It follows that, depending on contextual features, the list of relevant virtues that educators should foster in students change quite markedly. Katharina Stevens has argued that "there are in fact two sets of virtues an arguer has to master—and with them four sometimes very different roles" (2016, p. 375). Those different (and largely conflicting) sets of norms refer to two familiar ideal modes of argumentation, *adversarial* and *cooperative*, and arguers are required to use practical wisdom to establish what mode is appropriate to the current situation. As for the four roles, these are *knight* (reasonably defending one's position), *attacker* (questioning and criticizing a certain claim or argument), *teacher* (trying to spread true beliefs to others), and *student* (carefully considering someone else's position and reasons, to learn as much as possible from them), all of which are relevant to being a virtuous arguer, insofar as we also learn to correctly identify when it is appropriate to play each of them – again, by virtue of practical wisdom. Stevens' account has the great merit of providing quite a lot of structure to VAT, going beyond a mere list of virtues and trying to propose a principled order among them. How to translate such an account into educational practice, however, remains to be determined: this strikes me as a worthy endeavor for future research, thus I encourage VAT scholars (including Stevens herself) to pursue it.

A third concern about the standard list of argumentative virtues is that it does not seem interested in checking whether some virtues may be *derivative*, in the sense of being a predictable consequence of mastering other virtues. Since that list is very long and time in educational settings is extremely scarce (therefore valuable), this issue should take center stage in any proposal for using VAT to guide critical thinking instruction: we do not want to waste hours training students on a multitude of secondary virtues, if the same results would be better achieved by focusing instead on few primary ones, from which all the others would naturally follow. Here I do not mean "primary" and "secondary" as indicating relative importance, since I agree with Aberdein that argumentative virtues "overlap on multiple dimensions" and therefore cannot "be given a definitive species/genus classification" (2021, p. 1206): it is rather a matter of *psychological priority in their acquisition*. For instance, Aberdein's "big four", i.e. willingness to engage in argumentation, listen to others, modify one's own position, and question the obvious, seem all psychologically rooted in self-confidence, as long as it is complemented by a suitable respect for facts. This empirical suggestion is worth verifying: if correct, it would imply that virtue-based critical thinking education should focus mostly, or even exclusively, on these two pillars, i.e. building students' confidence as arguers and training them to respect facts and evidence, because doing so would result in the acquisition of all the other argumentative virtues on Aberdein's list.

This, however, brings us into contact with a more fundamental problem of using VAT as a blueprint for educational practice: *how to assess virtue*, both as a feature of an arguer (to what extent that person can be said to exhibit virtue X?) and as a characteristic of an act (how do we determine the virtuousness of a specific argumentation, with respect to virtue X?). Here the task is not to assign a specific number or grade to people or arguments, but rather to find suitable ways of verifying whether our educational efforts are producing the intended result, i.e. fostering certain argumentative virtues in behavior and character. This requires criteria that are clear and uncontroversial enough to elicit a sufficient level of intersubjective agreement among educators and students, so that they may all be on the same page in describing an argument or an arguer as virtuous (or vicious) in certain respects.

Once again, strong optimism on the universality of our intuitions may persuade VAT scholars to regard this problem as trivial: argumentative virtues, on this view, will assert themselves when present, and be equally manifest in their absence. True to my self-appointed role of the jiminy cricket of VAT, I must discourage virtue theorists from being too optimistic on that count: even assuming people to be fairly objective in assessing virtues and vices of their fellow arguers (which is unlikely, especially in adversarial situations; but see Mercier & Sperber, 2011, 2017, for discussion), humans are notoriously and demonstrably bad at judging themselves, in particular when the judgment in question is value-laden, as any talk of virtue necessarily is. Therefore, the idea that arguers may be even remotely impartial in assessing their own argumentative virtues strikes me as dangerously naïve: relying on intuition for agreeing on whether specific argumentative characters or acts of argument are virtuous is a pipe dream and, more likely than not, a recipe for disaster. How often have you encountered a closed-minded person that does not see themselves as very open-minded? Or a biased arguer that does not perceive their position as perfectly fair and even-handed?

Using virtues to scaffold the education of arguers requires moving beyond mere intuitions and towards verifiable criteria. Incidentally, this is not an unusual problem in assessing performance: many valuable traits and dispositions, including virtues, are naturally hard to assess, yet this does not make them any less valuable as educational goals. What is needed is a shift towards concreteness: for each proposed argumentative virtue, VAT scholars should specify a suitable *set of proxies*, i.e. actions and features that are observable and reliably connected with possession of that particular virtue in the appropriate contexts. A proxy, by definition, need not be a perfect substitute of the thing it stands for: there might be circumstances in which possession of the relevant virtue is not apparent in any proxy, and others in which a certain proxy is not indicative of the corresponding virtue. It suffices that the relationship is typically reliable, however, to qualify a proxy as a good approximation of the relevant virtue, and thus justify its use as a reasonable assessment tool.

To sum up, the quest to make VAT a true protagonist in the design of educational practices shows a lot of promise, yet it also entails a significant to-do list for VAT scholars. Firstly, they should work on converting mere lists of virtues, however informative they might have been in mapping the lay of the land at the onset of the approach (kudos to Aberdein for that, by the way), into more structured proposals: personally, I think the "two-by-four" model outlined by Stevens (2016) is an excellent first step in that direction, and I would love to see more work done in that spirit. Secondly, their effort to provide more principled accounts of how argumentative virtues stand in relation to one another should also be informed by empirical evidence on cognitive development and personality dynamics: not to derive from them an order of importance for virtues (that is clearly not an empirical matter), but rather to come up with optimal training solutions to foster effective acquisition of the required virtues in reasonable amounts of time. Finally, all virtues should be paired with a set of measurable proxies: observable aspects of argumentative behavior that provide reliable indication of an underlying virtue. These assessment tools are absolutely needed, not to assign meaningless "virtue points" or "critical thinking scores", but rather to track students' progress and, with that, the success (or lack thereof) of the proposed educational intervention.

5. Conclusions

In this paper, VAT was faced with three rather pointed questions: how did it fare in answering them? Overall, I would say it is a half full, half empty kind of scenario. The specificity challenge, originally posed by Goddu (2016), cannot be answered positively: at the end of the day, argumentative virtues do not seem to be specific of argumentation alone. This does not imply that the label "argumentative virtues" is meaningless, yet it forces us to look for a more precise definition of it, one that takes seriously the second question posed in this paper: how to systematically and demonstrably connect argumentative virtues with the quality of argumentation. Here some options do not seem to work (i.e., necessary and sufficient conditions), whereas others show more promise (i.e., probabilistic approaches), but developing them would require significant empirical work from VAT scholars, who unfortunately do not seem particularly keen on taking up this aspect of their theoretical enterprise. The need for an empirical turn is also relevant in facing the third challenge, which concerns how to make VAT a central inspiration for critical thinking education: here the paramount concern should be to provide instructors with actionable guidelines and observable proxies of virtuous argumentation, as opposed to abstract appeals to high-level virtues. Again, doing so would require leaving the ivory tower of philosophical speculation and going into the trenches of educational practice: in spite of the military metaphor, I believe this will provide a lot of insight and satisfaction to VAT scholars, hence I urge them to look upon this course of action with optimism and confidence.

Hopefully, these cursory remarks will be received by virtue theorists in the spirit in which they were intended: as suggestions for shoring up their theoretical approach, so that it might continue to evolve and thrive, and not as attacks meant to undermine its foundations and bring down the VAT edifice. Granted, some of the proposed revisions are quite radical, not only theoretically, but also in terms of job description: my frequent invitations to add empirical approaches to the VAT enterprise, for instance, might sound quite annoying and out of place to a philosophical audience. Nonetheless, if I am right (which remains to be seen, of course), VAT must become more used to make testable predictions and then take seriously the task of verifying them. This, incidentally, will also make it more relevant beyond the somewhat narrow scope of argumentation studies, turning argumentative virtues into a topic of interest for experimental psychologists and educational practitioners too. All considered, I think the future is bright for VAT, once some due maintenance of its core concepts has been successfully performed.

References

Aberdein, A. (2010). Virtue in argument. *Argumentation, 24(2)*, 165-179.
Aberdein, A. (2014). In defence of virtue: The legitimacy of agent-based argument appraisal. *Informal Logic, 34(1)*, 77-93.
Aberdein, A. (2016). The vices of argument. *Topoi, 35(2)*, 413-422.
Aberdein, A. (2021). Courageous arguments and deep disagreements. *Topoi, 40(5)*, 1205-1212.
Aberdein, A., & Cohen, D. H. (2016). Introduction: Virtues and arguments. *Topoi, 35(2)*, 339-343.
Bailin, S., & Battersby, M. (2016). Fostering the virtues of inquiry. *Topoi, 35(2)*, 367-374.
Bondy, P. (2015). Virtues, evidence, and ad hominem arguments. *Informal Logic, 35(4)*, 450-466.
Bowell, T., & Kingsbury, J. (2013). Virtue and argument: taking character into account. *Informal Logic, 33(1)*, 22-32.
Cohen, D. H. (2019). Argumentative virtues as conduits for reason's causal efficacy: Why the practice of giving reasons requires that we practice hearing reasons. *Topoi, 38(4)*, 711-718.
Cohen, D. H., & Miller, G. (2016). What virtue argumentation theory misses: The case of compathetic argumentation. *Topoi, 35(2)*, 451-460.
Dutilh Novaes, C. (2020). The role of trust in argumentation. *Informal Logic, 40(2)*, 205-236.
Finley, C. (2023). From virtue argumentation to virtue dialogue theory: how Aristotle shifts the conversation for virtue theory and education. *Educational Theory, 73(2)*, 153-173.

Godden, D. (2016). On the priority of agent-based argumentative norms. *Topoi, 35(2)*, 345-357.
Goddu, G. (2016). What (the hell) is virtue argumentation? In D. Mohammed & M. Lewinski (Eds.), *Argumentation and Reasoned Action. Proceedings of the 1st European Conference on Argumentation, Lisbon 2015, vol. II* (pp. 439-448). London: College Publications.
Kidd, I. J. (2016). Intellectual humility, confidence, and argumentation. *Topoi, 35(2)*, 395-402.
Knobe, J., Buckwalter, W., Nichols, S., Robbins, P., Sarkissian, H., & Sommers, T. (2012). Experimental philosophy. *Annual Review of Psychology, 63(1)*, 81-99.
Mäs, M., & Flache, A. (2013). Differentiation without distancing. Explaining bi-polarization of opinions without negative influence. *PLoS one, 8(11)*, e74516.
Mercier, H., & Sperber, D. (2011). Why do humans reason? Arguments for an argumentative theory. *Behavioral and Brain Sciences, 34*, 57–74.
Mercier, H., & Sperber, D. (2017). *The enigma of reason*. Cambridge: Harvard University Press.
Oliveira de Sousa, F. (2020). Other-regarding virtues and their place in virtue argumentation theory. *Informal Logic, 40(3)*, 317-357.
Paglieri, F. (2015). Bogency and goodacies: On argument quality in virtue argumentation theory. *Informal Logic, 35(1)*, 65-87.
Paglieri, F. (2022). Pandemic communication without argumentative strategy in the digital age: a cautionary tale and a call to arms. In S. Oswald, M. Lewiński, S. Greco & S. Villata (Eds.), *The pandemic of argumentation* (pp. 145-161). Cham: Springer.
Phillips, K. (2021). Deep disagreement and patience as an argumentative virtue. *Informal Logic, 41(1)*, 107-130.
Siegel, H. (2024). Arguing with arguments: argument quality, argumentative norms, and the strengths of the epistemic theory. In F. Paglieri, M. Marini & A. Ansani (Eds.), *The cognitive dimension of social argumentation. Proceedings of ECA 2022* (this volume). London: College Publications.
Stevens, K. (2016). The virtuous arguer: one person, four roles. *Topoi, 35(2)*, 375-383.
Stevens, K., & Cohen, D. H. (2021). Angelic devil's advocates and the forms of adversariality. *Topoi, 40(5)*, 899-912.
Tischer, J. F. (2022). Panmemic inoculation: how Taiwan is nerfing the pandemic with cute humour. *East Asian Journal of Popular Culture, 8(2)*, 183-204.
Vraga, E. K., Kim, S. C., & Cook, J. (2019). Testing logic-based and humor-based corrections for science, health, and political misinformation on social media. *Journal of Broadcasting & Electronic Media, 63(3)*, 393–414.

COMMENTARY ON FABIO PAGLIERI'S "ARGUMENTATIVE VIRTUES: BACK TO BASICS"

DANIEL COHEN
Colby College
dhcohen@colby.edu

Abstract

Fabio Paglieri's critical assessment of the first two decades of virtue-based approaches to argumentation theories identifies outstanding questions still in need of answers, central themes in need of further development, and paths for future applications. However, behind the guise of a modest and balanced critique is a radical re-visioning of large parts of the traditional agenda for argumentation theory.

1. Introduction

One thing that Fabio Paglieri clearly demonstrates with his characteristically scholarly, philosophically penetrating, and entertainingly insightful perspective on virtue theories is how much they have enriched the discourse of argumentation theory – at least insofar as the criticisms that have been raised against virtue theories, the challenges that have been posed for them, and the controversies that they have generated all count as "contributions". After all, without those contributions, we would have been deprived of Paglieri's own additions on all of these fronts.

Paglieri enters the discussions in the guise of a modest but well-meaning kibitzer but do not be fooled: he's a bomb thrower. His project is far from modest and he has contributed far too much to discussions about virtue theories to pass as an outsider. Because of his past contributions, however, he has at least earned the benefit of a doubt as to whether his radical proposals will turn out to be positive additions to virtue theories or occasions for retreat on the part of virtue theories.

And make no mistake, his proposal is radical: it contains elements that are far-reaching, it is ambitious, and it has the potential to be transformative. It goes right to the heart of the matter. Or maybe I should say it goes for the jugular. As it happens, I think that Andrew Aberdein and I both have pretty thick skins when it comes to criticisms, so we welcome his best shot. We will take your criticisms as criticisms of our position, not of us – as no doubt they were intended – so whether the criticisms hit their mark or not, our positions will improve one way or the other and we will all be better off.

By the way, the reference to the intentions behind Paglieri's criticisms and our ability to receive them appropriately is not adventitious. I have argued that having the right sort of thick skin is an argumentative virtue (Cohen 2017). Depending on how the qualifier "the right sort" gets cashed out – e.g., as being able to remain personally unmoved by criticisms of one's position and being argumentatively unmoved by criticisms of one's person – it might even be specific enough to satisfy Paglieri's challenge to provide an example of an argumentative virtue that is distinct from more general epistemic virtues and which makes a positive, causally effective contribution to argumentative engagements.

2. Paglieri's proposals and challenges

Let me elaborate on my characterization of Paglieri's comments as far-reaching, ambitious, and radical.

First, it is far-reaching because what he proposes has conceptual, doctrinal, and methodological implication for virtue argumentation theories. He is not suggesting minor tweaks or simple changes to approaching arguments with a focus on the characters of their participants. He starts with a modest and almost innocuous query about some details: "Is this the right list of argumentative virtues?" He then steps up a level with, "What exactly is an argumentative virtue?" By the end, he is taking on even bigger game: "What is an argument?" Every part of the project is up for grabs.

Second, it is radical and potentially transformative because there are a lot of moving pieces here and they move together. It might not be possible to make these changes is a serial, piecemeal, or partial fashion, so they would have to be adopted wholesale. Again, the seemingly modest kinds of conceptual change that are called for with respect to argumentative virtues – "redefinitions" – entails accompanying changes in what counts as arguing, what counts as an argument, and who counts as an arguer.

Above all, Paglieri's call to action is ambitious. The conceptual revisions that he identifies necessitate an equally revolutionary upheaval in methodology. He envisions long-term research projects that extend well beyond the borders of argumentation theory into other aspects of our intellectual lives including, notably, epistemology, pedagogy, and communicative practice.

I offer these comments as a commendation, not a criticism, because I think that at bottom, virtue-centric approaches to argumentation are themselves far-reaching, ambitious, and radical. Not every gets that. Paglieri does, and while that manifests itself in his proposals, by itself that still leaves it open to ask whether they are good, or viable, or likely deliver on their goals. For that matter, are the goals the right ones for argumentation theory? Here is where we need to pause. Once we redefine the concept of an argumentative virtue, we will need to make adjustments in how we understand the connection between good arguments and arguing well. Similar adjustments will be needed in the concept of an argument and, perforce, in argumentation theory. What matters is what we do with the new conceptual coming our way.

3. A new agenda

Let me lay my cards on the table: if all that we do when we create a brand-new set of conceptual tools is use them to build new answers to the old questions, then we have not grasped nor taken full advantage of what we have.

Here is the kind of situation that I have in mind. Virtue argumentation theory is arguer-centric, with more of a focus on arguing well than on good arguments as they have traditionally been understood, e.g., as measured by Johnson and Blair's (1977/2006) RSA test or standard definitions of cogency. Does VAT have the resources to explain relevancy, sufficiency, and acceptability or validity and soundness? I think so. Aberdein (2018) sketches out how to answer this question, so I am persuaded it can be done. However, I am not persuaded that it's the right question for VAT. Why recreate the right measure for the wrong kind of argument?

The same situation arises with respect Godden's (2016) challenges regarding the foundations of argumentation theory or Goddu's (2016) questions about how virtue theories apply in various areas. Are the virtues foundational or derivative, necessary or eliminable? Argumentation involves a large complex of concepts. Recognizing that that these are evolving concepts and that they have evolved together, takes the urgency out of questions about conceptual priority.

Paglieri's case studies exhibit this struggle between the old and new ways of thinking and the difficulties in effecting what amounts to a global *gestalt* switch. We can see evidence of some back-sliding in his discussion of the Galileo case. For example, at one point, Paglieri comments that Galileo's courage is irrelevant to what Paglieri refers to as "the merits of his argument." In context, it is clear that the phrase "his argument" whose merits are left unchanged refers to something like the illative core. This is the logician's concept of an argument as a premise-conclusion complex. It is independent of any arguers, context, or effects. He is right about that but there is more to be said. Galileo's courage certainly could have contributed to how well his argument <u>with the inquisitors</u> went. In the story that is told, the inquisitors, unlike the earth, were not moved during the trial, so his courage turned out to have been causally inefficacious. It could have been otherwise. His courage might have impressed some inquisitors and helped them engage with his reasons, in which case they – Galileo and the inquisitors – would have had a better <u>argument</u> – in the relevant sense of a give-and-take of reasons. Historically, that did not happen, so if we could go back in time and ask Galileo as he walked away from the engagement how his argument with the inquisitors went, he would have to say, "*Non bene*." Not great. Whatever the "merits of his argument," they did not transfer to their argument.

In fairness, Paglieri does acknowledge that argumentative virtues "increase the chances that a good argument will ensue," so arguments as more than just premise-inference-conclusion structures and arguments as engagements-among-arguers are on his radar. He does get it, so I suppose I am not being very virtuous in this argument by nit-picking, but I am doing so in the service of trying to show just how much of a challenge it will be to fully make the transitions he proposes.

The jury is still out on the changes Paglieri proposes, the questions he raises, and the research agenda he outlines. They are ambitious and challenging, so we might see them as daunting, but they are also exciting and promising, so we have reason to be bullish.

References

Abderdein, A. (2018). Inference and virtue. In S. Oswald & D, Maillat (Eds.). *Argumentation and Inference: Proceedings of the 2nd European Conference on Argumentation, Fribourg, 2017,* vol. 2, 1-9. London: College Publications.

Cohen, D. (2017). Virtue argumentation, value-argumentation, and the virtue of a thick skin. Conference on: Values in argumentation – Values of Argumentation, ArgLab, Nova Institute of Philosophy, Lisbon, 28-29 June, 2017.

Godden. D. (2016). On the priority of agent-based argumentative norms. *Topoi,* 35(2), 345-357.

Goddu, G. (2016). What (the hell) is virtue argumentation? In D. Mohammed & M. Lewinski (eds.), *Argumentation and Reasoned Action: Proceedings of the 1st European Conference on Argumentation, Lisbon 2015,* Vol. II, 439-488, London: College Publications.

Johnson, R. & Blair, J. A. (1977/2006). *Logical Self-Defense.* Toronto: McGraw-Hill.

ARGUMENTATIVE PATTERNS INITIATED BY CLOSED-LIST QUESTIONS IN ACCOUNTABILITY DIALOGUES. A CORPUS STUDY OF FINANCIAL CONFERENCE CALLS

ANDREA ROCCI
Università della Svizzera italiana (USI)
andrea.rocci@usi.ch

OLENA YASKORSKA-SHAH
Università della Svizzera italiana (USI)

GIULIA D'AGOSTINO
Università della Svizzera italiana (USI)

COSTANZA LUCCHINI
Università della Svizzera italiana (USI)

Abstract
To illustrate a research strategy aiming at the discovery of recurrent contextually significant patterns in dialogic argumentative activity types, we investigate *closed-list questions* as a minimal *argumentative pattern* and as component of broader dialogical argumentative patterns, in the Q&A phase of Earnings Conference Calls, a key financial communication activity type. The argumentative affordances of *closed-list questions* are discussed as well as their formalisation in the metalanguage of Inference Anchoring Theory. A small corpus study provides verification of the hypothesized occurrence of activity-relevant uses of closed list questions for *issue framing* and for *information elicitation* as well as of the question's potential in eliciting argumentative responses.

1. Introduction

In this paper, we investigate closed-list questions as a template for initial moves of dialogical *argumentative patterns*, which are functional to participant goals in a routinized financial communication activity type. In doing so, we demonstrate an approach to discovering and describing argumentative patterns characterizing argumentative activity types, which we are developing within the project *Mining argumentative patterns in context. A large scale corpus study of Earnings Conference Calls of listed companies*[1].

Earnings Conference Calls (henceforth ECCs) are public dialogical exchanges in the financial communications of listed companies. ECCs consist of teleconferences held by the top corporate executives (CEO, CFO, COO) for financial analysts, in connection with the announcement of quarterly results. During ECCs, managers present and interpret financial results, discuss major ongoing and planned business developments and discuss their own forecasts about the future earnings of the company. Much like the related activity type of the press conference, the genre template of the ECC includes a managerial presentation, followed by a Q&A where managers answer analysts' questions, as exemplified by the exchange in Example 1.

Example 1.
David Beckel (analyst): First one, just on Arena or MAGIC in general, I guess. Really impressive growth, obviously, from Arena in the quarter. I'm curious, do you have the data sets of... capable of giving you a holistic picture of your player base? I'm curious more specifically if that growth is coming at the expense of tabletop or if you're actually expanding the market base, and whether or not you expect mobile to further expand the market base.
Brian Goldner (Hasbro, CEO): Yes. Sure. So in fact, you're right. The Magic Arena had historically been expanding. It's accelerating in that effort. In fact, analog tabletop has performed incredibly well, and let me remind you that the analog and the tabletop business is performing incredibly well while people can't get together locally in their local favorite hobby shops or local gaming shops to play the game. [...] But we do expect an additional tailwind on the analog business when we're starting to see that as markets begin to reopen and people can begin to get back together

---

1 Mining argumentative patterns in context. A large scale corpus study of Earnings Conference Calls is a research project supported by the Swiss National Science Foundation (SNSF), Grant Number: 200857, main applicant Andrea Rocci (IALS, Institute of Argumentation, Linguistics and Semiotics, USI Lugano), co-applicant Chris Reed (ARG-tech, Centre for Argument Technology, University of Dundee).

again. [...] People get invited to come along and learn how to play MAGIC all the time. So yes, it's expansive. No, there is no cannibalization. And then in fact, Magic Arena is just allowing people to play at a distance who had never been able to be able reconnect with friends or family before and not necessarily be in their neighborhood. So all a net positive. (HAS Q1, 2021)

While these conferences add little "material information" to what the company has disclosed in the written earnings announcements published immediately before, research in finance (Matsumoto et al. 2011) has shown that they have a demonstrable impact on the market reaction to the quarterly results. The informative value of ECCs has been variously said to reside in the *sentiment* they express (Matsumoto et al. 2011), in the soft information or "color" they provide (García et al. 2023), and in the interaction itself (Jordan et al. 2018). In fact, Palmieri et al. (2015) have shown that this activity type is essentially argumentative and have hypothesized that in ECCs argumentation plays an important role in reducing analysts' uncertainty about managerial explanations, evaluations and forecasts, in particular "by pointing out that an already disclosed piece of information is relevant or not as a premise to support a standpoint on a price-related issue" (Palmieri et al. 2015: 123).

2. Argumentative patterns

While the argumentative dynamics of the ECC have been already described in small corpus studies (Palmieri et al. 2015, Rocci & Raimondo 2018b), testing the hypothesis that argumentation plays an important role in making ECC Q&As informative for the financial markets requires a large scale investigation of the correlation of features of argumentation in ECC with market data, or other measurable follow-ups such as analysts recommendation.

This involves a three-fold interdisciplinary challenge: the challenge of applying the proper *econometric methods* to investigate markets' reaction to argumentation in the ECC, the challenge of developing suitable computational *argumentation mining* techniques and applying them on a large scale and the challenge of choosing the *relevant features* pertaining to argumentation to be examined when investigating the effects of argumentation in ECC on the financial markets. This paper addresses the third challenge by turning to a context-specific notion of *argumentative pattern* (AP), and by making APs the main target of mining and key input of analytics.

2.1. Argumentative patterns as a pivotal notion for a large-scale investigation of argumentation in context

We use the term argumentative pattern, which we borrow from Pragma-Dialectical Theory (cf. van Eemeren 2017 and 2018), to mean a significant constellation of argumentative moves whose occurrence can be explained in view of the goals and constraints of the activity type.

The key assumption behind a pattern-based approach is that not all argumentation is equally worth "mining" and that we have a better chance of understanding the effects of argumentation on its context on a large scale if we choose to observe the distribution, not just of *any* feature of argumentation, but of features that we have principled theoretical reasons to relate to fundamental dynamics of the activity type.

In this paper we provide an illustration of the early stages of this approach by showing how we investigate a feature of question design (i.e., the closed list question) in a corpus of manually annotated ECC, in order to ascertain its contribution to the definition of significant patterns in the ECC activity type.

An approach based on APs involves a research cycle that starts from qualitative analyses of samples of argumentative discourse, where the logical, dialectical and rhetorical features of argumentative moves are put in relation with the institutionalised common *goals*, the normative *constraints* and the individual *incentives* that characterise a given activity type. This step will be immediately illustrated in relation to the ECC and the closed-list question design in Sections 2.2 and 3.

The following step involves working-out sufficiently precise definitions of candidate APs that need to be "translated" into a representation format usable for the structural annotation of a span of text or dialogue. In order to capture local patterns in the ECC Q&A, we adopted a representation format that is capable of "anchoring" the argumentative properties of the pattern to the dialogue acts performed by the participants. Inference Anchoring Theory (IAT, Budzynska and Reed 2011) provides the basis for such a representation. This step is illustrated in Section 4.2, in relation to closed-list question design and, generally, the modelling of analyst questions in ECCs.

Manual annotation of candidate patterns in a growing corpus of ECCs provides a first verification of the representation format and opens to the possibility of investigating the distribution of different candidate patterns first within meaningful case studies, then in larger corpora supporting cross-sectional and longitudinal studies. Section 5.2 presents a two steps process of annotation for a corpus of ECCs, which involves first the "coarse" annotation of discourse moves and question designs in the ECC Q&A, and then the full IAST reconstruction of extracted candidate patterns.

The results of two-steps process of annotation are exemplified for the case of the *closed-list question* in Sections 6 and 7. In particular, Section 7 reports a small scale study which involves the full argumentative analysis of 14 Q&A turn pairs initiated by closed list questions, to verify in the corpus the argumentative relevance of this question design and its hypothesized significance for the ECC activity types.

2.2. Argumentative patterns in the Earnings Conference Call activity type

Let us now go back to the question-answer pair in Example 1 and use it to illustrate how initial hypotheses about APs are developed by relating observed dialogical argumentative moves in the Q&A of the ECC with hypotheses on the overarching goals.

The exchange concerns the explanation of a *prima facie* positive business result: the growth of the number of players of the online card game MAGIC Arena. Analyst David Beckel proposes two alternative explanations for this result: either it comes at the expense of the tabletop version of the game, with MAGIC players migrating online during the global pandemic, or it is the result of an expanded market base. In doing so, the analyst evokes an argumentative confrontation on an issue of *explanation*, with two mutually exclusive standpoints on the issue, giving rise to what Pragma-Dialectical theory calls a mixed difference of opinion. In the same question turn, the analyst does three other things. First of all, he introduces the *explanandum*, the impressive growth of MAGIC Arena, with an assertive move we call *preface,* which serves to establish the background for the intelligibility of the question, but also to argue for the relevance of the issue. Furthermore, the analyst introduces two other questions in his question turn, asking for *data* that could allow to decide on the issue, and asking for a further predictive standpoint on additional growth, which is presented conditional to the expanded market base explanation being the case.

The argumentative significance of what the analyst does in the excerpt is clear if we understand the whole ECC as an argumentative discussion. In the ECC the ultimate issue addressed is a practical decision whether to buy, hold or sell the stock, which is based on an evaluation of the stock. In turn, the evaluation of the stock is based, by definition, on a forecast of future business performance of the company (cf. Palmieri 2018: 50). Future performance is predicted as resulting causally from the demand in the market and the company's ability to meet this demand profitably. Figure 1 displays this overarching inferential structure, which we call *prototypical argumentative pattern of firm valuation.*

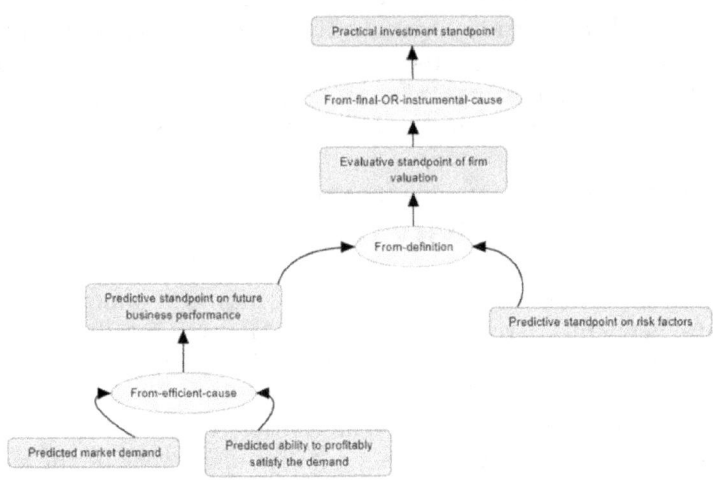

Figure 1. Prototypical argumentative pattern of firm valuation. The inferential structure representation follows the conventions of Inference Anchoring Theory (IAT, Budzynska & Reed 2011). The labels of the argument schemes in the inference nodes are those of the Argumentum Model of Topics (AMT, Rigotti & Greco 2019).

If we take the hierarchy of issues of the prototypical valuation pattern as a backdrop for the argumentative discussions in the ECC, it is clear that the explanatory issues of Q&A pair in Example 1 is relevant to establish the premises for predictions about future market demand which, in turn, contributes to establish predictions about future market performance. Only if the growth of online players turns out to be a genuine growth of the player base, it becomes relevant as a possible argument to predict further growth. This direction is clearly indicated by the sequencing of questions in the turn, where one of the possible answers to the *request of explanation* makes the following *request of predictive opinion* relevant.

Starting with Palmieri et al. (2015), it was observed that certain kinds of question design, such as the aforementioned *request of explanation* and *request of predictive opinion*, appear to be highly expectable in view of the hierarchy of issues implied by the argumentative pattern in Figure 1, and, to some extent conventionalised in the genre, with the use of specialised indirect question frames such as *could you... explain/clarify/confirm/provide more color on...* (cf. also Crawford Camiciottoli 2009).

Not only the recurring question designs in the ECC can often be explained as functional to the discussion of the *overarching issue* and to the testing of the prototypical argumentation pattern of the ECC, they can

ARGUMENTATIVE PATTERNS

also be transparently related to other *constraints* of the activity type and on *incentives* on participants.

At the level of dialectical commitments, financial analysts, bound to a questioner role during in the call, are committed to the antagonist's role of critically testing managerial standpoint, without committing to a standpoint of their own. At the level of their incentives for participation (cf. Yaskorska et al. 2022), analysts have a fundamental professional incentive to be right in their valuations, that's why they seek at the same time to enrich the information base for their valuation of the company (*information elicitation incentive*) and put managerial standpoints to the test by engaging in a sort of accountability dialogue with them on behalf of the investors (*critical incentive*). At the same time, they need to maintain an amicable relationship with managers (*relational incentive*).

Question design is the fundamental tool that analysts have at their disposal to pursue these incentives. This leads to the hypothesis that question designs are themselves to be seen as local APs, in that they are argumentatively *significant, conventionalised*, and *explainable* in terms of the goals and constraints of the activity type. Furthermore, they are the most immediate building blocks of broader APs in the ECC Q&A, spanning question-answer pairs. A detailed discussion of a typical question design of the ECC as a local argumentative pattern, namely the *request of confirmation of inference* is provided in Rocci & Raimondo (2018b).

3. The relevance of closed-list questions for an account of argumentative patterns in the ECC

If we examine closely the request of explanation from the Q&A pair (1) reproduced in Example 2 we can observe another interesting feature of its question design.

Example 2.
<u>David Beckel (analyst)</u>: *[...] I'm curious more specifically <if that growth is coming at the expense of tabletop> or <if you're actually expanding the market base>, [...].*

The question corresponds to what is generally called an *alternative question*, setting up a series of possible answers framed as mutually exclusive. *Alternative questions* represent one of the main question types, alongside with *open* (or *wh-* or *constituent*) and *polar* (or *yes/no*) *questions* (Biber et al. 1999, Quirk et al. 1985, Enfield et al. 2010, Stivers and Enfield 2010, Drake 2020). Here we will use the less common label *closed-list question* (Palmieri et al. 2015: 128) to refer specifically to alternative questions used by financial analysts in the ECC.

Alternative questions have been studied extensively within conversation analysis, interactional linguistics and semantics, in particular in their differences and similarities with polar (yes/no) questions. In contrast, in argumentation theory there are few studies (e.g., Hautli-Janisz et al. 2022) on question types, and none that investigates alternative questions in detail.

The question in example 2 raises an *issue* of explanation. As observed in Section 2.2, the issue is relevant as a sub-issue of the main issue of the ECC via the prototypical argumentative pattern of firm valuation (Figure 1).

As pointed out by Biezma and Rawlins (2012, 2015), alternative questions present a *disjunction*, that is to say, they propose a restricted set of alternative answers (Stivers and Enfield 2010), this set being presented as exhaustive (Biezma and Rawlins 2012). As demonstration of the *presupposition of exhaustivity*, Drake (2020) shows that answers that don't pick one of the proposed options are dispreferred turns. Answers can provide a different non-proposed alternative, but those are considered nonconforming answers resisting to the question's constraints.

In Example 2, analyst David Beckel frames the transfer of players at the expense of tabletop and the expansion of the market base as *mutually exclusive* and *exhaustive* alternatives. The question conveys as a presupposition the expectation that *they are not both true* and that *there are no other relevant alternatives*. It is not necessarily a presupposition of *strong* logical or ontological incompatibility and exhaustivity, it could simply mean that the listed options are exclusive and exhaustive *in practice*, for all relevant intents and purposes. It is interesting to observe that, in some cases, falsely presupposing an alternative framing can amount to a fallacy of *false dilemma* (cf. Lewiński and Aakhus 2023: 179 ff.).

These semantic features create affordances for argumentative moves that analysts exploit when they use closed-list questions in the ECC. They frame the issue in (2) in terms of what Pragma-Dialectics calls a mixed confrontation (see van Eemeren 2018: 36) where two supposedly alternative standpoints are advanced. More precisely, as a subtype of mixed confrontation where the alternative standpoints are treated as exhaustive, so that establishing one entails the refutation of the other (but see also Lewiński and Aakhus 2023: 184 ff.). The respondent reacting to such a framing is faced with a double choice: they can either reject the framing, assuming the burden of refuting the alternativity and/or the exhaustivity of the questioned propositions or accept the framing assuming the burden to defend one or the other proposition; see Figure 2. We will return on the pivotal aspect of *reaction to the pattern initiator* in Sections 4.2 as and 7.

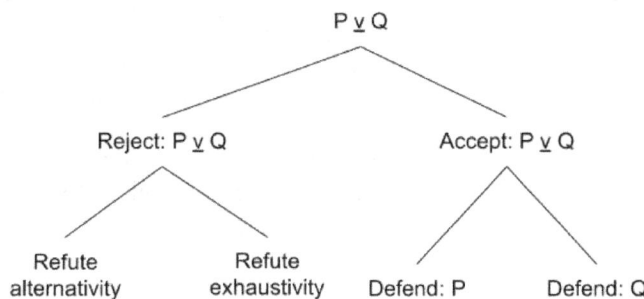

Figure 2. The dialectical choices in responding to alternative issue framing

In fact, closed-list questions allow analysts to provoke a confrontation by evoking it and to goad managers into a protagonist role, while maintaining neutrality and a purely antagonistic role. In view of the above examined features, it is natural to hypothesise that closed-list questions play a major role as part of an arsenal of pragmatic and rhetorical means that give ECC arguments a characteristic shape, and to assign them the status of APs.

Even if this is consistent with their antagonist role, analysts are generally loath to *directly* ask for a justification, using a challenging *why*-questions (Hautli-Janisz et al. 2021), as repeatedly confirmed by ECC corpus data (see Palmieri et al. 2015: 128,). This can be interpreted as avoidance of face threatening acts, consistent with their relational incentive. Closed-list questions can be an instrument indirectly trigger managers' argumentation. Consider the *request of predictive opinion* in Example 3, below:

Example 3.
Tami Zakaria (analyst): Do you expect that trend to continue for the rest of the year or should it be lower given you have announced price increases to your clients – consumers? (HAS Q1 2021)

By presenting two different developments as possible and, moreover, as *equally* possible, this type of question implicitly demands that the choice for one rather than the other is supported by argument. Likewise, if the *framing of the issue* is rejected and a "third" alternative is provided the answerer is prompted to argue for it. In Example 3 the framing of the issue is further reinforced by an argument ("given you have announced price increases to your clients – consumers?"), which operates at a meta-level, supporting the attribution of the second standpoint to the answerer. The analyst does not assume commitments about what will happen, yet she draws inferences about what managers might be thinking.

It has been observed that sometimes the second alternative is nothing more than the negation of the first (e.g., questions with the form ...*or not?*) so that the alternative question becomes logically equivalent to a yes/ no question. Interestingly, even in this case the alternative question is not pragmatically and rhetorically equivalent to the polar yes/no question, because alternatives are used "for seeking information when the addressee appears to be withholding it and the speaker wishes to close the issue". (Biezma 2009: 39). Consider Example 4 below:

Example 4.
<u>*Jaime Katz (analyst):*</u> *And then I think, originally, the 2023 outlook was for above 15.7% for operating margin, and that's been lifted a little bit. Is that primarily due to just the mix of the portfolio and where the returns are coming from? Or is there something else we should be thinking about? (HAS Q4 2021)*

The alternative questioning pattern *Just P... or P and something else* provides a powerful tool of information elicitation for analysts. While a manager can provide an incomplete explanation P and remain truthful in their disclosure, they cannot maintain that P is the *only* explanans without *being on record* asserting something false, which is likely to have serious consequences in the social and legal context of the ECC.

We hypothesize that *issue framing* and *information elicitation*, being two distinct activity relevant functions of the *closed-list* question can be seen as two recurrent and conventionalised sub-patterns of the AP.

4. Modelling local argumentative patterns of ECC question-answer turns

We have introduced above the notion of APs, as significant constellations of argumentative moves whose occurrence can be explained in view of the goals and rules of the activity type. While we take inspiration from Pragma-Dialectical Theory for this notion, we implement it in a different theoretical and methodological framework, where the formalism of Inference Anchoring Theory plays an important role. Our notion of AP is also separate from the one introduced by Musi and Aakhus (2018), even though we share with the authors the interest in a unit of analysis that can feed what they call a *macroscope* for argument mining.

4.1. General framework

The minimal requirements for recognizing an AP are (a) the presence of a constellation of discourse moves (i.e., speech acts, whose purpose in the activity type is recognizable by participants), (b) that are argumentatively

ARGUMENTATIVE PATTERNS

relevant (i.e., relevant for the resolution of an argumentative discussion), (c) and can be shown to fit (i.e., respond to) certain constraints of the activity type. Being specific to the activity type, APs offer interesting affordances both as the main target of the mining and as the basic units whose distribution is correlated with extra-discursive market data.

In the ECC analysis we focus on two main types of APs: (a) the *global APs of the ECCs* and (b) the *local dialogical APs of the Q&A*. The former takes the whole ECC as their relevant span, and result from the sum total of argumentation put forth by managers in anticipation of or as a reaction to analysts' questioning. For instance, Figure 1 presented a rough representation of one important global AP, which we called *prototypical argumentative pattern of firm valuation*, in terms of a IAT inferential structure, underspecified at the level of its pragmatic anchoring to discourse moves. Here we will not discuss them in detail, focusing instead on the *local dialogical APs of the Q&A*.

4.2. Using IAT to represent Q&A pairs in the ECC

Here we present elements of IAT analysis of our data using which we can further analyse our running Example 1. Being the most representative, such example is understandably also the most complex for analysis; therefore, for a comprehensive account, we will start the analysis drawing examples from the simpler cases of open and yes/no questions, which will provide a basic language for understanding complex cases of closed-list questions. In Example 5 an analyst (Jaime Katz) is asking for the explanation of the action, that was conducted by the company and a manager (Brian Goldner) provides the explanation. The joint exchange in which they engage can be visualised as an argument map, which we can see on the left-hand side of the diagram, where *explanandum* and *explanans* are connected through the relation of inference. On the right-hand side of the map, we also see the unfolding of a dialogue, which they are conducting in terms of a sequence of locutions connected via transition rules. The two phenomena (argumentation and conversation) have their common point in the communicative intentions with which participants conduct them. In the first locution, an analyst introduces the *explanandum* with the intention of requesting a justification from the manager. The manager introduces his explanation via asserting data and an opinion.

Example 5.
Do Jaime Katz (analyst): [...] *how have the retailers worked with you to accept that product?[...]*
Brian Goldner (Hasbro, CEO): [...] *We have been incredibly resourceful in finding several new ports and ways of bringing in product, working with our retailers. And the great news about our business and the categories where we're competing is they're in very high demand. (HAS Q2 2021)*

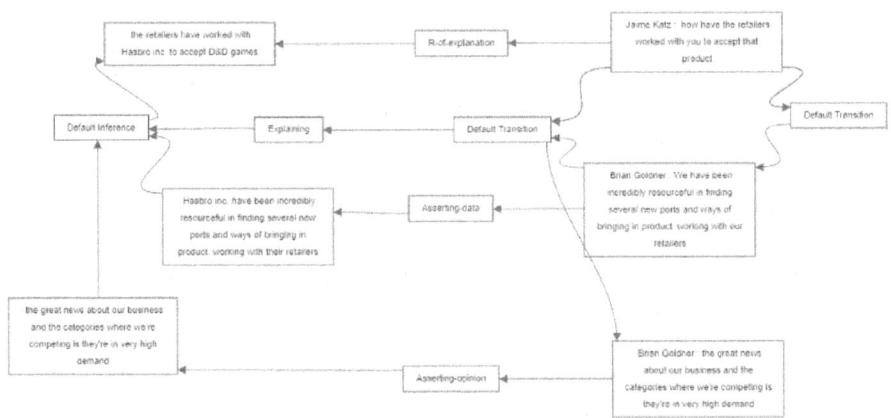

Figure 3. IAT representation of Example 5: open type of request of explanation

Looking closer to the request we can describe the way in which we model requests in our annotation. We take two dimensions of question parametrisation from the annotation scheme we will be discussing about in more detail in Section 5.2 in order to shape the question in IAT: *illocutionary force*, which in our case takes the value "r-of-explanation", and *logical structure*, which is in our case is "open". While the illocutionary force is modelled via tags on the illocutionary connections in the argument map (in the middle of the diagram), logical structure is captured via shaping the propositional content of the question in the left-hand side of the map. To show the structural difference between the types of questions, we could reshape the analyst's open r-of-explanation as a yes/no r-of-opinion, as in Example 6. Such a question would be represented as shown in Figure 4.

ARGUMENTATIVE PATTERNS

Example 6.
Do you think the retailers have been cooperative with you to accept that product?

Figure 4. IAT representation of Example 6: yes/no request of opinion

In contrast to open and yes/no questions, the logical structure of a closed-list question is highly complex for IAT representation. The discourse move is usually composed of at least two locutions that from an analytical point of view cannot be merged, since they are argumentative units presenting alternatives to which a manager is expected to refer to in his answer. Let's start from a request of opinion, during which an analyst asks for an opinion or a standpoint, as Michael McGovern does during his question in Example 7.

Example 7.
<u>Michael McGovern (analyst):</u> *And I guess one quick follow-up on order frequency. I was just wondering and on the 14% of customers that are now trying non-restaurant ordering for the first time or, excuse me, just using it. Do you expect that restaurants and non-restaurant can exhibit similar order frequency trends long-term or do you think that eating and ordering from restaurants is fundamentally a higher order frequency you kind of market? (DASH Q4 2021)*

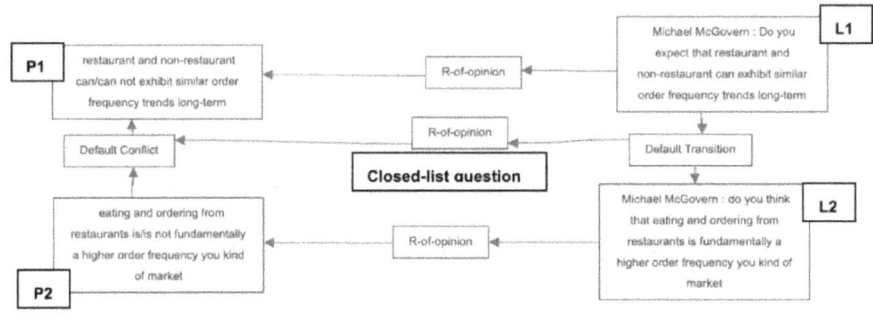

Figure 5. IAT representation of Example 7: closed-list question

The linguistic markers of both questions, i.e., "*Do you expect...*" and "*do you think that*" show that the analyst is asking about the manager's opinion, so in the corresponding IAT map we anchor such intention in the locutions L1 and L2, as shown in Figure 5. Also, we model those requests (similarly to the case of yes/no questions) shaping the propositional content of the locutions P1 and P2 accordingly. At the same time those propositional contents constitute a mutually exclusive, alternative view of the issue, so we can say that there is a relation of conflict between them. Yet, the relation of conflict is not introduced with the implicit intention of disagreement, but with the intention of pushing the manager towards providing which alternative is correct. The closed-list question is therefore annotated anchoring in the transition between the locutions L1; L2 in which the alternatives are suggested.

The illocutionary intentions characteristic for ECCs can be captured in two groups from the point of view of dialogical interaction (cf. Yaskorska-Shah et al. 2022). The first group comprises cases in which analysts ask for opinion, data, commitment or elaboration: they want managers to provide a standpoint or a premise, which are just a part of the argumentative structure. On the other hand, requests for justification, explanation or clarification are requests for the entire argumentative structure; this means that posing those types of questions an analyst wants a manager to state the standpoint, premises and the fact that there is a relation between them. Depending on those differences we model the propositional contents of questions differently; thus, the conceptual distinction between the two groups becomes visible in the analysis of examples such the one provided in Example 7, representing a request of opinion, and our running Example 1, representing a request of explanation, accordingly.

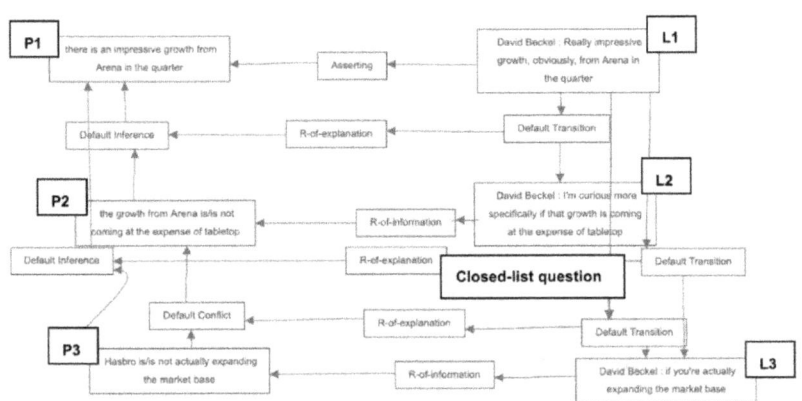

Figure 6. IAT representation of Example 1: closed-list type of request of explanation

In Figure 6 we see a sequence of locutions in which the analyst is first introducing a fact in L1, this move signalled via an "Asserting" illocutionary connection, and then provides in L2 and L3 two possible explanations for such data, which anchor an illocutionary connection of r-of-information with a logical structure of the yes/no type represented by P2 and P3. Explanation of a fact consists not only of *explanans*, but also of

ARGUMENTATIVE PATTERNS

the relation of inference between *explanans* and *explanandum*. The relations are anchored in the transitions between the locutions corresponding to the fact and its possible explanation, i.e., in the transitions L1;L2 and L1;L3 via illocutionary connection of r-of-explanation. This corresponds to the yes/no type of r-of-explanation questions. The principle of the closed-list question connection is exactly the same as in the request of opinion, i.e., two alternative explanations are proposed, with a relation of conflict between them.

Closed list questions are employed to frame an issue on the basis of conflicting alternatives, represented in an IAT map as two or more alternatives and at least one relation of conflict. Speaking in terms of dialogue dynamics, we can say that performing closed list questions, analysts are shaping an issue, as shown in the left-hand sides of Figure 5 and 6 and want managers to refer to such a framework in their answer. Therefore, as already stated in Section 3 aided by the scheme of Figure 2, we can expect a manager to react in either of the following ways: (1) [accept: $P \veebar Q$] accepting the framework, and by accepting one of the option he rejects all other alternatives, (2) [reject: $P \veebar Q$; refute exhaustivity] rejecting the whole framing, providing some new alternative answer, which was not suggested by an analyst or (3) [reject: $P \veebar Q$; refute alternativity] rejecting that there is a relation of conflict between the alternatives provided by an analyst and confirm some to all possible answers as not mutually exclusive. The dynamics of reaction to closed list questions are shown in practice in Section 7.

5. Corpus design and annotation

The present work is based on a small, annotated corpus of ECC transcripts, comprising the Q&A sections of 12 ECCs of three US listed companies during the four quarters of fiscal year 2021.

5.1. Corpus Collection

At an early stage of research on argumentative patterns in ECC Q&A, the purpose of corpus analysis not to test hypotheses on the significance of already established argumentative patterns under specific contextual conditions of the activity type. That's why initial corpus selection simply aimed at obtaining enough variety of moves, question designs and arguments and enough contextual information to be able to reconstruct arguments in-depth. To this aim, we chose three medium-large companies from different industrial sectors that were prominently featured in the business news at the time of corpus collection and which caught our attention for specific issues they were facing.

Table I. Issue based corpus selection

Company	Industry	End of year Market Cap. (2021)	Newsworthy issues between Q2 and Q3.
Door Dash (DASH)	Online food ordering	$51.08 B	End of Covid-19 restrictions. New regulation on delivery workers.
Hasbro (HAS)	Play and entertainment	$14.04 B	Death of CEO. Targeted by activist hedge fund.
Zillow (Z)	Internet technology / Real estate	$16.15 B	Losses due to failure of house trading AI system, subsequent layoff of 25% of workforce.

5.2. Corpus Annotation

For the purposes of the present paper, the annotation of the corpus can be described as a two-step pipeline.

Step 1 of the annotation, carried out using the INCEpTION platform (Klie et al. 2018), refines the analysis of dialogue moves in Q&A pairs and analyst question design in Palmieri et al. (2015). It features three annotation layers, which capture the following aspects of the dialogical exchange.

Layer 1 categorizes text segments in Q&A pairs as *question*, *reply*, *preface*, or *discourse regulator*. Questions are further categorized according to a threefold question type, distinguishing *open*, *yes/no* questions, and *closed-list*, which make the main focus of this paper.

Layer 2 captures fine grained question design by categorizing questions according to their *request type*, i.e. the type of speech act of the respondent they project. The remainder of this paper focuses on (closed-list) *requests of explanation* and *requests of opinion*.

Layer 3 groups together topically related questions (follow-ups, rephrases, specifications) forming *question spans* and is of no particular relevance here.

The inter-annotator agreement rate for *Step 1*, measured on 2 annotators by means of Cohen's Kappa (Cohen 1960), ranged from substantial to almost perfect for all features. Regarding the features of interest for this contribution, "Question type" (Layer 1) has κ=0.97 and "Request type" (Layer 2) has κ=0.80.

Step 2 involves the full reconstruction of the dialogical and inferential structure of Q&A pairs using the IAT theoretical framework, which has been discussed in detail in Section 4. Once a relevant corpus has undergone *Step 1* annotation, it is queried to extract Q&A pairs according to the features under investigation – in our case *closed-list* questions which realized either a *request of opinion* or a *request of confirmation* – which are selected for the full IAT reconstruction. This annotation is carried out using a dedicated tool, the OVA+[2] software (cf. Janier et al. 2014) and the resulting argument maps are stored and made publicly available through the AIFdb database (Lawrence and Reed 2014)[3]. For the purposes of this study OVA annotation was performed by one highly trained researcher only (one of the authors) and later discussed with another author. So, no agreement scores have been calculated.

6. Distribution of closed-list questions in the corpus

Before discussing the results of the full IAT analysis of closed-list questions in our corpus (*Step2*), it is worth briefly presenting some basic data on the distribution of closed-list questions in the corpus resulting from *Step1* annotation. Table II, below, shows that, consistently with the literature discussed in Section 4.1, closed-list questions represent least numerous question type. Zillow (Z), which has the smallest number of questions, also has the lowest number of closed-list questions (2) none of which is either a request of opinion or explanation, as shown in Table III. Interestingly, the request type most frequently associated with the closed-list question is the *request of confirmation*, briefly discussed in Section 2.2. Requests of confirmations, and especially the requests of confirmation of inference described by Rocci and Raimondo (2018b) are a complex question design, deserving a dedicated study of its own, which is forthcoming. Here to discuss the argumentative affordances of closed-list questions we will focus on the second and third most represented request types, namely requests of opinion and requests of explanation (see Table III).

Table II. Question types across each firm

	closed-list	open	yes/no	Total
DASH	20	118	34	172
HAS	19	98	29	146
Z	2	81	7	90
Total	*41*	*297*	*70*	*408*

---

2 http://ova.arg-tech.org/
3 http://corpora.aifdb.org/closedlist

Table III. Closed-list questions across requests types and companies

	request of clarification	request of commitment	request of confirmation	request of data	request of explanation	request of opinion	Total
DASH	3	1	6	4	3	3	20
HAS	3	0	7	1	3	5	19
Z	1	0	1	0	0	0	2
Total	7	1	14	5	6	8	41

7. A study of closed list requests of opinion and requests of explanations based on IAT

Request of opinion and requests of explanation are representative of each of the two groups of requests described in Section 4.2: the former inquiring the propositional content only, and the latter questioning an entire inferential structure. We used OVA+ to construct 14 argument maps, each representing one closed list question (posed by an analyst) and the related answer (from a manager) reconstructed in terms of the IAT analytical framework, as discussed and exemplified in Section 4.2. The analysis allowed to examine, for each Q&A pair, both the issue framing proposed by the analyst and manager's reaction to it.

In Table IV and Table V, we provide references to each IAT map in AIFdb[4] in terms of mapIDs making them retrievable for the reader.

Table IV. Summary of the analyses of closed-list requests of opinion.

#	mapID	Sub-pattern	Reaction	Argumentative response
1	25783	-none-	non-answer; accepting frame	yes
2	25784	*Information elicitation*	accepting frame	yes
3	25785	-none-	rejecting frame (alternativity)	yes
4	25800	*Confrontation framing*	accepting frame	yes
5	25787	-none-	accepting frame	yes
6	25788	-none-	accepting frame	no
7	25789	*Confrontation framing*	rejecting frame (alternativity)	yes
8	25790	*Confrontation framing*	rejecting frame (alternativity)	yes

---

4 http://corpora.aifdb.org/closedlist

ARGUMENTATIVE PATTERNS

Table V. Summary of the analyses of closed-list requests of explanation.

#	mapID	Sub-pattern	Reaction	Argumentative response
1	25801	-none-	accepting frame	yes
2	25802	Information elicitation	accepting frame	yes
3	25799	-none-	accepting frame	yes
4	25798	Confrontation framing	accepting frame	yes
5	25795	Information elicitation	rejecting frame, (exhaustivity)	yes
6	25796	Information elicitation	accepting frame	yes

The summary of the analysis presented in Table IV and V describes each of the 14 examples according to three parameters. The first parameter relates to the sub-patterns, as introduced in Section 3. The first, sub-pattern, which we will be calling *confrontation framing*, collects instances where analysts provide two options, one preferable to the other. The second, called *information eliciting*, describes cases when an analyst provides as one alternative a "basic" or standard" answer based on already disclosed information, which is opposed to the informatively empty alternative that something still unknown must be at work. In our corpus we found 4 clear instances of each sub-pattern.

The second parameter investigated was managers' reaction to the framing of the question. As described in Section 4.2, analysts leverage on the way the issue is presented to direct or shape managers' replies. In the case of locutions such as a requests of opinion, the framing would consist of two suggested propositional contents, holding a relation of conflict between them; in the case of a request of explanation, the framing consists of two inferential structures, each of which consisting of *explanans*, *explanandum* and an "explanatory" relation between them, plus a relation of conflict between the two structures. We have already discussed in Section 4.2 the types of reaction a manager can deploy in answering closed-list questions (cf. Figure 2). In the 14 examples we examined, most of the managers' reactions (11) accepted the proposed frame. In the case the frame is not accepted, however, two paths are available. The first path involves the rejection of *exhaustivity*; i.e. the manager rejects all the alternatives in the question by presenting a new hitherto unexamined option.

We have one case of exhaustivity-based rejection in the answers to requests of explanation (Example 8)). Here, however, the rejection is made expectable by the weakening of the alternative framing by the questioner himself: having provided a number of possible explanations of the weak performance during the previous quarter, the analyst turns to an open request of explanmation ("...*can you help me understand that?*") (Example 8). This allows the manager to provide his own explanation ignoring all three alternatives.

Example 8.
Eric Handler (analyst): Is that [...] reflecting animated program deliverables? Is that a timing issue? Is that consumer products, or can you help me understand that? (HAS Q2, 2021)

On the other hand, we say that a manager rejects the frame by refuting alternativity, when they state that more than one of the presumed alternatives is the case. This happens in three cases, with answers to requests of opinion.

All those strategies, however, are related to actually answering the question, which is not the only possible response. As noted in Palmieri et al. (2015) managers have also the option to explicitly refuse to answer, often providing arguments in support of their refusal. We found one justified refusal to answer in our data (mapID: 25783), reproduced as Example 9.

Example 9.
Prabir Adarkar (Door Dash, CEO): Secretary Walsh's comments actually suggest an openness to engage with the private sector to figure this stuff out. So it's a little early to signal <whether rideshare drivers will be grouped together with the broader gig economy or kept separate>. (DASH Q4 2021)

Clearly, here the manager accepts the frame by reproducing the proposed alternative in his answer (within angle brackets <...> in Example 9), but, at the same time, provides a reason for not actually answering yet.

Finally, the third parameter ("argumentative response") checked whether managers tend to support their answers with arguments. As we can see from both tables, answers were argumentatively justified in almost all cases, apart from one response to a request of opinion, in which a mere opinion was provided.

8. Conclusion and future work

In this paper we provided an outline of a strategy for the discovery of argumentative patterns in the Q&A phase of Earning Conference Calls activity type, exemplifying this strategy with closed-list questions, which are seen as a minimal AP in themselves and as the building block of significant APs spanning the Q&A turn initiated by the question.

In Section 2, having introduced a minimal notion of argumentative pattern as a significant constellation of argumentative moves whose occurrence can be explained in view of the goals and constraints of the activity type, we discussed its potential as a unit of analysis for large scale studies of argumentation in context, and, by introducing the prototypical argumentative pattern of firm valuation, we provided a concrete means for evaluating the potential relevance of patterns of moves in the Q&A as argumentative patterns of the activity.

Section 3 discusses the intrinsic argumentative affordances of closed-list (alternative) questions and makes a case on how they can potentially become relevant in an ECC Q&A context, through their use as means of issue framing and as information elicitation probes. The theoretically possible argumentative follow-ups are also mapped.

Section 4 introduces Inference Anchoring (IAT) Theory as the metalanguage for the description of potential argumentative patterns in ECCs and showcases how IAT can represent the argumentatively relevant features of closed-list questions with examples taken from the ECC corpus. Section 5 illustrates a two-step corpus-annotation process needed to empirically explore hypotheses about argumentative patterns such as those expounded in the previous sections, where Step 1 consist in a more coarse grained annotation of dialogue moves and question designs and Step 2 transforms Q&A pairs in fully analysed IAT argumentative map.

The results of this strategy as regards the distribution and argumentative functioning of closed-list questions in a small corpus of 12 ECCs are illustrated in Sections 6 and 7. Section 7, in particular, present the full IAT reconstruction, made available via AIFdb, of 14 closed-list questions in the corpus which adopt either a *request of opinion* or a *request of explanation* together with their follow-ups in the manager's response. This investigation shows that both frame rejecting and frame accepting responses happen, with all the theoretical possibilities represented. Both the *issue framing* and the *information elicitation* "sub-patterns" occurs in the corpus, with the second appearing only in the context of requests of explanation. As it was expected, the IAT analysis reveals that managerial follow-ups to closed-list questions are most often argumentative in nature and they invariably involve argumentation when the closed-list framing of the question is rejected. The analysis also showed that responses to closed-list questions can consist also in justified refusals to answer.

These results are suggestive and show a nice fit between theoretical hypotheses and in-depth qualitative analyses on the corpus, which encourage us to pursue the idea of closed-list questions as a key component of possibly several argumentative patterns in ECC Q&A, including in particular the *issue framing* and *information elicitation* sub-patterns. Together with the range of their possible answers they form a set of potentially significant patterns, worthy of further investigation.

Larger scale corpus analysis will be the main engine of future work. In fact, a new research cycle begins when the initial manually annotated corpus starts to be used as training data for algorithms for automatic *Step 1* and *Step 2* annotation. The training of machine learning algorithms for the recognition of basic dialogue moves in ECC is currently ongoing at IALS and we hope to be soon able to deploy *Step 1* annotation in a semi-automatic fashion, which will allow us to explore statistically significant correlations between dialogue moves, features of question design, and contextual variables, such as the financial circumstances of the company. This will allow us to select better targeted in-depth case studies in *Step 2*, refining hypotheses on candidate patterns. At both levels, exploring more in-depth the features of the answer turns will be critical.

References

Rahwan, I., Ramchurn, S. D., Jennings, N. R., McBurney, P., Parsons, S., & Sonenberg, L. (2003). Argumentation-based negotiation. *The Knowledge Engineering Review, 18(4)*, 343–375.

Tindale, C. W. (2007). *Fallacies and argument appraisal*. Cambridge: Cambridge University Press.

Biber, D., Johansson, S., Leech, G., Conrad, S., & Finegan, E. (1999). *Longman grammar of spoken and written English*. Edinburgh: Pearson Education Ltd.

Biezma, M. (2009). Alternative vs Polar Questions: the cornering effect. *Proceedings of SALT 19*.

Biezma, M., and Rawlins, K. (2012). Responding to alternative and polar questions. *Linguistics and Philosophy, 35(5)*, 361–406.

Biezma, M., and Rawlins, K. (2015). Alternative Questions. *Language and Linguistics Compass, 9(11)*, 450–468.

Budzynska, K., Janier, M., Reed, C., Saint-Dizier, P., Stede, M., & Yakorska, O. (2014). A Model for Processing Illocutionary Structures and Argumentation in Debates. *Proceedings of the Ninth International Conference on Language Resources and Evaluation (LREC'14)*, 917–924.

Budzynska, K., & Reed, C. (2011) Speech acts of argumentation: inference anchors and peripheral cues in dialogue. *Computational models of natural argument: papers from the 2011 AAAI Workshop*, 3–10.

Clayman S. E., & Heritage, J. (2002). Questioning presidents: Journalistic deference and adversarialness in the press conferences of U.S. Presidents Eisenhower and Reagan. *Journal of Communication, 52(4)*, 749–775

Cohen, J. (1960). A Coefficient of Agreement for Nominal Scales. *Educational and Psychological Measurement, 20(1)*, 37–46.

Crawford Camiciottoli, B. (2009). 'Just wondering if you could comment on that': Indirect requests for information in corporate earnings calls. *Text and Talk, 29(6)*, 661–681.

Drake, V. (2020). Alternative questions and their responses in English interaction. *PRAG, 31(1)*, 62–86.

Enfield, N. J., Stivers, T., & Levinson, S. C. (2010). Question–response sequences in conversation across ten languages: An introduction. *Journal of Pragmatics, 42(10)*, 2615–2619

García, D., Hu, X., & Rohrer, M. (2023). The colour of finance words. *Journal of Financial Economics, 147*(3), 525–549.

Hautli-Janisz, A., Budzynska, K., McKillop, C., Plüss, B., Gold, V., & Reed, C. (2022). Questions in argumentative dialogue. *Journal of Pragmatics, 188*, 56–79.

Janier, M., Lawrence, J., & Reed, C. (2014). OVA+: an Argument Analysis Interface. *Computational Models of Argument*, 463–464.

Jason V, C., Venky, N., & Jordan, S. (2018). Manager-analyst conversations in earnings conference calls. *Review of Accounting Studies, 23*, 1315–1354.

Lewiński, M. & Aakhus, M. (2023). Argumentation in Complex Communication. Managing Disagreement in a Polylogue. Cambridge: Cambridge University Press

Klie, J.-C., Bugert, M., Boullosa, B., Eckart de Castilho, R., & Gurevych, I. (2018). The INCEpTION Platform: Machine-Assisted and Knowledge-Oriented Interactive Annotation. *Proceedings of the 27th International Conference on Computational Linguistics: System Demonstrations*, 5–9.

Lawrence, J., & Reed, C. (2014). AIFdb Corpora. *Computational Models of Argument: Proceedings of COMMA 2014*, 465–466.

Matsumoto, D., Pronk, M., & Roelofsen, E. (2011). What Makes Conference Calls Useful? The Information Content of Managers' Presentations and Analysts' Discussion Sessions. *The Accounting Review, 86*(4), 1383–1414. https://doi.org/10.2308/accr-10034

Musi, E., & Aakhus, M. (2018). Discovering Argumentative Patterns in Energy Polylogues: A Macroscope for Argument Mining. *Argumentation*, 1–34.

Palmieri, R. (2018). The Role of Argumentation in Financial Communication and Investor Relations. In Laskin, A. V. (Ed), *The Handbook of Financial Communication and Investor Relations* (pp. 45–60). New York, NY: John Wiley & Sons.

Palmieri, R., Rocci, A., & Kudrautsava, N. (2015). Argumentation in earnings conference calls. Corporate standpoints and analysts' challenges. *Studies in communication sciences, 15(1)*, 120–132.

Quirk, R., Greenbaum, S., Leech, G., & Svartvik, J. (1985). *A Comprehensive Grammar of the English Language*. London:Longman.

Rigotti, E., & Greco, S. (2019). *Inference in Argumentation: A Topics-Based Approach to Argument Schemes*. Cham: Springer International Publishing: Imprint: Springer.

Rocci, A., & Raimondo, C. (2018a). Conference calls: A communication perspective. In Laskin, A. V. (Ed), *The Handbook of Financial Communication and Investor Relations* (pp. 293–308). New York, NY: John Wiley & Sons.

Rocci, A., & Raimondo, C. (2018b). Dialogical Argumentation in Financial Conference Calls: The Request of Confirmation of Inference (ROCOI). In S. Oswald & D. Maillat (Eds.), *Argumentation and Inference: Proceedings of the 2nd European Conference on Argumentation* (pp. 699–715). College Publications.

Stivers, T., & Enfield, N. J. (2010). A coding scheme for question–response sequences in conversation. *Journal of Pragmatics, 42(10)*, 2620–2626 van Eemeren, F. H. (2017). Context-dependency of argumentative patterns in discourse. *Journal of Argumentation in Context*, 6(1), 3–26. https://doi.org/10.1075/jaic.6.1.01eem

van Eemeren, F. H. (2018). Argumentation Theory: A Pragma-Dialectical Perspective. Cham: Springer.

Yaskorska-Shah, O., Rocci, A., & Reed, C. (2022). Conversation shaping questions: a taxonomy used for mapping argumentative dialogues in financial discourse. *Workshop on Computational Models of Natural Argument (CMNA 22)* https://ceur-ws.org/Vol-3205/.

DILEMMATIC ARGUMENTATION: A PRAGMA-DIALECTICAL APPROACH TO A CLASSICAL TOPOS

CHARLOTTE VAN DER VOORT
Leiden University Centre for Linguistics
c.van.der.voort@hum.leidenuniv.nl

Abstract

This paper examines dilemmatic argumentation in argumentation theoretical terms. With a dilemmatic argument, one presents a disjunction of two alternatives combined with two arguments that lead to one conclusion. Ancient rhetoricians claimed it to be an extremely powerful argument, while formal logicians have pointed out its fallacious nature. In this paper, I define the main form of dilemmatic arguments and introduce a threefold typology of dilemmatic arguments based on their combination of the type of standpoint and used argument schemes. For each type this paper develops an argumentative pattern with its associated critical questions to differentiate, analyze and evaluate dilemmatic arguments. A first exploration of the strategic function of dilemmatic argumentation in the final section shows that it often may have a whiff of unreasonableness but that it is also a persuasive strategic maneuver that allows a discussant to put forward one's argument while anticipating the possible refutation or critique of an opponent at the same time.

1. The 'invention' of rhetoric and the dilemmatic argument

In antiquity, the invention of the art of rhetoric is ascribed to two legendary 'inventors', the Syracusans Corax and Tisias (e.g., Hinks, 1940; Cole, 1991:22-27; Schiappa, 2013:49-54).[1] This legendary starting point of rhetoric is accompanied by a fascinating but puzzling type of argument:

---

[1] For ancient versions of this story, e.g., Sextus Empiricus' Adversus Mathematicos 2.96-99 and the Byzantine Prolegomena in Rabe (1931:26-27, 52-53, 67). Maximus Planudes (Rabe 1931:67) remarks that Corax uses a dilemma ('dilemmatoi grèsmenos').

the dilemmatic argument. The story goes as follows. After democracy was established in Syracuse, Corax began to teach the art of oratory. Tisias became his first pupil. They agreed that Tisias would only have to pay tuition if he would win his first court case; only then the lessons would have been successful. After their course, however, Tisias decided not to put his rhetorical skills into practice and did not pursue any lawsuit. Corax grew impatient and sued his former pupil to demand the tuition. In court, Corax argued that he should get paid if he wins, but also if he loses: should Corax win, Tisias must grant Corax' current demand to pay, and should he lose, Tisias wins and must pay following their earlier agreement. For a moment there seemed no way out, but Tisias proved to be a good student: he argued that he does not have to pay, for if he wins, Corax' claim does not need to be met, and if he loses, he has lost his first case and does not have to pay for the apparently failed lessons. When the jury had heard these arguments, they expelled Corax and Tisias from the courtroom with the verdict 'from a bad crow ('*corax*'), a bad egg'.

The type of argument used by Corax and Tisias is a dilemmatic argument (also known as 'Morton's fork' or by the description 'damned if you do and damned if you don't'). It presents two contradictory alternatives (winning or losing the case) and connect both to their desired outcome (tuition or no tuition). Although this type of argument does not necessarily appear rare in classical nor modern discourse, the literature that deals with dilemmatic argumentation is 'scarce, casual and not always reliable' as Nuchelmans remarks in his historical study of dilemmatic arguments in classical and Renaissance rhetoric (1991:5). The existing literature is mainly limited to brief discussions by the ancient rhetoricians (Aristotle, *Rhetorica* 1399a.17-27; Anonymous, *Rhetorica Ad Herennium* II.38-42; Cicero, *De Inventione* I.44-45, I.83-84; Quintilian, *Institutio Oratioria* V.10.69-71; pseudo-Hermogenes, *De Inventione* IV.6; see Nuchelmans 1991)[2] and modern formal logicians (e.g., Smith, 1938:148-156; Copi & Cohen, 1990:245-252; Parry & Hacker, 1991:395-406; and from a more informal approach: Tomić, 2013, 2021), who focus on respectively the definition and examples or on the formal form of dilemmatic argumentation. Whereas the rhetoricians characterized dilemmatic argumentation as a powerful instrument for persuasion 'from which, if it's true, there is no escape' (Cicero, *De Inventione* I.83; cf. pseudo-Hermogenes, *De Inventione* IV.6.1), logicians have remarked that '[d]ilemmatic arguments are [...] more often fallacious than not' (Jevons, 1883:159; Keynes, 1906:366). Both aspects we recognize from the story of Corax and Tisias: they use a seemingly strong argument that brings their

---

[2] The rhetorical sources treat dilemmatic argumentation under different terms: 'topos of two opposites' (Aristotle), 'duplex conclusio' (Rhetorica Ad Herrenium), 'divisio' (Rhetorica Ad Herrenium, Quintilian), 'complexio' (Cicero), 'tò dilêmmaton' (pseudo-Hermogenes).

opponent into a catch-22-situation, but the jurors note that the reasonableness falls short.

How dilemmatic argumentation 'in the wild' is used as a persuasive argument or when such an argument is fallacious or not, is not discussed thoroughly, if at all, by the ancient rhetoricians and modern logicians. The perspective of argumentation theory, by contrast, can shed light on these questions. Modern argumentation theory focuses namely on the actual use and function of argumentation to analyze and evaluate how discussants resolve their difference of opinion (Van Eemeren et al., 1987:94-111). However, to date, argumentation theorists have left dilemmatic argumentation largely unaddressed, and those who have touched upon it (Perelman & Olbrechts-Tyteca, 1969:236-240; Schellens, 1985:162-163; Kienpointer, 1992:314; Gerlofs, 2009:155-160) have in common that they are concise, quite unsystematic, and often superficial by citing only artificial examples.

This paper hence studies dilemmatic argumentation in argumentation theoretical terms to examine its possible variants, develop an analytical and assessment tool, and explore why this type of argument is so seemingly persuasive. The next paragraph starts with the main form of dilemmatic argumentation. In the subsequent paragraphs 3 and 4, I use the pragma-dialectical theory of argumentation to introduce a threefold typology of dilemmatic arguments and develop for each type an argumentative pattern and its associated critical questions. In paragraph 5, I explore the strategic function of dilemmatic argumentation and show how this argument can be used to put one's argument forward while anticipating the opponent's critique.

2. The main form of dilemmatic argumentation

The ancient rhetorical treatments show that dilemmatic argumentation is characterized by two features: a dilemmatic argument makes use of two alternatives, of which one necessarily must be the case (often because the two alternatives are logical opposites; Aristotle, *Rhetorica* 1399a19-21; pseudo-Hermogenes, *De Inventione* IV.6.3), and both alternatives lead to the same (for the opponent often negative) result. The most complete definition is provided by Quintilian who stresses both aspects: 'One can [...] give one's opponent the choice between two propositions, one of which must be true, and ensure that whichever he chooses does his case harm' (V.10.69; tr. Russell, 2002).

Pseudo-Hermogenes remarks that there are two types of dilemmatic constructions (*De Inventione* IV.6.15-19): two alternatives that lead to the same conclusion, or two alternatives that lead to two different consequents. The former can be illustrated with an example of Quintilian (V.10.69; tr. Russell, 2002):

Example 1. *The man who can bear pain will lie under torture, the man who cannot will also lie.*

Here both alternatives lead to one conclusion (a tortured man will lie). The latter kind of dilemmatic construction, the one with two different consequents, can be illustrated with the example Aristotle discusses (*Rhetorica* 1399a.22-24; tr. Waterfield, 2018):

Example 2. *The priestess [attempted] to dissuade her son from becoming a public speaker. 'For', she said, 'if your policies are just, men will hate you, and if they are unjust the gods will hate you.'*

The mother presents here two alternatives (just or unjust policies) that lead to two different consequents (being hated by men or by the gods). We must note, however, that the consequents still lead to one final conclusion: the advice not to become an orator. Both examples thus work in a funamentally similar way, but the latter has an intermediate step before reaching the 'inevitable' conclusion.

Through classical rhetoric, this argumentative phenomenon came to be known as 'the dilemma' in logic handbooks (e.g., Whately, 1853:131-137; Jevons, 1883:158-160; Keynes, 1906:363-366; Smith, 1938:148-156; Copi & Cohen, 1990:245-252; Parry & Hacker, 1991:395-406; Cook, 2009:46-47). The logicians describe the dilemma as a disjunction of two hypothetical syllogisms, i.e. a conjunction of two hypothetical syllogisms ('if p, then q' and 'if r, then q') and a disjunction of the elements in the antecedents ('p or r'). The dilemma thus differs from the false dilemma fallacy that only involves a (false) disjunctive claim (p or q; not-p, thus q; Boone, 1999:6). Just like pseudo-Hermogenes, they distinguish between what they call a simple dilemma (with one consequent) and the complex dilemma (with two different consequents):[3]

---

[3] The formal logicians usually also distinguish a simple and complex destructive type of the dilemma. The destructive type makes use of a modus tollens while the constructive types make use of the modus ponens. This distinction is not so relevant in argumentation theoretical terms, since both modi differ only in the reconstruction of the (often left implicit) bridging premise – compare:

Modus tollens reconstruction	Modus ponens reconstruction
1. Not-p	1. Not-p
1.1 Not-q	1.1 Not-q
1.1' If p, then q	1.1' If not-q, then not-p

Simple constructive dilemma	Complex constructive dilemma
If p, then q and if r, then q	*If p, then q and if r, then s*
p or r	*p or r*
Therefore q	*Therefore q or s*

Two remarks must be made about these logical forms. Firstly, as said, a dilemma is characterized by two alternatives leading to one conclusion. The form of the complex constructive dilemma above, however, represents just two *modi ponentes* that are connected by a disjunction. Each modus ponens has now its own consequent (q or s) and has, therefore, not the very persuasive power that characterizes the dilemmatic argument (e.g.: 'if it is hot, I like ice cream and if it is cold, I like hot chocolate; it is hot or cold; thus I like ice cream or hot chocolate'). Only when both consequents are combined in an overarching conclusion one can call the argumentation dilemmatic, as is, for instance, the case with example 2 where the two consequents, hated by men (q) or by gods (s), are connected in the conclusion that one should not become an orator.

Another remark must be made about the disjunction in both schemes (p or r). Jevons (1883:159) argued that '[d]ilemmatic arguments are [...] more often fallacious than not, because it is seldom possible to find instances where two alternatives exhaust all the possible cases, unless indeed one of them be the simple negative of the other in accordance with the law of the excluded middle'. According to the definition by Quintilian, dilemmatic arguments are defined by 'two propositions, one of which must be true'. Only when one of both alternatives in a dilemma is the case ('true' in logical terms), the conclusion is compelling. Hence the persuasive power of the dilemma lies in the comprehensiveness of the disjunctive premise. Jevons' remark is, therefore, not so much about the form of the dilemma, but about the content of the disjunctive premise. There are two ways that the disjunction includes all possible options: when the two alternatives are logical oppositions (p or not-p), or when the two alternatives are no logical opposites (p or r) but in their context relate to all possible options ('the red pill or the blue pill') between one must choose. The disjunction thus must be exclusive. Since in the latter case the negation of one of the options ('not the blue pill') is a confirmation of the other option ('the red pill'), the notation 'p or not-p' for the disjunctive premise is sufficient and, moreover, it emphasizes the comprehensiveness of the disjunction in a valid dilemma. The main form of the dilemma therefore is (with the possible intermediate step in the case of a complex dilemma between brackets):

Main form of the dilemma
If p (then r,), then q and if not-p (then s,), then q
p or not-p
Therefore q

3. Argumentative patterns and reasonableness criteria

By no means are all dilemmatic arguments reasonable. The judges, for instance, did not consider Corax' and Tisias' dilemmatic arguments sound, and Jevons already remarked that a dilemmatic argument is fallacious if it allows for a third alternative. The possible fallaciousness of dilemmatic arguments is, however, not a formal matter (as Tomić, 2013, 2021 takes as the starting point), but a matter of content: problems with dilemmatic argumentations lie in the propositional content of the disjunctive premise that may allow for other alternatives (Keynes, 1906:366; Lewínski, 2014:104 *contra* Tomić, 2013) or in the hypothetical premises. Formal logic focuses only on the (in)valid forms of arguments and, therefore, cannot help us to evaluate the possible content of the premises.

Unlike practice-based rhetoric and structure-based logic, modern argumentation theories focus on the content, real use, and function of argumentation. The pragma-dialectical theory of argumentation is perhaps the most systematic of these in integrating the normative and descriptive dimension of argumentation (Van Eemeren & Grootendorst, 1984, 1992, 2004). Its method is grounded in the ideal model of a critical discussion that functions, on the one hand, as an analytical tool to reconstruct argumentation and, on the other hand, results in a set of rules to critically evaluate argumentation (Van Eemeren & Grootendorst, 19992:208-209).

One of these rules is the 'argument scheme rule' that assesses whether an argument is an appropriate defense of a standpoint. The pragma-dialectical theory of argumentation distinguishes between three main types of argument schemes: argumentation based on a causal relation, symptomatic/sign relation, or analogy (Van Eemeren & Grootendorst, 1992:97). The type of justification that connects argument and standpoint is expressed in the bridging premise. To assess whether the justificatory relation and thus the argument scheme is applied correctly, a critical question can be linked to this bridging premise. As an illustration, the scheme of causal argumentation:

1.   For X holds Y
1.1 For X holds Z
1.1' Z leads to Y   Q. Does Z lead to Y?

Most arguments – like dilemmatic arguments – however, consist of a more complex set of argumentative steps. Such arguments cannot be described merely by the three main types of argument schemes, but require a reconstruction that reflects the different argumentative moves.

This is possible with the pragma-dialectical notion of an argumentative pattern: 'a constellation of argumentative moves in which, [...], in defence of a particular type of standpoint a particular argument scheme or combination of argument schemes is used in a particular kind of argumentation structure' (Van Eemeren, 2017:19-20). For example, pragmatic argumentation can be described in a (very brief) pattern:[4]

1. Action X must be carried out
1.1a X leads to Y
1.1b Y is desirable
1.1a-b' If action X leads to a desirable result such as Y, X must be carried out

In an argumentative pattern, the different arguments anticipate critical responses of the other discussant(s) and from these critical questions can be derived. For pragmatic argumentation, these questions are (Van Eemeren & Garssen, 2020:17-18):[5]

Q1. Does X lead to Y? (1.1a)
Q2. Is Y desirable? (1.1b)
Q3. Must actions that lead to a desirable result Y always be carried out? (1.1a-b')
Q4. Could Y not be achieved more easily or economically by other actions? (1.1a-b')
Q5. Does X not have unavoidable undesirable side-effects? (1.1a-b')

Reconstructing an argumentative pattern and its associated critical questions for dilemmatic argumentation is a suitable tool for its analysis and evaluation.

4. Reconstruction and assessment of dilemmatic arguments

No satisfactory reconstruction of dilemmatic argumentation exists yet. Classical rhetoric has given extensive consideration to the production and refutation of such arguments, but gives anything but a systematic

---

[4] Based on Feteris (2002:21); Van Eemeren & Garssen (2020:17) describe pragmatic argumentation as an argument scheme with argument 1.1 Action X leads to desirable result Y. Since this contains two propositions, I prefer to divide both into two coordinative arguments (1.1a-1.1b).

[5] My order. I deleted their fourth question ('would another result be not more desirable than Y?') since it is in my opinion irrelevant for the conclusion 'X must be carried out' whether there is an even more desirable consequence.

overview. Modern argumentation theorists do pursue this systematic treatment, but dilemmatic argumentation has received little attention by them so far: there are four argumentation theoretical treatments of dilemmatic argumentation (Perelman & Olbrechts-Tyteca, 1969:236-240; Schellens, 1985:162-163; Kienpointer, 1992:314; Gerlofs, 2009:155-160), but they have in common that they are concise, quite unspecific, and mostly rehearsals of the earlier rhetorical or logical insights. Moreover, these contributions stay all on the surface by citing some (new or classical) artificial examples, and neither systematically discuss the different types, nor the reasonableness criteria of dilemmatic argumentation. This is a hiatus for who is studying real examples of dilemmatic arguments. When I was collecting and analyzing the dilemmatic arguments, I noticed crucial differences between the examples which showed that only distinguishing the main form of the dilemma is not enough for a thorough understanding. For that reason, I developed a typology of dilemmatic arguments that can form the starting point for more in-depth research on this type of argumentation.

By bringing together the different (logical, rhetorical, and argumentation) theoretical treatments, I have distinguished three main types of dilemmatic arguments. These differ not in form but in content. This distinction can be made on the basis of the type of standpoint and the type of argument schemes that are used. Pragma-dialectics distinguishes between three types of standpoints: descriptive ('X is fact Y'), evaluative ('X is evaluation Y') and prescriptive ('X must be carried out') standpoints (Van Eemeren & Grootendorst, 1992:159). Considering these and the type of argument schemes used, we can distinguish three types of dilemmatic arguments: the 'pragmatic dilemmatic argument' that combines a prescriptive standpoint with two pragmatic arguments, the 'assertive dilemmatic argument' that posits a statement or judgement (descriptive or evaluative standpoint) based on two symptomatic or causal arguments, and the 'impossible-choice-dilemma' that combines two pragmatic arguments with a descriptive standpoint.

Below I develop for each of the three dilemmatic types an argumentative pattern and a set of critical questions, and illustrate these with an example. Since dilemmatic argumentation is characterized by a clear main form, the three argumentative patterns overlap partly with regard to their essential argumentative moves: the presentation of a disjunction and two alternative arguments that converge into one conclusion (i.e. standpoint). The differences between the three dilemmatic types, however, affect the argumentative pattern of each type and thus its critical questions; each type needs therefore its own reconstruction and evaluation.

4.1. Pragmatic dilemmatic argument

In their *New Rhetoric*, Perelman and Olbrechts-Tyteca (1969:236) remark that the conclusion of a dilemma 'is a statement, [or] a course of action'. Although this distinction and their discussion of examples are little precise, their remark is useful to distinguish two different dilemma's that we can encounter in the rhetorical theories: the dilemma that calls for action and the dilemma that asserts something. Here I start with the first type; the latter I analyze in the next section.

Perelman's and Olbrecht-Tyteca's statement derives probably from Aristotle's definition of dilemmatic argumentation (*Rhetorica* 1399a.17ff.; tr. Waterstone 2018): 'a speaker who must persuade or dissuade an audience about two courses of action that are mutually exclusive'. Aristotle describes here what we nowadays know as pragmatic argumentation. He illustrates this with the abovementioned example 2 of a mother who dissuades her son to become an orator. With this example, an action is discouraged by presenting two (contradictory) conditions of the action ('making just or unjust policies') that result in undesirable effects ('being hated by men or by the gods').

This type of dilemmatic argumentation advises (against) an action based on two pragmatic arguments, which I hence call a pragmatic dilemmatic argument. The argumentative pattern of this type consists of three coordinative arguments that support the prescriptive standpoint: one disjunctive premise with two conditions under which the action can be carried out (1.1a), and two pragmatic arguments (1.1b and 1.1c). The bridging premise expresses the relation between those three arguments and the standpoint. The negative variant (recommendation against something), such as Aristotle's example, can be reconstructed as follows:

The argumentative pattern of pragmatic dilemmatic argumentation
1. Action X should not be carried out
1.1a Performing X presupposes either X1 or X2
1.1b X1 should not be carried out
1.1b.1a X1 leads to Y
1.1b.1b Y is undesirable
1.1c X2 should not be carried out
1.1c.1a X2 leads to Z
1.1c.1b Z is undesirable
1.1a-c' If X either leads to undesirable result Y or undesirable result Z, X should not be carried out

For the evaluation of a pragmatic dilemmatic argument, each of the argumentative moves in this pattern can be assessed with one or more critical questions. As the logicians (and the rhetoricians as well) remarked, the disjunction of a dilemma should not allow for a third alternative. Therefore, for 1.1a it must be asked whether X1 and X2 are the only two possibilities (Q1). If not, the dilemmatic argument can be refuted by presenting a third alternative that leads to a different conclusion. Regarding the pragmatic arguments 1.1b and 1.1c, it must be assessed whether X1 and X2 result in the mentioned effects Y and Z (Q2-3) and whether the effects are indeed undesirable (Q4). These (Q2/3 and Q4) correspond to the first two critical questions for pragmatic argumentation that Van Eemeren & Garssen distinguished (paragraph 3 *supra*). Also their third ('must actions that lead to a (un)desirable result always (not) be carried out?') and fourth question ('does action X not have unavoidable (un)desirable side-effects?') are relevant for pragmatic dilemmatic argumentation (Q5-6 and Q7). Not relevant is their fifth question ('could result Y not be achieved more easily or more economically by other actions?'), because dilemmatic argumentation aims to describe all the possible actions in a situation. This results in the following set of critical questions:

Q1. Is there, besides X1 and X2, a third condition X3 possible that allows X to be carried out?
Q2. Does X1 lead to Y?
Q3. Does X2 lead to Z?
Q4. Are Y and Z undesirable?
Q5. Does X1 not lead to another result (Y') which makes it preferable to carry out X, since the result Y does not outweigh result Y'?
Q6. Does X2 not lead to another result (Z') which makes it preferable to carry out X, since the result Y does not outweigh result Z'?
Q7. Must actions that lead to an undesirable result (Y or Z) always be discouraged?

Critical questions Q5 and Q6 require some further clarification. These assess if the dilemma can be 'reversed' into a more plausible and reasonable dilemma. The possibility to turn around a dilemma is a fascinating aspect that the classical rhetoricians have already examined extensively (Aristotle, *Rhetorica* 1399a.25-27, *Rhetorica Ad Herrenium* II.39, Cicero, *De Inventione* I.83, Aulus Gellius, *Noctes Atticae* V.11.3-8; also Montefusco, 2010). To illustrate this, we can look at the counter-advise in Aristotle's example. With the same antecedents one can defend the opposite standpoint (*Rhetorica* 1399a.25-26):

Example 4. *[The mother] might equally have advised him to take up public speaking on the grounds that 'if your policies are just, the gods will love you, and if they are unjust men will love you.'*

In this response, the standpoint is 'reversed' and the desirable results are connected to the same antecedents in the pragmatic arguments:

1. I must become an orator
1.1a An orator has just or unjust policies
(1.1b One must become an orator with just policies)
1.1b.1a Just policies lead to loving gods
(1.1b.1b Being loved by gods is desirable)
(1.1c One must become an orator with unjust policies)
1.1c.1a Unjust policies lead to loving men
(1.1c.1b Being loved by men is desirable)
(1.1a-c' If becoming an orator either leads to being loved by gods or men, one should become an orator)

This reversed dilemma is, however, weaker than the original dilemma since the reversed dilemma can be refuted by the objection that the discussant does not take into account the undesirable side-effects: if the policies are unjust, you may be loved by the people – although this can be doubted as well, i.e. Q3 – but the question is whether that outweighs the hatred of the gods (of course only if you empathize with the religious context of the time). Critical questions Q5 and Q6 hence assess whether possible side-effects nullify the standpoint of the pragmatic dilemma.

4.2. Assertive dilemmatic argument

Many dilemmatic arguments take a different form than the pragmatic type. Most examples in classical as well as modern theoretical discussions result in 'a statement' as Perelman and Olbrechts-Tyteca have remarked. This statement can be descriptive or evaluative, for which reason I name this type of dilemmatic argumentation the assertive dilemmatic argument. Example 1 (of Quintilian) is an instance of an assertive dilemmatic argument with a (implicit) descriptive standpoint ('torture will result in lying') combined with two alternative scenarios ('enduring pain or not') – note that when used in a discourse, such descriptive dilemma probably will used for an evaluative standpoint (e.g., torture is useless, because it always results in lying). A similar kind of dilemmatic argument can also be formed with an evaluative standpoint. For example, in this Twitter message where the leadership qualities of Boris Johnson, at that time the Prime Minister of the UK, are evaluated after it became known that Johnson appointed Chris Pincher as Member of Parliament while there were allegations of inappropriate sexual behavior against Pincher:

Example 5. *Look either Boris Johnson knew about abuse allegations against Chris Pincher & still put him in charge of MP welfare or he didn't* *bother doing vetting checks on him before putting him in charge of MP welfare[.] Either way, Johnson was incompetent & didn't care[.]* [6]

In this example, the dilemmatic argument has an evaluative standpoint ('Johnson was incompetent') supported by two alternative scenarios.

Descriptive or evaluative standpoints are both commonly supported by either symptomatic or causal arguments (Wagemans, 2014:27). Whereas the pragmatic dilemma uses pragmatic arguments to support the prescriptive standpoint, the assertive dilemma thus uses symptomatic or causal arguments to present two possible scenarios that are typical of or lead to the assertion in the standpoint. This results in the following argumentative pattern (cf. Gerlofs' reconstruction 2009:159):[7]

The argumentative pattern of assertive dilemmatic argumentation
1.   For X holds Y
1.1a   For X holds either Z1 or Z2
1.1b   Z1 is characteristic of / leads to Y
1.1b.1a For Z1 holds A
1.1b.1b A is characteristic of / leads to Y
1.1c   Z2 is characteristic of / leads to Y
1.1c.1a For Z2 holds B
1.1c.1b B is characteristic of / leads to Y
1.1a-c' If either Z1 or Z2 is true of X and both are symptoms of / leads to Y, Y is true of X

Like the argumentative pattern of pragmatic dilemmatic argumentation, the pattern of assertive dilemmatic argumentation contains a disjunctive premise (1.1a) and two arguments (1.1b and 1.1c). The two arguments can be supported (but are often left implicit) with extra subordinate arguments (1.1b.1a-b and 1.1c.1a-b) that anticipate critique on the propositions in 1.1b or 1.1c. For instance, example 5 can be reconstructed as:

1.   Boris Johnson was incompetent
1.1a   Johnson either knew about the allegations of appropriate behavior against Chris Pincher or he didn't bother doing vetting checks [and thus didn't know] before he put Pincher in charge of MP welfare

---

[6] Twitter message from @nazirafzal (July 4th, 2022). Retrieved from https://twitter.com/nazirafzal/status/1543890513169883137 on July 6th, 2022.

[7] Gerlofs discusses and reconstructs an example of what I call an assertive dilemma, but does not develop a general argumentative pattern. In her reconstruction (that was helpful for reconstructing my pattern here), she distinguishes a disjunctive premise and two arguments as well.

(1.1b   If Johnson knew about the allegations and appointed Pincher, he was incompetent)
(1.1b.1a   Knowing about the allegations and appointing Pincher is a sign of amorality)
(1.1b.1b   Amorality is a characteristic of incompetence)
1.1c   If Johnson didn't bother doing vetting checks [and thus didn't know about the allegations], he was incompetent
(1.1c.1a   Not doing vetting checks is a sign of negligence)
(1.1c.1b   Negligence is a characteristic of incompetence)
(1.1a-c' If Johnson either didn't care about the allegations or didn't care about doing vetting checks are signs of incompetence, Johnson was incompetent)

The critical questions of the assertive dilemma correspond partially to the questions of the pragmatic dilemma because of the similar main form of the dilemma. Again, one must assess whether the alternatives Z1 and Z2 do not allow for a third possibility (Q1). Relevant for arguments 1.1b and 1.1c is only the main critical question for causal or symptomatic argumentation (Q2 and Q3). Finally, as with the pragmatic dilemma, it must be ensured that a reversal of the dilemma does not lead to a more plausible defense of the opposite standpoint (Q4):

Q1. Is there, besides Z1 and Z2, a third alternative Z3 possible?
Q2. Does Z1 result in / Is Z1 symptomatic for (case/evaluation) Y?
Q3. Does Z2 result in / Is Z2 symptomatic for (case/evaluation) Y?
Q4. Is it not more plausible that Z1 and Z2 result in / are symptomatic for not-Y than for Y?

When we assess example 5, we can doubt whether the said alternatives are the only two (Q1): there was at that time, for instance, a possibility that vetting checks were done but that Johnson was not briefed about these. A discussant can bring in this third alternative and try to argue for another outcome (not-Y). Subsequently, questions Q2 and Q3 about the two symptomatic arguments can be answered positively at first: deliberately appointing someone with a questionable background and not doing vetting checks are probable signs of an incompetent leader. However, argument 1.1b is based on the presupposition that the allegations against Pincher are such that he should not be appointed anymore to a high office. One can question this and try to argue against the dilemma that Johnson did know about the allegations but did not consider the allegations serious enough to deny Pincher the office. Lastly, the dilemma cannot be reversed to a more reasonable dilemma (Q4): arguing that Johnson was competent since he knew about the allegations (e.g., and gave Pincher a second chance, what can be a sign of competency) or he did not do vetting checks (e.g., which shows his trust in his colleagues), does not provide a more reasonable dilemma. Before it seems as if no dilemma can be reversed to a plausible alternative, a plausible reversal can be illustrated with the assertive dilemma of example 1: 'torture makes one speak the truth, for either one can bear the pain and will know that speaking the truth is the only way out, or one cannot and speaks the truth out of panic'.

4.3. Impossible-choice-dilemma

Schellens (1985:162-163), Kienpointer (1992:314), and Gerlofs (2009:156-157) describe a type of dilemmatic argument that does not occur in the ancient rhetorical treatments (but does in the rhetorical practice, see Craig, 1993:23). This type of dilemma states that a choice between two alternatives cannot be made because both are equally (un)desirable. Gerlofs calls this the 'colloquial dilemma', since it comes close to the meaning of 'dilemma' as we use it in everyday language. Although all three understand this dilemma similarly and give a reconstruction, none is fully satisfactory. Kienpointer's scheme distinguishes three arguments: a choice between two options has to be made; one option is appropriate; the other option is appropriate. The scheme is unsatisfactory, because its conclusion that both options thus are appropriate seem not to be the thing argued for. Schellens' scheme (on which Kienpointer bases himself) is clearer in this regard. Schellens emphasizes that in the case of a dilemma a choice between two options must be necessary, and reconstruct his scheme as follows:

> A choice between action A1 and A2 is necessary
> Action A1 results in B
> Action A2 results in C
> B and C are equally undesirable
> Therefore: A1 and A2 are equally undesirable

Schellens shows that there are two possible alternatives that both lead to equally undesirable consequents. He categorizes this scheme as pragmatic argumentation since this dilemma makes use of two pragmatic arguments. Schellens' scheme is, however, not fully satisfactory as well: his first premise ('a choice is necessary') is unrelated to the conclusion. When we take into account this first premise, the (now implicit) overarching conclusion would be that a reasonable choice between both is impossible.

Gerlofs' representation (2009:157) of the 'colloquial dilemma' does reflect the impossibility of a reasonable choice between two equally undesirable alternatives. This is represented in the standpoint of her reconstruction:

1. Whether we should do X or not, cannot be decided
1.1a Either we do X and Y will happen, or we don't do X and Z will happen
1.1a.1a If we do X, then Y
1.1a.1b If we don't do X, then Z
1.1a.1c Either we do X or we don't do X
1.1b Y and Z are equally undesirable

Gerlofs combines here a descriptive standpoint with two coordinative arguments. The first (complex) argument 1.1a shows that there are two alternatives that both lead to their own result (1.1a.1a-1b), and that a choice between both is inevitable (1.1a.1c). The second coordinative argument 1.1b explicates that both consequents are equally undesirable and correspond, therefore, with Schellens' scheme.

Gerlofs' reconstruction is a bit convoluted: 1.1a.1a with 1.1a.1b and 1.1b form together two pragmatic arguments but are not transparently represented as such. This also results in a compound (disjunctive) causal premise 1.1a and a subordinative disjunctive premise 1.1a.1c that could be more conveniently reconstructed as coordinative to the pragmatic arguments. Therefore, I introduce a more orderly argumentative pattern with as coordinative arguments the choice between the two alternatives, the two pragmatic arguments, and the fact that both are equally (un)desirable (like Schellens). These arguments support a descriptive standpoint, namely that a choice is impossible (like Gerlofs). By analogy to the other argumentative patterns of dilemmatic argumentation, my argumentative pattern of the impossible-choice-dilemma reads as follows:

The argumentative pattern of the impossible-choice-dilemma
1. A reasonable choice regarding X is impossible
1.1a A choice must be made between X1 or X2
1.1b X1 must (not) be chosen
1.1b.1a X1 results in Y
1.1b.1b Y is (un)desirable
1.1c X2 must (not) be chosen
1.1c.1a X2 results in Z
1.1c.1b Z is (un)desirable
1.1d Y and Z are equally (un)desirable
1.1a-d' If X1 and X2 lead to either Y or Z, and Y and Z are equally undesirable, no reasonable choice regarding X can be made

Because the impossible-choice-dilemma contains a disjunctive premise and two pragmatic arguments like the pragmatic dilemma, the critical questions of both are partly similar: it must be questioned whether a third alternative is not possible (Q1), whether the alternatives result in Y and Z (Q2 and Q3), and whether these results are (un)desirable (Q4). However, different from the pragmatic dilemma – where the degree of the undesirability is not relevant but where it is relevant whether there are no side-effects that make action X more desirable than undesirable or vice versa – it must be assessed if both alternatives have *equally* (un)desirable effects. The impossible-choice-dilemma cannot be reversed like the other two dilemmatic types of argument: the standpoint 'a reasonable choice is possible' is already reached when one of the alternatives is less (un)desirable than the other. Therefore, the fifth and sixth critical questions are different from the pragmatic dilemma. This also holds for the seventh question, since we do not have a prescriptive but descriptive standpoint in the impossible-choice-dilemma:

Q1. Is there, besides X1 and X2, a third condition X3 possible?
Q2. Does X1 lead to Y?
Q3. Does X2 lead to Z?
Q4. Are Y and Z (un)desirable?
Q5. Does X1 not lead to another result (Y') which makes X1 more preferable to X2?
Q6. Does X2 not lead to another result (Z') which makes X2 more preferable to X1?
Q7. Can the choice be made on other grounds than the effects?

An example of the impossible-choice-dilemma is used by Dutch populist columnist Jan Dijkgraaf in a cynical opinion letter to the Dutch Minister of Health at that time, who was in charge of the vaccination strategy during the beginning of the COVID-19-pandemic:

Example 6. *We don't have to vaccinate, but if we don't, we are denied certain (fundamental) rights. Either we get vaccinated with a vaccine of which the long-term side-effects are completely unknown and are allowed 'approved by the Dutch state' to enter restaurants, attend events and celebrate Christmas with family. Or we are second-class citizens who refuse to be blackmailed and are punished for it by the Dutch State.*[8]

In his letter, Dijkgraaf argues that the Dutch government pretends to give its citizens a choice, while that it is an impossible choice between two evils. Although the government advises to vaccinate, 'critical' citizens like Dijkgraaf are caught between a rock and a hard place:

(1. A reasonable choice regarding vaccination is impossible)
1.1a You can get a vaccine or not
(1.1b Being vaccinated is undesirable)
1.1b.1a Vaccination can result in long-term side-effects

---

[8] Dijkgraaf, J. (November 19th, 2020). 'Briefje van Jan – aan Hugo de Jonge', *The Post Online*, retrieved from (https://tpo.nl/2020/11/11/briefje-van-jan-aan-hugo-de-jonge-17/) on April 22th, 2021. My translation.

(1.1b.1b   Long-term side-effects are undesirable)
(1.1c   Not being vaccinated is undesirable)
1.1c.1a   Not being vaccinated results in being treated as a second-class citizen and punishment by the Dutch State
1.1c.1a.1   You are not allowed to visit restaurants, events, and family
(1.1c.1b   Being treated as a second-class citizen is undesirable)
(1.1d   Long-term side-effects and being treated as a second-class citizen are equally undesirable)
(1.1a-d' If vaccination or no vaccination both lead to equally undesirable effects, no reasonable choice regarding vaccination can be made)

Whereas getting vaccinated or not are the only two possibilities (Q1), all the other critical questions may encounter criticism depending on the type of audience evaluating this argument. A skeptic citizen would evaluate this argument probably as quite reasonable and hence differently from a citizen who trusts medical science and, therefore, does not expect long-term side-effects (Q2) or does not consider the restrictions for health care purposes to be undesirable (Q3). Moreover, some would find it more important to undertake as much as possible in the short-term (Q5), while for others the long-term side-effects outweigh that (Q6). Critical question Q7 is not relevant for this context but would have been relevant only if another party would make the decision (e.g., when the government makes vaccination mandatory).

5. Dilemmatic argumentation as strategic maneuver

In the argumentative reality, no discussant merely pursues the goal of being reasonable. Discussants also want to convince their opponent. However, when the ideal of reasonableness is lost sight of and effectiveness prevails, argumentation will become fallacious. Discussants, therefore, should attempt to keep a balance between being reasonable (the rhetorical goal) and being effective (the dialectical goal). The pragma-dialectical theory of argumentation refers to this balancing between one's rhetorical and dialectical goal as 'strategic maneuvering' (Van Eemeren, 2010). How one strategically maneuvers is reflected in three aspects: one's opportune choice of the topical potential, one's adaptation to the audience demands, and one's presentational devices (Van Eemeren, 2010:94-95).

Strategic maneuvering is a useful analytical instrument to gain a better understanding of the use and function of dilemmatic argumentation. As we have seen in the previous paragraph, almost all examples of dilemmatic argumentation evoke some critical response. Those who look at the critical questions of the different dilemmatic argumentative patterns may ask whether it is possible at all that a dilemmatic argument is fully reasonable: is it possible that no third alternative is imaginable, that there are no other side-effects that weaken the argument (in the pragmatic dilemma), or that both effects are equal (in the impossible-choice-dilemma)? If dilemmatic argumentation has a whiff of unreasonableness, why would one use a dilemmatic argument, and what makes this such a powerful argument as the ancient rhetoricians have claimed it to be? Strategic maneuvering can provide an answer to these questions. In the next paragraph, I explore what the strategic function of dilemmatic arguments can be in order to introduce a possible line of further research on dilemmatic argumentation.

The three aspects of dilemmatic argumentation make it possible to conceive dilemmatic argumentation more broadly than only a logical form or scheme as is done in recent literature. The presentation of two (contradictory) alternative scenarios is, first and foremost, a stylistic choice that contributes to the rhetorical power of this argument – for this reason, some rhetoricians see dilemmatic argumentation not as an argumentative topos but rather as a stylistic device (*Rhetorica Ad Herrenium* IV.52, pseudo-Hermagoras, *De Inventione* IV.6.1). How these arguments are connected to the standpoint – and thus which type of dilemmatic argument is used – is part of the topical repertoire a discussant has. Finally, one also adapts one's topical and stylistic choices to the wishes and expectations of the audience. Below I examine what the strategic function of a dilemmatic maneuver can be. Again I want to emphasize the exploratory nature of this section: based one case study, I show how a dilemmatic argument can be strategically used in an argumentative discourse.

5.1. A case study: the strategic function of dilemmatic argumentation

Responding to the trail of atrocities that was left by Russian soldiers in Ukraine, Fiona Hill, a former senior director for European and Russian Affairs on the United States National Security Council, argued:

Example 7. *'This clearly isn't a special military operation, is it? [...] If there was genuinely a special military operation to "liberate" a fraternal country from you know what Putin was describing as "Nazis", you would not expect this kind of conduct. So either this is a complete breakdown of command and control, or it is actually being sanctioned in some way to teach the Ukrainians a lesson. Either way, this is actually pretty disastrous* *and obviously requires some serious response in the international community'.*⁹

Hill, who held no political office at the time, gave her view on the situation in Ukraine after images came out of the Bucha massacre in March 2022. She considers the cruel conduct of the Russian army a reason for the international community to respond since it no longer matches with what you would expect of a "special military operation". With an assertive dilemmatic argument (in bold), she makes clear that Ukraine awaits a catastrophe no matter how (the interpretation of the implicit elements is mine):

1. The Russian invasion of Ukraine requires some sort of response of the international community
1.1 There clearly is no "special military operation" going on
1.1.1 The conduct of the Russian army is (disproportionally) disastrous
1.1.1.1a There is either a complete breakdown of command and control, or it [i.e. the Bucha massacre] is actually being sanctioned in some way to teach the Ukrainians a lesson
(1.1.1.1b A complete breakdown of command and control leads to an extremely disastrous conduct)
(1.1.1.1b.1a If there a complete breakdown of command and control, the Russian army is uncontrollably on a rampage)
(1.1.1.1b.1b An uncontrollable army on rampage is a sign of an extremely disastrous conduct)
(1.1.1.1c Atrocities being sanctioned to teach Ukrainians a lesson leads to an extremely disastrous conduct)
(1.1.1.1c.1a Atrocities being sanctioned to teach Ukrainians a lesson means that there is a deliberate plan to destroy a country)
(1.1.1.1c.1b Deliberately planning to destroy a country is a sign of an extremely disastrous conduct)

With the assertive dilemma, Hill argues that the conduct of the Russian army is, in any case, disastrous (1.1.1) and thus not what one would expect from a "special military operation" (1.1). When we assess the reasonableness of this dilemma, critical questions Q2-4 probably can be answered positively: a complete breakdown of control (Q2) and deliberate sanctions lead to a for Ukraine disastrous conduct of the Russian army (Q3), and the opposite standpoint ('there is no disastrous conduct') cannot be defended more plausibly with the same scenarios (Q4). One can however criticize Hill's reduction of the situation to either a complete loss of control or a full command of the atrocities (Q1): there was at that time a possibility

---

9 Face the Nation (2022, April 3rd). Fiona Hill on alleged Russian atrocities in Ukraine and Putin's future [online video]. Retrieved from https://www.youtube.com/watch?v=u-BPEUjkkfI&t=36s on April 5th, 2022.

that the Kremlin was (largely) in command of the military but that the atrocities were committed by rogue individuals.

Why would Hill use a precarious disjunction to support a quite obvious statement: that the conduct of the Russian army is disastrous? The two alternatives in the disjunction presuppose either a "special military operation" which has become uncontrollable, or no "special military operation" but a war mission to teach Ukrainians a lesson. Hill makes clear that she does not believe that the Russian invasion is a "special military operation". That means that she could also have used a monolemmatic argument instead, for example: 'there is no "special military operation", for the disastrous conduct of the Russian army suggests that the Kremlin is whipping out the Ukrainian nation'. What is then the strategic function of presenting her argument as a dilemmatic argument instead?

The use of a dilemmatic argument here is strategic for two reasons. First of all, for a long time it was – and perhaps still is – not clear what the exact intentions of the Kremlin were: was it a mission to "liberate" the pro-Russian separatists in the Donbas region, or is it a (slowly progressing) war to annex Ukraine? With her dilemmatic argument Hill sidesteps this speculation: regardless of the Kremlin's true intentions, Hill considers the situation disastrous enough to respond internationally. But more importantly, Hill's use of a dilemmatic argument shows how she strategically maneuvers in the polarized political arena of 'the international community' that she calls to action, which is the second strategic function of this dilemma. Some people and some governments refused to speak of a war in Ukraine. If Hill would have brought forward her argument monolemmatically (i.e. only current 1.1.1c), some possibly would have criticized her argumentation because it disregards the possibility that just some Russian soldiers committed the atrocities. With the dilemmatic presentation of her argument, she anticipates this possible refutation. Also the phrasing of the two scenarios can be seen as strategic; Hill adapts her argumentation to both types of audience: by calling it a 'complete' loss of control she responds to the feelings of those who fully condemn the Russian conduct, and with 'sanctions to teach the Ukrainians a lesson' she avoids calling it a total and unjust war for those who do not fully condemn the invasion. By using the disjunction – even if it is not completely exclusive and thus not fully reasonable – it seems that rejecting one scenario leads inevitably to the other scenario, which makes the conclusion was compelling. Dilemmatic argumentation, therefore, can be seen as a topical and stylistic choice to present two alternatives that respond to and anticipate the presumptions of different audiences in order to 'immunize' a particular conjecture or controversy.

6. Conclusion

In this paper, I have brought dilemmatic argumentation from classical rhetoric and formal logic to the perspective of the pragma-dialectical theory of argumentation. In classical rhetoric, two main features of dilemmatic argumentation were defined: it presents two alternatives of which one seemingly must be the case, and which both lead to the same conclusion. The logical main form shows that a dilemmatic argument consists of, at least, a disjunctive premise that combines two syllogisms. Based on pragma-dialectical insights, I have introduced three types of dilemmatic arguments that differ from each other in the combination of the type of standpoint and argument schemes used: the pragmatic dilemmatic argument, the assertive dilemmatic argument, and the impossible-choice-dilemma. The argumentative pattern that this paper has developed for each type makes it possible to differentiate the types, analyze examples, and evaluate their reasonableness with the associated critical questions. With these, an ancient judge could have argued that Corax' and Tisias' assertive dilemmatic arguments fall short concerning Q2 and Q3: one cannot arbitrarily and advantageously enforce either the law or a previously made agreement.

In his historical overview of dilemmatic argumentation in rhetorical treatises, Nuchelmans (1991:5) expressed the wish 'may the result [of this monograph] incite others to corrections and supplementations'. With this paper, I attempted to meet his wish by introducing a threefold typology for the analysis and evaluation of dilemmatic argumentation, and by a hopefully inspiring first exploration of its strategic function. Dilemmatic argumentation is an intriguing strategic maneuver: it is a topical and stylistic device to adapt one's argumentation to different audiences. It allows a discussant to put forward one's argument while anticipating a possible refutation or critique of an opponent. With these insights, I aim to 'incite others to corrections and supplementations' in my turn, since the use of dilemmatic argumentation deserves more attention than it has received so far.

References

Boone, D.N. (1999). The Cogent Reasoning Model of Informal Fallacies. *Informal Logic 19(1)*, 1-39.
Cole, T. (1991). *The Origins of Rhetoric in Ancient Greece*. Baltimore: Johns Hopkins University Press.
Cook, R. T. (2009). *A Dictionary of Philosophical Logic*. Edinburgh: Edinburgh University Press.
Copi, I.M. & Cohen, C. (1990). *Introduction to Logic (8th edition)*. New York & London: Macmillan Publishing Company.
Craig, C.P. (1993). Form as Argument in Cicero's Speeches: A Study of Dilemma. Atlanta: Scholars Press.
Eemeren, F.H. van (2017). Prototypical Argumentative Patterns: Exploring the Relationship Between Argumentative Discourse and Institutional Context. Amsterdam & Philadelphia: John Benjamins Publishing.
Eemeren, F. H. van. (2010). Strategic Maneuvering in Argumentative Discourse Extending the Pragma-Dialectical Theory of Argumentation. Amsterdam & Philadelphia: John Benjamins Publishing.
Eemeren, F. H. van & Grootendorst, R. (1984). Speech Acts in Argumentative Discussions: A Theoretical Model for the Analysis of Discussions Directed towards Solving Conflicts of Opinion. Berlijn & Dordrecht: Foris Publications/De Gruyter.
Eemeren, F.H. van, Grootendorst, R. & Kruiger, T. (1987). Handbook of Argumentation Theory: A Critical Survey of Classical Backgrounds and Modern Studies. Dordrecht: De Gruyter.
Eemeren, F. H. van & Grootendorst, R. (1992). *Argumentation, Communication, and Fallacies: A Pragma-Dialectical Perspective*. New York: Lawrence Erlbaum Associates.
Eemeren, F.H. van & Grootendorst, R. (2004). *A Systematic Theory of Argumentation: The Pragma-Dialectical Approach*. New York: Cambridge University Press.
Eemeren, F.H. van, Garssen, B., Krabbe, E.C.W., Snoeck Henkemans, A.F., Verheij, B. & Wagemans, J.H.M. (2014). *Handbook of Argumentation Theory*. Dordrecht, Heidelberg, Londen & New York: Springer.
Eemeren, F.H. van, Garssen, B. (2020). Argument Schemes: Extending the Pragma-Dialectical Approach. In F.H van Eemeren & B. Garssen, (Eds.), *From Argument Schemes to Argumentatieve Relations in the Wild* (pp.11-23). Chaam: Springer.
Feteris, E.T. (2002). Filosofische Achtergronden Van Een Pragma-Dialectisch Instrumentarium Voor De Analyse En Beoordeling Van Pragmatische Argumentatie. *Tijdschrift voor Taalbeheersing 24(1)*, 14-31.
Gerlofs, J.M. (2009). The Use of Conditionals in Argumentation: A Proposal for the Analysis and Evaluation of Argumentatively used conditionals. Dissertation, University of Amsterdam.
Hinks, D.A.G. (1940). Tisias and Corax and the Invention of Rhetoric. *Classical Quarterly 34(1-2)*, 61-69.
Jevons, W.S. (1883). *The Elements of Logic: A Text-book*. New York & Chicago: Sheldon and Company.

Keynes, J.N. (1906 [1884]). *Studies and Exercises in Formal Logic (fourth edition)*. London: Macmillan Company.
Kienpointner, M. (1992). Alltagslogik. Struktur und Funktion vom Argumentationsmustern. Stuttgart: Frommann-Holzboog.
Lewiński, M. (2014). Argumentative Polylogues: Beyond Dialectical Understanding of Fallacies. *Studies in Logic, Grammar and Rhetoric 36*, 193-218.
Montefusco, L.C. (2010). Rhetorical Use of Dilemmatic Arguments. *Rhetorica 28(4)*, 363-383.
Nuchelmans, G. (1991). *Dilemmatic Arguments: Towards a History of their Logic and Rhetoric*. (Verhandelingen der Koninklijke Nederlandse Akademie van Wetenschappen, afdeling Letterkunde, Nieuwe Reeks, deel 145). Amsterdam, New York, Oxford & Tokyo.
Parry, W.T. & Hacker, E.A. (1991). *Aristotelian Logic*. New York: State University Press of New York.
Perelman, C., & Olbrechts-Tyteca, L. (1969). *The New Rhetoric: A Treatise on Argumentation*. (transls. Wilkinson, J. & Weaver, P.). London: University of Notre Dame Press.
Rabe, H. (1931) (Ed.). *Prolegomenon Sylloge (Rhetores Graeci, XIV)*. Leipzig: Teubner.
Russell, D.A. (2002). *Quintilian: The Orator's Education, Volume I-V*. Cambridge, MA & London: Harvard University Press.
Schellens, P. J. (1985). Redelijke Argumenten. Een Onderzoek Naar Normen Voor Kritische Lezers. Dordrecht: Foris Publications.
Schiappa, E. (2003). *Protagoras and Logos: A Study in Greek Philosophy and Rhetoric*. Columbia, S.C.: University of South Carolina Press.
Smith, H.B. (1938). *A First Book in Logic*. New York: Harper & Brothers.
Tomić, T. (2013). False Dilemma: A Systematic Exposition. *Argumentation 27*, 347-368.
Tomić, T. (2021). The Distinction Between False Dilemma and False Disjunctive Syllogism. *Informal Logic 41(4)*, 607-639.
Wagemans, J.H.W. (2014). Een systematische catalogus van argumenten. *Tijdschrift voor Taalbeheersing 36(1)*, 11-30.
Waterfield, R. (2018). *Aristotle: The Art of Rhetoric*. Oxford: Oxford University Press.
Whately, R. (1853 [1826]). Elements of Logic Comprising The Substance of The Article in The Encyclopaedia Metropolitana. New York: Harper & Brothers.

The Concept of Argumentation in Chinese Writing Education and the Modern Transformation of Chinese Thinking and Reasoning

HAILONG WANG
Wuhan University
spwhl99999@163.com

Abstract

From the beginning of the 20th century to the present, Chinese thinking and reasoning, with thousands of years of historical tradition, has undergone a modern transformation in face of westernization. The concept of argumentation in Chinese writing education can be taken as a research object to explain the transformation. Through the dissemination and reception of the concept of argumentation, Chinese people had undergone significant changes in their consciousness of independent thinking and abilities of logical reasoning from 1900s to 1940s. However, the concept of argumentation was under control by strong political ideology from the 1950s to 1970s and the process of the transformation of Chinese thinking and reasoning was interrupted during that time. The discussion on the theory of three elements of argumentation from 1980s to 2020s promotes Chinese people's understanding of the concept of argumentation and is conducive to the modern transformation of Chinese thinking and reasoning. It is hoped that this historical type of research can introduce the dissemination and reception of western argumentation concepts in China and present the modern transformation of Chinese thinking and reasoning.

KEYWORDS: argumentation, rhetoric, Chinese writing education, thinking and reasoning

From the beginning of the 20th century to the present, Chinese thinking and reasoning, with thousands of years of historical tradition, has undergone a modern transformation in face of westernization. The transformation which has gone through more than 100 years can be divided into the following three stages according to the two historical nodes in China: 1900s-1940s, 1950s-1970s, and 1980s-2020s. The concept of argumentation in Chinese writing education can be taken as a research object in order to study the modern transformation of Chinese thinking and reasoning. Recently, Christian Plantin (2018) provided a very detailed explanation for the concept in Dictionary of Argumentation: An Introduction to Argumentation Studies. It is also mainly understood as a synonym of "argument" in China. But if we trace back to China in the early 20th century, we will find that the western concept argumentation was firstly translated in Chinese writing textbooks in face of westernization, originally understood by many Chinese scholars as a mode of discourse and frequently juxtaposed with description, narration and explanation. This meaning of the concept of argumentation is mentioned in Encyclopedia of Rhetoric and Composition edited by Theresa Enos (1996), which seems to be ignored by Chinese argumentation researchers.

1. The concept with enlightening function and the transformation from the 1900s to 1940s

China has a tradition of argumentative writing for thousands of years, and there are also many concepts of argumentative genre. The concept of argumentation in Chinese writing education since the 20th century is fundamentally different from the traditional Chinese concept of argumentative genre. The argumentative genre in ancient China mainly has two characteristics. One is following the words of saints, which means that views expressed in the traditional argumentative genre should consistent with the words of the saints, especially Confucian sages and men of virtue. Once the views different from those of predecessors are put forward, they are often denounced as heresy. The other is that there is no definite method of argument in writing, that is, there is no fixed rule in the writing education of argumentative genre. The ancient Chinese emphasized the natural expression of cultivating literary spirit, with a certain degree of mysticism. These two characteristics which were incompatible with modern society were concentrated in the writing of the eight-legged essay in the Ming and Qing Dynasties. They would obviously change as Chinese society entered the modern era.

At the beginning of the 20th century, the traditional Chinese concept of argumentation was in a very embarrassing situation because the concept of the eight legged essay was declining with the abolition of the imperial examination system. In addition, due to the rise of the vernacular movement, the traditional argumentative writing lost its language carrier, and the traditional concept of argumentation relying on the classical Chinese language cannot adapt to vernacular argumentative writing. At that time, some famous intellectuals in Chinese society, including Wang Tao and Liang Qichao, used modern Japanese vocabulary, grammar and western knowledge of logical reasoning and rhetorical methods to carried out new argumentative writing in newspapers and periodicals, which was called a new writing practice. What concept to describe the new argumentative writing practice became a problem that the Chinese intellectual community needed to solve. The establishment of new schools in the late Qing Dynasty provided convenient conditions for the introduction of new knowledge into China, and the western concept of argumentation was therefore introduced into China in the 1900s.

The western concept of argumentation was indirectly introduced into China through Japanese rhetoric. After the China-Japan War of 1894-1895, China began to learn from Japan. Many Chinese young people chose to study abroad in Japan at that time and brought back a lot of modern knowledge. It is in this background that Japanese rhetoric knowledge was introduced into China. In 1905, two books published in China, namely, Tang Zhenchang's *A Textbook of Rhetoric* and Long Bochun's *A Textbook of Words: The Rhetorical Volume*, were currently considered to be the initial works to introduce Japanese rhetoric (Huo, 2019). The copyright consciousness of Chinese people in the 1900s is not clear. The contents of these two books were almost the same as some Japanese rhetorical works actually. Both of them mentioned the concept of argumentation in the context of writing education, which made this concept not translated into *Lunzheng* (argument) in Chinese at first, but into *Yilunwen* (argumentative writing). By the 1920s, Chen Wangdao, Gao Yuhan and others who had studied abroad in Japan also compiled some writing textbooks, which made the concept of argumentation in the writing education widely spread in China (You, 2005).

In fact, Japanese rhetoric was just a copy of western rhetoric. The concept of argumentation in Japanese rhetoric originated from the modes of discourse theory which dominated western rhetoric in the 19th century (Tomasi, 2004). In the *Philosophy of Rhetoric* published in 1776, George Campbell (1776), a Scottish rhetorician, summed up four purposes of discourse according to the four faculties of human, namely, enlightening the understanding, pleasing the imagination, moving the passions and influencing the will. Alexander Bain (1867), another Scottish rhetorician, divides compositions into five categories in his book *English Composition and Rhetoric: A Manual*: narration, description, exposition, oratory, and poetry. As Scottish rhetoric influenced the United States, American rhetoricians perfected the division of the modes of discourse from Samuel Newman's *A Practical System of Rhetoric* (1827) to Adams Hill's *The Principles of Rhetoric* (1895) and John Genung's *Outline of Rhetoric* (1896). Hill and Genung both divided the discourse into narration, description, exposition and argumentation (Connors, 1981). The modes of discourse theory was based on the psychology of faculty and the philosophy of rhetoric in the 18th century and developed by American rhetoricians in the 19th century (D'Angelo, 1984). With the theory was introduced into Japan at the end of the 19th century, it was also introduced into Chinese writing education by translation at the beginning of the 20th century. The concept of argumentation in the theory of modes/forms of discourse then as a carrier fundamentally changed the Chinese traditional ways of thinking and reasoning, popularized the consciousness of self-expression and the method of logical reasoning in the whole society.

It can be seen that the western concept of argumentation in the first half of the 20th century meant a mode of discourse or a genre of composition in Chinese writing education. But the western concept of argumentation as a western term was essentially different from the concept of Chinese traditional argumentative writing. The first difference was in nature. The concept of Chinese traditional argumentative writing was mysterious and appealed to inspiration, while the concept of argumentation was scientific and had a theoretical basis. The second difference was in purpose. The concept of Chinese traditional argumentative writing was to describe as a form of writing practice, while the concept of argumentation was to teach as an embodiment of pedagogy. The third difference was in formation. The concept of Chinese traditional argumentative writing was derived from the characteristics of composition by induction, while the concept of argumentation was derived from the faculty in psychology by deduction. These three differences won't be directly reflected in the text writing, but they reflected the modern transformation of Chinese thinking and reasoning. Through the dissemination and reception of this western concept of argumentation, Chinese people had undergone significant changes in their consciousness of independent thinking and abilities of logical reasoning.

The consciousness of independent thinking was very insufficiency in Chinese traditional argumentative writing. Most argumentative writing in ancient China required following the words of saints. The western concept of argumentation introduced in the early 20th century promoted the independent thinking of Chinese people in two aspects. One is to spread the sense of independence by introducing the purpose and value of argumentation. With the development of New Culture Movement in the 1920s, almost every writing textbook will take expressing one's own thoughts as the core to introduce the concept of argumentation. For example, in 1922, Chen Wangdao (1922) said that the intention of argumentation lies in making your own words believable. In 1924, Sun Lianggong (1924) said that the function of argumentation was to express your own thoughts and change the views of others. Ye Shengtao (1924) said that the argumentation was a text that expounded one's own opinions and criticized other's opinions. These three scholars were all famous writing educators in the first half of the 20th century and their writing textbooks had a wide influence in China. Even a hundred years later, Chinese teachers still refer to their works to teach students to write. Their expressions of the western concept of argumentation undoubtedly promoted the modern transformation of Chinese thinking.

The abilities of logical reasoning were also very insufficiency in Chinese traditional argumentative writing. Although Mozi, a thinker of the pre-Qin period, once wrote down some knowledge of logic, he lost his voice in Chinese tradition with the decline of Mohism. Compared with deduction and induction, the method of analogy was more widely used in argumentative writing in ancient China. It was not until the beginning of the twentieth century that the western concept of argumentation in Chinese writing education emphasized the important role of deductive and inductive logical knowledge. By using the knowledge of deduction and induction, people had the ability to reasonably express their views on the basis of independent thinking. The deduction and induction also replaced analogy and quotation as the main methods of argument in argumentative writing. Such a modern transformation of Chinese reasoning should be attributed to the introduction of the western concept of argumentation into China. If the independent thinking makes modern Chinese people want to argue, then logical reasoning makes modern Chinese people good at arguing.

Immanuel Kant (1996) once said that enlightenment was man's leaving his self-caused immaturity and immaturity was the incapacity to use one's intelligence without the guidance of others. The dissemination and acceptance of the western concept of argumentation in China led to the modern transformation of Chinese thinking and reasoning, which is the embodiment of the enlightenment function of the argumentation. The consciousness of independent thinking and the abilities of logical reasoning not only made Chinese people get rid of the traditional way of argument but also made the concept of science and rationality deeply rooted in the hearts of Chinese people. However, the transformation only lasted from the 1900s to the 1940s. With the change of political power and social nature, the modern transformation of Chinese thinking and reasoning entered the next stage. And the concept of argumentation with enlightenment function was under control by strong political ideology from the 1950s to 1970s.

2. The concept under political ideology and the transformation from the 1950s to 1970s

In the late 1940s, the People's Republic of China was founded. The relationship between China and the Soviet Union was very friendly under the international background at that time, which to a large extent led to China's rejection of western culture and its full acceptance of the influence of the Soviet Union. On December 30th, 1949, Qian Junrui, the Chinese Minister of Education, said in the summary report at the first National Conference on Education that Chinese education should be reformed and advocated drawing on the experience of the Soviet Union (Gu, 2001). Since then, China learned from the Soviet Union in an all-round way, and the transformation process in the first half of the 20th century undergone a complete change, which was directly reflected in the change of the concept of argumentation in Chinese writing education. At the normative level of the concept, the argumentation was gradually considered to be composed of "three elements" including the "thesis", the "proof", and the "method of argument". The so-called "three elements of argumentation" is still widely applied in China. At the descriptive level of the concept, the argumentation was transformed by the powerful political ideology of Marxism and Leninism, and described as a tool of political struggle and proletarian revolution.

Under the theory of "the three elements", only things including the correct "thesis", the abundant "proof", and the appropriate "method of argument" can be regarded as a complete good argumentation. This kind of argumentation theory occupied a dominant position in Chinese writing education in the second half of the 20th century. It also appeared in the Soviet Union's writing education. For example, in *The System of Composition Teaching in Russian Class (Grades 4-8)*, Ravensky, a famous writing teacher of the Soviet Union, stated in Chapter 8 for introducing the ability of argument that a good argument, generally speaking, always includes three elements: the "thesis", the "proof", and the "method of argument" (Wu, 1982). The logic of the Soviet Union may be the source of it. The Chinese version of the Soviet Union's logical work The Logical Theory of Argument and Refutation first appeared the formulation of "the three elements of argumentation". There are at least two Chinese versions of this logic work written by Soviet logician Asmus. In the chapter "The Structure of Argument" in this book, Asmus said that in all arguments (no matter what is to be proved), there are always: 1. the thesis, 2. the basis of argument (proof), 3. the methods of argument. Before his statement, the theory of three elements of argumentation did not appear in China (Asmus, 1955).

The expression of the theory of three elements of argumentation gradually appeared in various writing textbooks around 1960 in China including *How to write an argumentative article* written by Zhang Fushen (1957) in 1957, *Lectures on Writing* compiled by the Department of Chinese literature of Hunan Normal University in 1958, *Basic Knowledge of Writing* edited by Hu Wenshu (1961) in 1961, and Basic Knowledge of Argumentative Writing edited by Ou Lequn (1963) in 1963. These writing textbooks which popularized the theory of three elements in China all mentioned that the element of "thesis" was used to refer to the object of argument, the element of "proof" was used to refer to the evidence of argument, and the element of "method of argument" was used to refer to the process of argument. At that time, people engaged in writing education believed that these three elements were sufficient to define the concept of argumentation from the normative level. By the 1970s, almost all textbooks in China introduced the theory of three elements of argumentation when teaching the knowledge of writing. In short, the theory of three elements, as the embodiment of the normative concept of argumentation, originated in the 1950s under the influence of the Soviet Union's logical theory and composition teaching, and was widely accepted by the public in the 1970s after its dissemination in the 1960s.

Perhaps from today's point of view, the theory of three elements of argumentation had some defects, but it made at least two aspects of contribution from the 1950s to the 1970s. One is that compared with the argumentative writing education in the first half of the 20th century, the theory of three elements had deepened people's understanding of argumentation and made people systematically acquire the knowledge of argument. Another one is that with the popularization of the theory of three elements in China, the basic knowledge of logic, including deduction and induction, was widely spread and accepted in China, because the element of "method of argument" in the theory was actually the basic knowledge of logical reasoning. In other words, the theory of three elements of argumentation should have a certain degree of promoting effect on the transformation of Chinese thinking and reasoning. However, the influence of political ideology on the concept of argumentation in Chinese writing education from the 1950s to 1970s unfortunately weakened the practical effect of the theory of three elements. What's worse, the theory of three elements was no longer a tool for thinking and reasoning, but a tool for the propaganda of the political ideology in the 1970s.

In fact, political ideology had penetrated into the concept of argumentation in the 1950s under the influence of the Soviet Union. At that time, most writing textbooks described the concept of argumentation as a cultural tool of the proletarian revolution. People believed that the argumentative writing should conform to Marxism and Leninism, make good comments, and create good public opinion conditions for the proletariat. The descriptive concept of argumentation was greatly influenced by the political ideology. At the same time, the theory of three elements, as a normative concept of argumentation, was also influenced by political ideology. The element of a correct "thesis" should conform to Mao Zedong's thought, the element of a strong "proof" should be Mao Zedong's expression, and the element of "methods of argument" should be used to support Mao Zedong's thought. During the Chinese Cultural Revolution (1966-1976), the concept of argumentation was completely under the control of political ideology. The concept of argumentation under the control of political ideology obviously weakened the consciousness of independent thinking and the ability of logical reasoning.

In general, the transformation of Chinese thinking and reasoning in the first half of the 20th century was closely related to the concept of argumentation with enlightening function. However, with the close relationship between China and the Soviet Union in the 1950s, the argumentation gradually became a concept under political ideology because of the impact of the Soviet Union. The concept of argumentation temporarily lost the function of enlightening and the process of the transformation of Chinese thinking and reasoning from the 1900s to the 1940s was interrupted during the 1950s to 1970s.

3. The concept for rational expression and the transformation from the 1980s to 2020s

China's reform and opening-up advocated emancipating the mind in the late 1970s, which gave the concept of argumentation an opportunity to get rid of the control of political ideology since the 1980s, and the transformation of Chinese thinking and reasoning had the opportunity to return to the path of rationality. The theory of three elements of argumentation was questioned and criticized a lot after China's reform and opening up, because it was the recognition of the concept of argumentation under the past political ideology. In 1980, Lu Jiansan (1980), a very famous Chinese teacher, published a paper called "A Brief Introduction to Argumentative Writing". On the one hand, he proposed that the concept of argumentation, which was generally composed of the "thesis", the "proof" and the "methods of argument", should also add the element of "topic". On the other hand, he thought that some argumentative writing may not be fully applicable to the theory of three elements. Lu Jiansan's discussion was of guiding significance, because he opened two major directions for the academic circles to question the theory of three elements of argumentation for more than 40 years: one was why the concept of argumentation consisted only of the three elements, and the other was why the three elements were necessary for the concept of argumentation.

After Lu Jiansan, many scholars proposed that the concept of argumentation was not just consisted only of the "thesis", the "proof" and the "method of argument". They pointed out the existence of the element of "topic" by questioning the element of "thesis", and point out the existence of the element of "conclusion" by questioning the element of "methods of argument". When the theory of three elements of argumentation first appeared in China, scholars did not clearly realize the distinction between the element of "topic" and the element of "thesis". But in the 1980s, they gradually understood that a "thesis" should be a judgment which was the viewpoint put forward by the author, and a "topic" was a main problem discussed in an argumentative writing. Based on this new cognition, some scholars put forward the theory of four elements of argumentation including the "topic", the "thesis", the "proof" and the "method of argument". As the element of "thesis" was questioned, the element of "methods of argument" was also challenged. Some scholars thought that the "method of argument" was a tool for argumentative writing or a means of expressing one's view rather than an element of the concept of argumentation. They proposed to take the element of "conclusion" as a new element of argumentation, even to replace the "method of argument" with the element of "conclusion". Even if new elements such as the "topic" and the "conclusion" had been put forward, some scholars were used to adhering to the original theory of three elements of argumentation. Although Xie Zhili (1999), a Chinese writing theorist, put forward the theory of five elements of argumentation at two levels in 1999, that is, the "topic", the "thesis", the "proof" and the "conclusion" as the elements of the first level and the "method of argument" as the element of the second level, the original theory of three elements of argumentation was still widely accepted in Chinese writing education at the end of the 20th century.

Compared with discussing what elements constituted the concept of argumentation, more scholars began to attack the disadvantages of the theory of three elements. The first disadvantage is its rigid in form. Some scholars criticized that argumentative composition teaching was too modular, which made the concept of argumentation lack of vitality. They thought that the cognition of the concept of argumentation should be diversified. The second disadvantage is that it restricts the writer's thoughts. Some scholars believed that although there were three elements in the concept of argumentation, it was easy to cause the problems of narrow vision and limited thinking if it was applied to all argumentative writing. They believed that the theory of three elements would have a negative impact on the richness of argumentative writing practice. The third disadvantage is that its effectiveness is very limited. Many scholars found that There were lots of good argumentative composition that did not include the elements of "thesis", the "proof" or the "method of argument". They claimed that students should not analyze all the argumentative writing with the theory of three elements. The above three criticisms all pointed to a conclusion that the theory of three elements was not necessary for the concept of argumentation.

In the 21st century, more and more scholars proposed to completely abandon the theory of three elements of argumentation since it was unfavorable and unnecessary. In fact, this view was first put forward by Yu Shaoqiu (1991) in the 1990s, but it was not supported by scholars at that time. Later, although the viewpoint of abolishing was put forward occasionally, it did not have much impact, and the theory of three elements was still very mainstream in the first decade of the 21st century. In the 2010s, the voice of abolishing the theory of three elements of argumentation gradually became widespread because Pan Xinhe, a famous professor of Fujian Normal University, wrote three papers in succession to oppose the three elements theory of argumentation around 2012. The papers were "Criticisms on the Theory of Three Elements of Argumentation" (2011), "On the Disadvantages of the Theory of Three Elements of Argumentation" (2012), and "A Reconstruction of the Elements of Argumentation" (2012). The attack of these three papers on the theory of three elements did not exceed the doubts of the academic circles at the end of the 20th century, but they did have a strong influence on Chinese writing education. People began to notice the limitations of the theory and seriously consider the concept of argumentation that they had been aware of.

Although the voice against the theory of three elements was growing, many scholars still believed that the theory had its irreplaceable advantages. The first advantage is that the theory of three elements is convenient for teaching and it can help students quickly understand how to write argumentative essays. The second advantage is that the theory makes it easy for people to grasp the concept of argumentation and it is a useful tool for people to analyze an argumentative essay. The third advantage is that the theory of three elements is the most accepted knowledge in the field of Chinese writing education at present, and it is difficult for other theories to replace it immediately. Scholars who support the theory of three elements generally believe that even if some opposing views are reasonable, the role of the "thesis", the "proof" and the "method of argument" should not be ignored from the perspective of reality. They claim that the theory of three elements on argumentation has sustained vitality and is still not outdated.

At present, the discussion on the theory of three elements is still a hot topic in Chinese writing education, which promotes Chinese people's understanding of the concept of argumentation and is conducive to the modern transformation of Chinese thinking and reasoning. The discussion not only made the concept of argumentation more widely disseminated and accepted, but also restored the weakened consciousness of independent thinking and the abilities of logical reasoning in Chinese people. The modern transformation of Chinese thinking and reasoning returns to the right direction of development. But we must realize the limitation of the theory of three elements because it is essentially based on the knowledge of western classical logic. Although the knowledge of deduction and induction contained in the theory of three elements are important, the development of Chinese thinking and reasoning cannot be satisfied with them. Due to the limitation of the age during the 1950s to 1970s, some western theories on argumentation, such as Toulmin's model, were not introduced into China in time. Now, the academic circles should learn from the argumentation theories put forward since the second half of the 20th century for the modern transformation of Chinese thinking and reasoning.

In fact, many argumentation theories, including pragma-dialectical theory, have been spread in China. But the spreading still stays at the theoretical level and is not combined with the practice of Chinese thinking and reasoning. A hundred years ago, the western knowledge of classical logic was applied to Chinese writing education with the spread of the concept of argumentation in China. However, the pragma-dialectical theory that has been introduced into China for some time has not been used in the practice of Chinese writing education. In this sense, the transformation from 1980s to 2020s was not enough for Chinese thinking and reasoning. It is hoped that this historical type of research can enlighten the dissemination of the western concept of argumentation and the transformation of Chinese thinking and reasoning. Chinese academic circles on argumentation need to combine theories with Chinese practice in the future.

References

Asmus. (1955). *Logical theory of argument and refutation*. Beijing: SDX Joint Publishing Company.
Bain, A. (1867). *English Composition and Rhetoric: A Manual*. New York: D. Appleton and Company.
Campbell, G. (1776). *The Philosophy of Rhetoric*. London: Printed for W. Strahan; and T. Cadell, in the Strand; and W. Creech at Edinburgh.
Chen, W. (1922). Zuowen Fa Jiangyi [Lectures on Written Composition]. Shanghai: Minzhi Press.
Connors, R. (1981). The Rise and Fall of the Modes of Discourse. *College Composition and Communication, 32(4)*, 444-455.
D'Angelo, F. (1984). Nineteenth-Century Forms/Modes of Discourse: A Critical Inquiry. *College Composition and Communication, 35(1)*, 31-42.
Enos, T. (1996). Encyclopedia of Rhetoric and Composition: Communication from Ancient Times to the Information Age. New York and London: Routledge.
Genung, J F. (1896). *Outline of Rhetoric*. Boston: Ginn & Company.
Gu, H. (2001). *Centennial Events of Modern Chinese Education in China*. Shanghai: Shanghai Education Publishing House.
Hill, A S. (1895). *The Principles of Rhetoric*. New York: American Book Company.
Hu, W. (1961). *Basic Knowledge of Writing*. Shanghai: Shanghai Education Publishing House.
Huo, S. (2019). *Selected Works of Chinese Modern Rhetoric*. Shanghai: Shanghai Education Publishing House.
Kant, I. (1996). *Practical Philosophy*. Cambridge: Cambridge University Press.
Lu, J. (1980). On Argumentation. *Chinese Frontier, (9)*, 5-7.
Newman, S. (1827). A Practical System of Rhetoric or The Principles & Rules of Style. Portland: Shirley & Hyde.
Ou, L. (1963). *Basic Knowledge of Argumentative Writing*. Lanzhou: Gansu People's Publishing House.
Plantin, C. (2018). Dictionary of Argumentation: An Introduction to Argumentation Studies. London: College Publications.
Pan, X. (2011). Criticisms on the Theory of Three Elements of Argumentation, *Modern Chinese Language, (7)*, 10-13.

Pan, X. (2012). On the Disadvantages of the Theory of Three Elements of Argumentation, *Chinese Planning, (1)*, 17-22.

Pan, X. (2012). A Reconstruction of the Elements of Argumentation, *Chinese Planning, (6)*, 19-23.

Sun, L. (1924). Lunshuowen Zuofa Jiangyi [Lectures on the Argumentative Writing]. Shanghai: The Commercial Press.

Tomasi, M. (2004). Rhetoric in Modern Japan: Western Influences on the Development of Narrative and Oratorical Style. Honolulu: University of Hawaii Press.

Wu, L. (1982). *Composition Teaching in the Soviet Union*. Beijing: Educational Science Publishing House.

Xie Z. (1999). A New Understanding on the Elements of Argumentation. *Journal of Shanxi Normal University, (3)*, 92-95.

Ye, S. (1924). *Zuowen Lun [On Written Composition]*. Shanghai: The Commercial Press.

You, X. (2005). Conflation of Rhetorical Traditions: The Formation of Modern Chinese Writing Instruction. *Rhetoric Review, 24(2)*, 150-169.

Yu, S. (1991). Against the Theory of Three Elements of Argumentation, *Chinese learning, (9)*, 10-13.

Zhang, F. (1958). *How to Write an Argumentative Article*. Taiyuan: Shanxi People's Publishing House.

COMMENTARY ON: THE CONCEPT OF ARGUMENTATION IN CHINESE WRITING EDUCATION AND THE MODERN TRANSFORMATION OF CHINESE THINKING AND REASONING BY HAILONG WANG

FRANK ZENKER
College of Philosophy, Nankai University, Tianjin, China

Using 'argumentation' as a synonym for 'thinking and reasoning' on account of the former activity being an outward expression of the latter, Hailong Wang traces the 20th-century transformation in China of the concept of argumentation as a "mode of discourse [sic] frequently juxtaposed with description, narration and explanation." This transformation spans three periods: 1900-1940s (roughly from the end of the Qing Dynasty to the founding of the People's Republic of China), 1950-1970s (from erecting a planned economy modeled on Soviet Communism to the end of the Cultural Revolution), and 1980-2020s (the "Reform and Opening Up" period). Each period Wang associates with a single keyword: enlightenment, political ideology, and rational expression, respectively.

Wang describes how the historical process of Westernization first led Chinese scholars to understand argumentation along Kantian lines as a tool to promote independent thinking. Under the influence of Japanese rhetorical scholarship—described as a copy of 19th-century Western rhetoric—the traditionally mystical and intuitive aspects of the *form* of argumentation were downplayed, leading to a more scientific, logical understanding of it. During the second period, by contrast, under the influence of the Soviet Union the scholarly understanding of argumentation in China aligned with the *content* of Communist ideology and propaganda. During the third period, finally, this understanding "return[ed] to the path of rationality," indeed "to the right direction of development," by incorporating mid to late 20th-century Western scholarship on argumentation. The second period thus appears as a *misdirected* intellectual development, a description that broadly agrees with how China's then-political development is evaluated by many Chinese intellectuals today.

Wang's discussion of the third period devotes significant space to the "theory of three elements"— a normative product of Soviet scholarship according to which arguments are composed of a *thesis*, a *proof*, and a *method* of argument—, recent criticisms thereof for being neither necessary, nor sufficient for an analysis of *good* argumentation, as well as the theory's proposed extensions into four or five elements (including the *topic* and the *conclusion*). While all elements appear readily identifiable in a Western scholarly tradition, the theory and its (ongoing) impact on scholarship and education in China are news to this commentator. A fuller discussion of the theory's intellectual origin, together with application examples of it, would complement this notable contribution to writing a history of argumentation in China.

LIST OF CONTRIBUTORS TO ALL VOLUMES
(IN ALPHABETICAL ORDER BY SURNAME)

SCOTT AIKIN
Vanderbilt University

JOSÉ ALHAMBRA
Autonomous University of Madrid

JOSE M. ALONSO-MORAL
Universidade de Santiago de Compostela, Spain

KATIE ATKINSON
University of Liverpool

SHARON BAILIN
Faculty of Education, Simon Fraser University, Vancouver Canada

MARK BATTERSBY
Department of Philosophy, Capilano University, Vancouver Canada

TREVOR BENCH-CAPON
University of Liverpool

SARAH BIGI
University of the Sacred Heart, Milan, Italy

PETAR BODLOVIĆ
NOVA Institute of Philosophy (ArgLab),
FCSH, Nova University of Lisbon

MIEKE BOON
University of Twente

ELENA CABRIO
Université CôteD'Azur, CNRS, Inria, I3S

JOHN CASEY
Northeastern Illinois University

ALEJANDRO CATALA
Universidade de Santiago de Compostela, Spain

LIST OF CONTRIBUTORS

DORIANA CIMMINO
Independent researcher

DANIEL COHEN
Colby College

FEDERICA COMINETTI
Università dell'Aquila

CLAUDIA COPPOLA
Università Roma Tre, La Sapienza Università di Roma

MARÍA INÉS CORBALÁN
ArgLab-IFILNOVA, NOVA Universidade de Lisboa

HÉDI VIRÁG CSORDÁS
Assistant Lecturer at Budapest University of Technology and Economics

GIULIA D'AGOSTINO
Università della Svizzera italiana (USI)

DANIEL DE OLIVEIRA FERNANDES
University of Fribourg, Switzerland

EMMANUELLE DIETZ
Airbus Central R&T, Germany

ÁLVARO DOMÍNGUEZ-ARMAS
NOVA Institute of Philosophy, NOVA University of Lisbon

GONEN DORI-HACOHEN
University of Massachusetts Amherst

IOVAN DREHE
Technical University of Cluj-Napoca

LUCIJA DUDA
University of Manchester

MICHEL DUFOUR
University Sorbonne-Nouvelle

ALINA DURRANI
University of Massachusetts Amherst

List of Contributors

CATARINA DUTILH NOVAES
Department of Philosophy, Vrije Universiteit Amsterdam

ISABELA FAIRCLOUGH
University of Central Lancashire

LOGAN FIELDS
Advancing Machine and Human Reasoning (AMHR) Lab, University of South Florida

JOSÉ ÁNGEL GASCÓN
Departamento de Filosofía, Universidad de Murcia

GIULIA GIUNTA
University of Neuchâtel

GEOFFREY C. GODDU
University of Richmond

SARA GRECO
USI-Università della Svizzera italiana

MARCELLO GUARINI
University of Windsor, Canada

PASCAL GYGAX
University of Fribourg, Switzerland

DALE HAMPLE
Western Illinois University

AMALIA HARO MARCHAL
University of Granada

ANNETTE HAUTLI-JANISZ
University of Passau

BITA HESHMATI
University of Groningen

MIKA HIETANEN
Lund University

MICHAEL J. HOPPMANN
Northeastern University, Boston

List of Contributors

BROOKE HUBSCH
The Pennsylvania State University

BETH INNOCENTI
University of Kansas

CHIARA JERMINI-MARTINEZ SORIA
Università della Svizzera italiana

ANTONIS KAKAS
Dept. Computer Science, University of Cyprus, Cyprus

ALEXANDRA KARAKAS
Assistant Lecturer at Budapest University of Technology and Economics

IRYNA KHOMENKO
Taras Shevchenko National University of Kyiv

ZLATA KIKTEVA
University of Passau

KONRAD KILJAN
University of Warsaw
Laboratory of The New Ethos, Warsaw University of Technology

GABRIJELA KIŠIČEK
University of Zagreb

MARCIN KOSZOWY
Laboratory of The New Ethos, Warsaw University of Technology

ADAMOS KOUMI
Dept. Computer Science, University of Cyprus, Cyprus

MANFRED KRAUS
University of Tübingen

LEONARD KUPŚ
Faculty of Psychology and Cognitive Science
Adam Mickiewicz University, Poznań, Poland

NIILO LAHTI
The University of Eastern Finland, School of Theology

JOHN LAWRENCE
Centre for Argument Technology, University of Dundee, UK

LAWRENCE LENGBEYER
United Stated Naval Academy

MARCIN LEWIŃSKI
Nova Institute of Philosophy
Nova University Lisbon, Portugal

JIAXING LI
Nankai University

YAN-LIN LIAO
Department of Philosophy, Sun Yat-sen University, China

JOHN LICATO
Advancing Machine and Human Reasoning (AMHR) Lab, University of South Florida

DAVIDE LIGA
University of Luxembourg

EDOARDO LOMBARDI VALLAURI
Università Roma Tre

COSTANZA LUCCHINI
Università della Svizzera italiana (USI)

CHRISTOPH LUMER
University of Siena, Italy

GIORGIA MANNAIOLI
Università Roma Tre, La Sapienza Università di Roma

MAURIZIO MANZIN
University of Trento

ZAID MARJI
Advancing Machine and Human Reasoning (AMHR) Lab, University of South Florida

HUBERT MARRAUD
Universidad Autónoma de Madrid (Spain)

LIST OF CONTRIBUTORS

SANTIAGO MARRO
Université CôteD'Azur, CNRS, Inria, I3S

VIVIANA MASIA
Università Roma Tre

DAVIDE MAZZI
University of Modena and Reggio Emilia (Italy)

GUIDO MELCHIOR
University of Graz

CHIARA MERCURI
Università della Svizzera Italiana

DIMA MOHAMMED
Institute of Philosophy, Faculty of Social and Human Sciences, NOVA University of Lisbon, Portugal

ELENA MUSI
University of Liverpool

HENRI MÜTSCHELE
Heinrich- Heinrich Heine University Düsseldorf, Germany

ZI-HAN NIU
Department of Philosophy, Sun Yat-sen University, China

PAULA OLMOS
Universidad Autónoma de Madrid

MARIANA OROZCO
University of Twente

RAHMI ORUÇ
Ibn Haldun University, Comparative Literature, ArguMunazara Research Center

STEVE OSWALD
University of Fribourg, Switzerland

WENQI OUYANG
Department of Philosophy, Sun Yat-sen University, China

LIST OF CONTRIBUTORS

FABIO PAGLIERI
Istituto di Scienze e Tecnologie della Cognizione, Consiglio Nazionale delle Ricerche (ISTC-CNR), Italy

ROOSMARYN PILGRAM
Leiden University Centre for Linguistics

FEDERICO PUPPO
University of Trento

MENNO H. REIJVEN
University of Amsterdam

THÉOPHILE ROBINEAU
Université Paris Cité

ANDREA ROCCI
Università della Svizzera italiana (USI)

MARIA GRAZIA ROSSI
Institute of Philosophy, Faculty of Social and Human Sciences, NOVA University of Lisbon, Portugal

LUCIA SALVATO
Università Cattolica del Sacro Cuore

CRISTIÁN SANTIBÁNEZ,
Universidad Católica de la Santísima Concepción

MENASHE SCHWED
Ashkelon Academic College, Israel

BLAKE D. SCOTT
Institute of Philosophy, KU Leuven

HARVEY SIEGEL
University of Miami

ILIA STEPIN
Universidade de Santiago de Compostela, Spain

JÁNOS TANÁCS
Department of Argumentation Theory and Marketing, ELTE, Budapest

List of Contributors

GIULIA TERZIAN
ArgLab-IFILNOVA, NOVA Universidade de Lisboa

CHRISTOPHER W. TINDALE
Department of Philosophy, University of Windsor, Ontario, Canada

SERENA TOMASI
University of Trento

MARIUSZ URBAŃSKI
Faculty of Psychology and Cognitive Science
Adam Mickiewicz University, Poznań, Poland

MEHMET ALÌ ÜZELGÜN
IFILNOVA, Universidade Nova de Lisboa
CIES-ISCTE, Instituto Universitário de Lisboa

CHARLOTTE VAN DER VOORT
Leiden University Centre for Linguistics
JAN ALBERT VAN LAAR
University of Groningen

LOTTE VAN POPPEL
Center for Language and Cognition Groningen

SERENA VILLATA
Université CôteD'Azur, CNRS, Inria, I3S

JACKY VISSER
Centre for Argument Technology, University of Dundee, UK

JEAN H.M. WAGEMANS
University of Amsterdam

HAILONG WANG
Wuhan University

JIANFENG WANG
Fujian Normal University, China

MARK WEINSTEIN
Montclair State University

HARALD R. WOHLRAPP
Universität Hamburg

LIST OF CONTRIBUTORS

MING-HUI XIONG
Guanghua Law School, Zhejiang University, China

OLENA YASKORSKA-SHAH
Università della Svizzera italiana (USI)

SHIYANG YU
College of Philosophy, Nankai University

GÁBOR Á. ZEMPLÉN
Eötvös Loránd University-Faculty of Economics, ELTE, Budapest

FRANK ZENKER
College of Philosophy, Nankai University, Tianjin, China

www.ingramcontent.com/pod-product-compliance
Lightning Source LLC
Chambersburg PA
CBHW071941220426
43662CB00009B/942